Meta-Analysis

Meta-Analysis

A Structural Equation Modeling Approach

Mike W. -L. Cheung

National University of Singapore, Singapore

This edition first published 2015

Registered office
John Wiley & Sons Ltd, The Atrium, Southern Gate, Chichester, West Sussex, PO19 8SQ, United Kingdom

For details of our global editorial offices, for customer services and for information about how to apply for permission to reuse the copyright material in this book please see our website at www.wiley.com.

Library of Congress Cataloging-in-Publication Data applied for

A catalogue record for this book is available from the British Library.

ISBN: 9781119993438

Set in 10/12pt TimesRoman by Laserwords Private Limited, Chennai, India.
Printed and bound in Singapore by Markono Print Media Pte Ltd.

1 2015

For my family—my wife Maggie,
my daughter little Ching Ching, and my parents

Contents

Preface xiii

Acknowledgments xv

List of abbreviations xvii

List of figures xix

List of tables xxi

1 Introduction **1**
1.1 What is meta-analysis? 1
1.2 What is structural equation modeling? 2
1.3 Reasons for writing a book on meta-analysis and structural equation modeling 3
1.3.1 Benefits to users of structural equation modeling and meta-analysis 6
1.4 Outline of the following chapters 6
1.4.1 Computer examples and data sets used in this book 8
1.5 Concluding remarks and further readings 8
References 9

2 Brief review of structural equation modeling **13**
2.1 Introduction 13
2.2 Model specification 14
2.2.1 Equations 14
2.2.2 Path diagram 15
2.2.3 Matrix representation 15
2.3 Common structural equation models 18
2.3.1 Path analysis 18
2.3.2 Confirmatory factor analysis 19
2.3.3 Structural equation model 21
2.3.4 Latent growth model 22
2.3.5 Multiple-group analysis 23

2.4 Estimation methods, test statistics, and goodness-of-fit indices 25
2.4.1 Maximum likelihood estimation 25
2.4.2 Weighted least squares 26
2.4.3 Multiple-group analysis 28
2.4.4 Likelihood ratio test and Wald test 28
2.4.5 Confidence intervals on parameter estimates 29
2.4.6 Test statistics versus goodness-of-fit indices 34
2.5 Extensions on structural equation modeling 38
2.5.1 Phantom variables 38
2.5.2 Definition variables 39
2.5.3 Full information maximum likelihood estimation 41
2.6 Concluding remarks and further readings 42
References 42

3 Computing effect sizes for meta-analysis 48
3.1 Introduction 48
3.2 Effect sizes for univariate meta-analysis 50
3.2.1 Mean differences 50
3.2.2 Correlation coefficient and its Fisher's z transformation 55
3.2.3 Binary variables 56
3.3 Effect sizes for multivariate meta-analysis 57
3.3.1 Mean differences 57
3.3.2 Correlation matrix and its Fisher's z transformation 59
3.3.3 Odds ratio 60
3.4 General approach to estimating the sampling variances and covariances 60
3.4.1 Delta method 61
3.4.2 Computation with structural equation modeling 64
3.5 Illustrations Using R 68
3.5.1 Repeated measures 69
3.5.2 Multiple treatment studies 71
3.5.3 Multiple-endpoint studies 73
3.5.4 Multiple treatment with multiple-endpoint studies 75
3.5.5 Correlation matrix 77
3.6 Concluding remarks and further readings 78
References 78

4 Univariate meta-analysis 81
4.1 Introduction 81
4.2 Fixed-effects model 83
4.2.1 Estimation and hypotheses testing 83
4.2.2 Testing the homogeneity of effect sizes 85
4.2.3 Treating the sampling variance as known versus as estimated 85

4.3 Random-effects model 87
4.3.1 Estimation and hypothesis testing 88
4.3.2 Testing the variance component 90
4.3.3 Quantifying the degree of the heterogeneity of effect sizes 92
4.4 Comparisons between the fixed- and the random-effects models 93
4.4.1 Conceptual differences 93
4.4.2 Statistical differences 94
4.5 Mixed-effects model 96
4.5.1 Estimation and hypotheses testing 97
4.5.2 Explained variance 98
4.5.3 A cautionary note 99
4.6 Structural equation modeling approach 100
4.6.1 Fixed-effects model 100
4.6.2 Random-effects model 101
4.6.3 Mixed-effects model 102
4.7 Illustrations using R 105
4.7.1 Odds ratio of atrial fibrillation between bisphosphonate and non-bisphosphonate users 105
4.7.2 Correlation between organizational commitment and salesperson job performance 108
4.8 Concluding remarks and further readings 116
References 117

5 Multivariate meta-analysis 121
5.1 Introduction 121
5.1.1 Types of dependence 121
5.1.2 Univariate meta-analysis versus multivariate meta-analysis 122
5.2 Fixed-effects model 124
5.2.1 Testing the homogeneity of effect sizes 125
5.2.2 Estimation and hypotheses testing 126
5.3 Random-effects model 127
5.3.1 Structure of the variance component of random effects 128
5.3.2 Nonnegative definite of the variance component of random effects 129
5.3.3 Estimation and hypotheses testing 131
5.3.4 Quantifying the degree of heterogeneity of effect sizes 132
5.3.5 When the sampling covariances are not known 133
5.4 Mixed-effects model 134
5.4.1 Explained variance 135
5.5 Structural equation modeling approach 136
5.5.1 Fixed-effects model 136
5.5.2 Random-effects model 137
5.5.3 Mixed-effects model 138

5.6 Extensions: mediation and moderation models on the effect sizes 140
5.6.1 Regression model 141
5.6.2 Mediating model 143
5.6.3 Moderating model 144
5.7 Illustrations using R 145
5.7.1 BCG vaccine for preventing tuberculosis 146
5.7.2 Standardized mean differences between males and females on life satisfaction and life control 156
5.7.3 Mediation and moderation models 161
5.8 Concluding remarks and further readings 174
References 174

6 Three-level meta-analysis 179
6.1 Introduction 179
6.1.1 Examples of dependent effect sizes with unknown degree of dependence 180
6.1.2 Common methods to handling dependent effect sizes 180
6.2 Three-level model 183
6.2.1 Random-effects model 183
6.2.2 Mixed-effects model 187
6.3 Structural equation modeling approach 188
6.3.1 Two representations of the same model 189
6.3.2 Random-effects model 191
6.3.3 Mixed-effects model 193
6.4 Relationship between the multivariate and the three-level meta-analyses 195
6.4.1 Three-level meta-analysis as a special case of the multivariate meta-analysis 195
6.4.2 Approximating a multivariate meta-analysis with a three-level meta-analysis 196
6.4.3 Three-level multivariate meta-analysis 198
6.5 Illustrations using R 200
6.5.1 Inspecting the data 201
6.5.2 Fitting a random-effects model 202
6.5.3 Obtaining the likelihood-based confidence interval 203
6.5.4 Testing $\tau^2_{(3)} = 0$ 204
6.5.5 Testing $\tau^2_{(2)} = 0$ 205
6.5.6 Testing $\tau^2_{(2)} = \tau^2_{(3)}$ 205
6.5.7 Testing types of proposals (grant versus fellowship) 206
6.5.8 Testing the effect of the year of application 207
6.5.9 Testing the country effect 209
6.6 Concluding remarks and further readings 210
References 211

7 Meta-analytic structural equation modeling **214**
7.1 Introduction 214
7.1.1 Meta-analytic structural equation modeling as a possible solution for conflicting research findings 215
7.1.2 Basic steps for conducting a meta-analytic structural equation modeling 217
7.2 Conventional approaches 218
7.2.1 Univariate approaches 218
7.2.2 Generalized least squares approach 221
7.3 Two-stage structural equation modeling: fixed-effects models 223
7.3.1 Stage 1 of the analysis: pooling correlation matrices 224
7.3.2 Stage 2 of the analysis: fitting structural models 227
7.3.3 Subgroup analysis 233
7.4 Two-stage structural equation modeling: random-effects models 233
7.4.1 Stage 1 of the analysis: pooling correlation matrices 234
7.4.2 Stage 2 of the analysis: fitting structural models 235
7.5 Related issues 235
7.5.1 Multiple-group structural equation modeling versus meta-analytic structural equation modeling 236
7.5.2 Fixed-effects model: two-stage structural equation modeling versus generalized least squares 237
7.5.3 Alternative random-effects models 239
7.5.4 Maximum likelihood estimation versus restricted (or residual) maximum likelihood estimation 242
7.5.5 Correlation coefficient versus Fisher's z score 242
7.5.6 Correction for unreliability 243
7.6 Illustrations using R 244
7.6.1 A higher-order confirmatory factor analytic model for the Big Five model 244
7.6.2 A regression model on SAT (Math) 258
7.6.3 A path model for cognitive ability to supervisor rating 266
7.7 Concluding remarks and further readings 273
References 274

8 Advanced topics in SEM-based meta-analysis **279**
8.1 Restricted (or residual) maximum likelihood estimation 279
8.1.1 Reasons for and against the maximum likelihood estimation 280
8.1.2 Applying the restricted (or residual) maximum likelihood estimation in SEM-based meta-analysis 281
8.1.3 Implementation in structural equation modeling 283
8.2 Missing values in the moderators 289
8.2.1 Types of missing mechanisms 289

8.2.2 Common methods to handling missing data 290
8.2.3 Maximum likelihood estimation 291
8.3 Illustrations using R 294
8.3.1 Restricted (or residual) maximum likelihood estimation 295
8.3.2 Missing values in the moderators 300
8.4 Concluding remarks and further readings 309
References 310

9 Conducting meta-analysis with M*plus* 313
9.1 Introduction 313
9.2 Univariate meta-analysis 314
9.2.1 Fixed-effects model 314
9.2.2 Random-effects model 317
9.2.3 Mixed-effects model 322
9.2.4 Handling missing values in moderators 325
9.3 Multivariate meta-analysis 327
9.3.1 Fixed-effects model 328
9.3.2 Random-effects model 333
9.3.3 Mixed-effects model 337
9.3.4 Mediation and moderation models on the effect sizes 340
9.4 Three-level meta-analysis 346
9.4.1 Random-effects model 346
9.4.2 Mixed-effects model 351
9.5 Concluding remarks and further readings 353
References 354

A A brief introduction to R, `OpenMx`, and `metaSEM` packages 356
A.1 R 357
A.2 `OpenMx` 362
A.3 `metaSEM` 364
References 368

Index 369

Preface

"If all you have is a hammer, everything looks like a nail."
—Maslow's hammer

Purpose of This Book

There were two purposes of writing this book. One was personal and the other was more "formal." I will give the personal one first. The primary motivation for writing this book was to document my own journey in learning structural equation modeling (SEM) and meta-analysis. The journey began when I was a undergraduate student. I first learned SEM from Wai Chan, my former supervisor. After learning a bit from the giants in SEM, such as Karl Jöreskog, Peter Bentler, Bengt Muthén, Kenneth Bollen, Michael Browne, Michael Neale, and Roderick McDonald, among others, I found SEM fascinating. It seems that SEM is *the* statistical framework for all data analysis. Nearly all statistical techniques I learned can be formulated as structural equation models.

In my graduate study, I came across a different technique—meta-analysis. I learned meta-analysis by reading the classic book by Larry Hedges and Ingram Olkin. I was impressed that a simple yet elegant statistical model could be used to synthesize findings across studies. It seems that meta-analysis is the key to advance knowledge by combining results from different studies. As I was trained with the SEM background, everything looks like a structural equation model to me. I asked the question, "could a meta-analysis be a structural equation model?" This book summarized my journey to answer this question in the past one and a half decades.

Now, I will give a more formal purpose of this book. With the advances in statistics and computing, researchers have more statistical tools to answer their research questions. SEM and meta-analysis are two powerful statistical techniques in the social, educational, behavioral, and medical sciences. SEM is a popular tool to test hypothesized models by modeling the latent and observed variables in primary research, while meta-analysis is a *de facto* tool to synthesize research findings from a pool of empirical studies. These two techniques are usually treated as two unrelated topics in the literature. They have their own strengths, weaknesses, assumptions, models, terminologies, software packages, audiences, and even journals (*Structural Equation Modeling: A Multidisciplinary Journal* and *Research Synthesis Methods*). Researchers working in one area rarely refer to the work in the other area. Advances in one area have basically no impact on the other area.

There were two primary goals for this book. The first one was to present the recent methodological advances on integrating meta-analysis and SEM—the *SEM-based meta-analysis* (using SEM to conducting meta-analysis) and meta-analytic structural equation modeling (conducting meta-analysis on correlation matrices for the purpose of fitting structural equation models on the pooled correlation matrix). It is my hope that a unified framework will be made available to researchers conducting both primary data analysis and meta-analysis. A single framework can easily translate advances from one field to the other fields. Researchers do not need to reinvent the wheels again.

The second goal was to provide accessible computational tools for researchers conducting meta-analyses. The `metaSEM` package in the R statistical environment, which is available at http://courses.nus.edu.sg/course/psycwlm/Internet/metaSEM/, was developed to fill this gap. Using the `OpenMx` package as the workhorse, the `metaSEM` package implemented most of the methods discussed in this book. Complete examples in R code are provided to guide readers to fit various meta-analytic models. Besides the R code, M*plus* was also used to illustrate some of the examples in this book. `R` (3.1.1), `OpenMx` (2.0.0-3654), `metaSEM` (0.9-0), `metafor` (1.9-3), `lavaan` (0.5-17.698), and M*plus* (7.2) were used in writing this book. The output format may be slightly different from the versions that you are using.

Level and Prerequisites

Readers are expected to have some basic knowledge of SEM. This level is similar to the first year of research methods covered in most graduate programs. Knowledge of meta-analysis is preferable though not required. We will go through the meta-analytic models in this book. It will also be useful if readers have some knowledge in R because R is the main statistical environment to implement the methods introduced in this book. Readers may refer to Appendix at the end of this book for a quick introduction to R. For readers who are more familiar with M*plus*, they may use M*plus* to implement some of the methods discussed in this book.

MIKE W.-L. CHEUNG
Singapore

Acknowledgments

I thank Wai Chan, my former supervisor, for introducing me to the exciting field of structural equation modeling (SEM). He also suggested me to explore meta-analytic structural equation modeling in my graduate studies. I acknowledge the suggestions and comments made by many people: Shu-fai Cheung, Adam Hafdahl, Suzanne Jak, Yonghao Lim, Iris Sun, and Wolfgang Viechtbauer. All remaining errors are mine. I especially thank my wife for her support and patience. My daughter was born during the preparation of this book. I enjoyed my daughter's company when I was writing this book. Part of the book was completed during my sabbatical leave supported by the Faculty of Arts & Social Sciences, the National University of Singapore. I also appreciate the funding provided by the Faculty to facilitate the production of this book. I thank Heather Kay, Richard Davies, Jo Taylor, and Prachi Sinha Sahay from Wiley. They are very supportive and professional. It has been a pleasure working with them.

The `metaSEM` package could not be written without R and `OpenMx`. Contributions by the R Development Core Team and the OpenMx Core Development Team are highly appreciated. Their excellent work makes it possible to implement the techniques discussed in this book. I have to specially thank the members of the OpenMx Core Development Team for their quick and helpful responses in addressing issues related to `OpenMx`. I also thank Yves Rosseel for answering questions related to the `lavaan` package. Finally, the preparation of this book was mainly based on the open-source software. This includes LaTeX for typesetting this book, R for the analyses, `Sweave` for mixing R and LaTeX, Graphviz and dot2tex for preparing the figures, GNU make for automatically building files, Git for revision control, Emacs for editing files, and finally, Linux as the platform for writing.

List of abbreviations

Abbreviation	Full name
CFA	confirmatory factor analysis
CFI	comparative fit index
CI	confidence interval
FIML	full information maximum likelihood
GLS	generalized least squares
LBCI	likelihood-based confidence interval
LL	log likelihood
LR	likelihood ratio
MASEM	meta-analytic structural equation modeling
ML	maximum likelihood
NNFI	non-normed fit index
OR	odds ratio
OLS	ordinary least squares
RAM	reticular action model
REML	restricted (or residual) maximum likelihood estimation
RMD	raw mean difference
RMSEA	root mean square error of approximation
SE	standard error
SEM	structural equation modeling
SMD	standardized mean difference
SRMS	standardized root mean square residual
TLI	Tucker–Lewis index
TSSEM	two-stage structural equation modeling
UMM	unweighted method of moments
WLS	weighted least squares
WMM	weighted method of moments

List of figures

1.1	Publications using meta-analysis and structural equation modeling. (a) Actual number of publications per year and (b) percentage of publications.	4
2.1	Two graphical model representations of a simple regression.	16
2.2	A path model with one mediator.	19
2.3	A confirmatory factor analytic model.	20
2.4	A structural equation model.	22
2.5	A latent growth model.	24
2.6	A moderated mediation model. (a) Group 1 and (b) Group 2.	24
2.7	Relationship between a likelihood ratio and a Wald tests.	32
2.8	Binomial distribution with $n = 10$ and $x = 8$ (a) and $n = 100$ and $x = 80$ (b) and with solid lines (true log-likelihood) and dashed lines (quadratic approximation of log-likelihood).	32
2.9	A regression model with a phantom variable.	39
2.10	A regression model with and without a definition variable.	40
3.1	A structural equation model for repeated measures.	65
3.2	A structural equation model for multiple treatment study.	66
3.3	A structural equation model for multiple-endpoint study.	67
3.4	A structural equation model for a correlation matrix.	69
4.1	A model with one variable y.	82
4.2	Distributions of $\chi^2_{\text{df}=1}$ and a 50:50 mixture of $\chi^2_{\text{df}=0}$ and $\chi^2_{\text{df}=1}$.	91
4.3	(a) Fixed-effects versus (b) random-effects meta-analyses.	95
4.4	A univariate fixed-effects meta-analytic model.	100
4.5	A univariate random-effects meta-analytic model.	102
4.6	A univariate mixed-effects meta-analytic model treating the moderator as a variable.	103
4.7	A univariate mixed-effects meta-analytic model treating the moderator as a design matrix.	104
5.1	A multivariate fixed-effects model with two effect sizes.	136
5.2	A multivariate random-effects model with two effect sizes.	138
5.3	A multivariate mixed-effects model with two effect sizes and one moderator.	139
5.4	A structural equation model to correct for unreliabilities.	141
5.5	A regression model between two effect sizes.	142

5.6 A mediation model with two effect sizes. 144
5.7 A moderation model with two effect sizes. 146
5.8 Confidence ellipses on the vaccinated and the nonvaccinated groups. 154
5.9 Confidence ellipses on the two effect sizes with forest plots. 155
5.10 Confidence ellipses on the SMDs of life satisfaction and life control. 159
6.1 Two representations of a random-effects model. 190
6.2 A three-level random-effects model with two studies in the *j*th cluster. 192
6.3 A three-level mixed-effects model with two studies and one moderator in the *j*th cluster. 194
7.1 A fixed-effects model of the first stage of the TSSEM approach. 226
7.2 A higher-order model for the Big Five model. 245
7.3 A regression model of math aptitude with spatial ability and verbal ability as predictors. 257
7.4 A path model of cognitive ability, job knowledge, work sample, and supervisor rating. 266
8.1 A univariate random-effects meta-analysis with three studies. 285
8.2 A univariate random-effects meta-analysis with three studies using REML estimation. 286
8.3 A univariate random-effects meta-analysis with a moderator. 293
8.4 A multivariate random-effects meta-analysis with two effect sizes per study and a moderator. 294
8.5 A three-level meta-analysis with two studies and a level-3 moderator in the *j*th cluster. 295
8.6 A three-level meta-analysis with two studies and a level-2 moderator in the *j*th cluster. 296
9.1 A univariate fixed-effects meta-analytic model in M*plus*. 315
9.2 A univariate random-effects meta-analytic model in M*plus*. 318
9.3 A univariate mixed-effects meta-analytic model in M*plus*. 322
9.4 Multivariate fixed-effects meta-analytic models in M*plus*. 329
9.5 A multivariate random-effects meta-analytic model in M*plus*. 333
9.6 A multivariate mixed-effects meta-analytic model in M*plus*. 337
9.7 A mediation model with two effect sizes in M*plus*. 340
9.8 A moderation model with two effect sizes in M*plus*. 343
A.1 A plot of y against x with the best fitted regression line. 360

List of tables

1.1 Datasets used in this book. 7
5.1 Long format data for a multivariate meta-analysis. 135
5.2 Wide format data for a multivariate meta-analysis. 135
6.1 Long format data for a two-level meta-analysis. 190
6.2 Wide format data for a two-level meta-analysis. 191
6.3 Wide format data for a three-level meta-analysis. 191
6.4 Two effect sizes nested within *k* clusters 199

1

Introduction

This chapter gives an overview of this book. It first briefly reviews the history and applications of meta-analysis and structural equation modeling (SEM). The importance of using meta-analysis and SEM to advancing scientific research is discussed. This chapter then addresses the needs and advantages of integrating meta-analysis and SEM. It further outlines the remaining chapters and the data sets used in the book. We close this chapter by addressing topics that will not be further discussed in this book.

1.1 What is meta-analysis?

Pearson (1904) was often credited as one of the earliest researchers applying ideas of meta-analysis (e.g., Chalmers et al., 2002; Cooper and Hedges, 2009; National Research Council, 1992; O'Rourke, 2007). He tried to determine the relationship between mortality and inoculation with a vaccine for enteric fever by averaging correlation coefficients across 11 small-sample studies. The idea of combining and pooling studies has been widely used in the physical and social sciences. There are many successful stories as documented in, for example, National Research Council (1992) and Hunt (1997). The term *meta-analysis* was coined by Gene Glass in educational psychology to represent "the statistical analysis of a large collection of analysis results from individual studies for the purpose of integrating the findings" (Glass 1976, p.3).

Validity generalization, another technique with similar objectives, was independently developed by Schmidt and Hunter (1977) in industrial and organizational psychology in nearly the same period. Later, Hedges and Olkin (1985) wrote a classic text that provides the statistical foundation of meta-analysis. These techniques have been expanded, refined, and adopted in many disciplines.

Meta-Analysis: A Structural Equation Modeling Approach, First Edition. Mike W. -L. Cheung.

Companion Website: www.wiley.com/go/cheung/meta_analysis

Meta-analysis is now a popular statistical technique to synthesizing research findings in many disciplines including educational, social, and medical sciences.

A meta-analysis begins by conceptualizing the research questions. The research questions must be empirically testable based on the published studies. The published studies should be able to provide enough information to calculate the effect sizes, the ingredients for a meta-analysis. Detailed inclusion and exclusion criteria are developed to guide which studies are eligible to be included in the meta-analysis. After extracting the effect sizes and the study characteristics, the data can be subjected to a statistical analysis. The next step is to interpret the results and prepare reports to disseminate the findings.

This book mainly focuses on the statistical issues in a meta-analysis. Generally speaking, the statistical models discussed in this book fall into three dimensions:

(i) fixed-effects versus random-effects models;

(ii) independent versus nonindependent effect sizes; and

(iii) models with or without structural models on the averaged effect sizes.

The first dimension is fixed-effects versus random-effects models. Fixed-effects models provide conditional inferences on the studies included in the meta-analysis, while random-effects models attempt to generalize the inferences beyond the studies used in the meta-analysis. Statistically speaking, the fixed-effects models, also known as the *common effects models*, are special cases of the random-effects models.

The second dimension focuses on whether the effect sizes are independent or nonindependent. Most meta-analytic models, such as the univariate meta-analysis introduced in this book, assume independence on the effect sizes. When there is more than one effect size reported per study, the effect sizes are likely nonindependent. Both the multivariate and three-level meta-analyses are introduced to handle the nonindependent effect sizes depending on the assumptions of the data. The last dimension is whether the research questions are related to the averaged effect sizes themselves or some forms of structural models on the averaged effect sizes. If researchers are only interested in the effect sizes, conventional univariate, multivariate, and three-level meta-analyses are sufficient. Sometimes, researchers are interested in testing proposed structures on the effect size. This type of research questions can be addressed by testing the mediation and moderation models on the effect sizes (Section 5.6) or the meta-analytic structural equation modeling (MASEM; Chapter 7).

1.2 What is structural equation modeling?

SEM is a flexible modeling technique to test proposed models. The proposed models can be specified as path diagrams, equations, or matrices. SEM integrates several statistical techniques into a single framework—path analysis in biology and sociology, factor analysis in psychology, and simultaneous equation and errors-in-variables models in economics (e.g., Matsueda 2012). Jöreskog (1969, 1970, 1978) was usually credited as the one who first integrated these techniques into a single framework. He further proposed computational feasible approaches to conduct the analysis. These algorithms were implemented in `LISREL` (Jöreskog and Sörbom, 1996), the first SEM package in the market. At nearly the same time, Bentler contributed a lot in the methodological development of SEM (e.g., Bentler 1986, 1990; Bentler and Weeks, 1980). He also wrote a user friendly program called `EQS` (Bentler, 2006) to conduct SEM. The availability of `LISREL` and `EQS` popularized applications of SEM in various fields. Both Jöreskog and Bentler received the Award for Distinguished Scientific Applications of Psychology (American Psychological Association, 2007a, b) "[f]or [their] development of models, statistical procedures, and a computer algorithm for structural equation modeling (SEM) that changed the way in which inferences are made from observational data; namely, SEM permits hypotheses derived from theory to be tested."

Many recent methodological advances have been developed and integrated into M*plus*, a popular and powerful SEM package (Muthén and Muthén, 2012). SEM is now widely used as a statistical model to test research hypotheses. Readers may refer to, for example, MacCallum and Austin (2000) and Bollen (2002) for some applications in the social sciences.

1.3 Reasons for writing a book on meta-analysis and structural equation modeling

There are already many good books on the topic of meta-analysis (e.g., Borenstein et al., 2010; Card, 2012; Hedges and Olkin, 1985; Lipsey and Wilson, 2000; Schmidt and Hunter, 2015; Whitehead, 2002). Moreover, meta-analysis has also been covered as special cases of mixed-effects or multilevel models (e.g., Demidenko, 2013, Goldstein, 2011, Hox, 2010; Raudenbush and Bryk, 2002). It seems that there is no need to write another book on meta-analysis. On the other hand, this book did not aim to be a comprehensive introduction to SEM neither. Before answering this question, let us first review the current state of applications of meta-analysis and SEM in academic research.

Figure 1.1 shows two figures on the numbers of publications using meta-analysis and SEM in Web of Science. The figures were averaged over 5 years. For example, the number for 2010 was calculated by averaging from 1998 to 2012. Figure 1.1a depicts the actual numbers of publications, while Figure 1.1b converts the numbers to percentages by dividing the numbers by the total numbers of publications. The trends in both figures are nearly identical in terms of actual numbers and percentages. One speculation why the numbers on meta-analysis are higher than

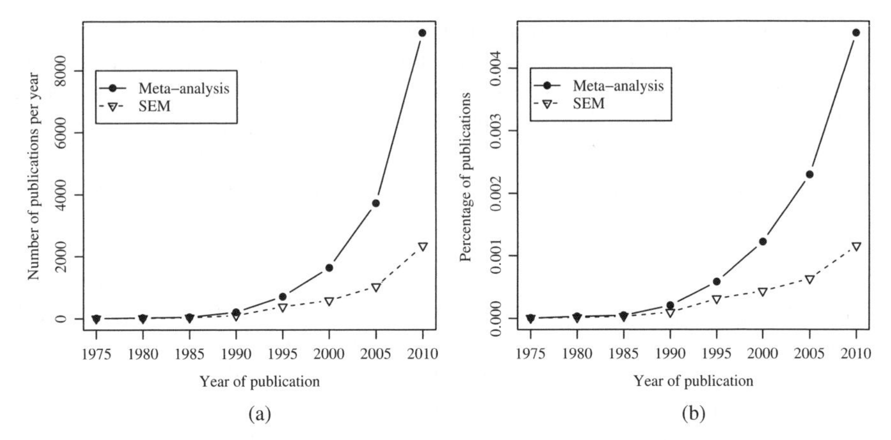

Figure 1.1 Publications using meta-analysis and structural equation modeling. (a) Actual number of publications per year and (b) percentage of publications.

those on SEM is that meta-analysis is very popular in medical research, whereas SEM is rarely used in medical research (cf. Song and Lee, 2012). Anyway, it is clear that both techniques are getting more and more popular over time.

Although both SEM and meta-analysis are very popular in the educational, social, behavioral, and medical sciences, both techniques are treated as two unrelated techniques in the literature. They have their own assumptions, models, terminologies, software packages, communities, and even journals (*Structural Equation Modeling: A Multidisciplinary Journal* and *Research Synthesis Methods*). These two techniques are also considered as separate topics in doctoral training in psychology (Aiken et al., 2008). Users of SEM are mainly interested in primary research, while users of meta-analysis only conduct research synthesis on the literature. Researchers working in one area rarely refer to the work in the other area. Users of SEM seldom have the motivation to learn meta-analysis and vice versa. Advances in one area have basically no impact on the other area.

There were some attempts to bring these two techniques together. One such topic is known as MASEM (e.g., Cheung and Chan, 2005b; Viswesvaran and Ones, 1995). There are two stages involved in an MASEM. Meta-analysis is usually used to pool correlation matrices together in the stage 1 analysis. The pooled correlation matrix is used to fit structural equation models in the stage 2 analysis. As researchers usually apply ad hoc procedures to fit structural equation models, some of these procedures are not statistically defensible from an SEM perspective. Therefore, one of the goals of this book (Chapter 7) was to provide a statistically defensible approach to conduct MASEM.

Another reason for writing this book was to integrate meta-analysis into the general SEM framework. This helps to advance the methodological development in both areas. There are many such examples in the literature. Consider the classic example of analysis of variance (ANOVA) and multiple regression. Before the seminal work of Cohen (1968) and Cohen and Cohen (1975), "[t]he textbooks in 'psychological' statistics treat [multiple regression, ANOVA, and ANCOVA] quite separately, with wholly different algorithms, nomenclature, output, and examples" (Cohen 1968, p. 426). Understanding the mathematical equivalence between an ANOVA (and analysis of covariance (ANCOVA)) and a multiple regression helps us to comprehend the details behind the general linear model that plays an important role in modern statistics.

SEM is another successful story in the literature. The general linear model, path analysis, and confirmatory factor analysis (CFA) are some well-known special cases of SEM. It has been shown that many models used in the social and behavioral sciences are indeed special cases of SEM. For example, many item response theory (IRT) models can be analyzed as structural equation models with binary or categorical variables as indicators (e.g., Takane and Deleeuw, 1987). The main advantage of analyzing IRT models as structural equation models is that many of the SEM techniques can be directly applied to address research questions that are challenging in traditional IRT framework. For example, researchers may test IRT models with multiple traits (multiple factor models in SEM), with covariates as predictors (multiple indicators multiple causes in SEM), with missing data (full information maximum likelihood (FIML) estimation in SEM), and with nested structures (multilevel SEM) (Muthén and Asparouhov, 2013).

Another recent example is the recognition of multilevel models as structural equation models (Bauer, 2003; Curran, 2003; Mehta and Neale, 2005; Mehta and West, 2000; Rovine and Molenaar, 2000). Understanding the similarities between multilevel models and structural equation models helps to develop the multilevel SEM (e.g., Mehta and Neale, 2005; Muthén, 1994; Preacher et al., 2010). There are at least two methodological advances of integrating multilevel models and SEM. First, graphical models, which are popular in SEM, have been developed to represent multilevel models (Curran and Bauer, 2007). Another advance is that various goodness-of-fit indices in SEM have been *exported* to multilevel models (Wu et al., 2009). Readers may refer to, for example, Bollen et al. (2010), Matsueda (2012), and Kaplan (2009) for the recent methodological advances in SEM.

The current SEM framework is far beyond the original SEM developed by Jöreskog and Bentler. Modern SEM framework integrates techniques and models from several disciplines. For example, M*plus* (Muthén and Muthén, 2012) combines traditional SEM, multilevel models, complex survey analysis, mixture modeling, survival analysis, latent class models, some IRT models, and even Bayesian inferences into a single statistical modeling framework. Another general framework is the generalized linear latent and mixed models (GLLAMM) (Skrondal and Rabe-Hesketh, 2004) that integrate SEM, generalized linear models, multilevel models, latent class models, and IRT models.

This book provides the foundation of integrating meta-analysis into the SEM framework. Latent variables in a structural equation model are used to represent the *true* effect sizes in a meta-analysis. Meta-analytic models can then be analyzed as structural equation models. This approach is termed *SEM-based meta-analysis* in this book. Many state-of-the-art techniques in SEM are available to researchers doing meta-analysis by using the SEM-based meta-analysis.

1.3.1 Benefits to users of structural equation modeling and meta-analysis

There are several advantages of integrating meta-analysis into the SEM framework. For the SEM users, the SEM-based meta-analysis extends their statistical tools to conduct research with meta-analysis. Suppose that their primary research interests are in studying the training effectiveness with SEM; the SEM-based meta-analysis allows them to conduct a meta-analysis on the same topic without leaving the SEM framework. Many of the terminologies in meta-analysis can be translated into the terminologies in SEM. Software developers may explore the possibilities to develop an integrated SEM framework for researchers doing primary and meta-analysis. For example, M*plus* can be used to implement many of the SEM-based meta-analysis introduced in this book (see Chapter 9).

For the meta-analysis users, the SEM-based meta-analysis provides some new research tools to address research questions in meta-analysis. For example, users may apply the SEM-based meta-analysis to conduct univariate, multivariate, and three-level meta-analyses that handle missing values in moderators in the same SEM framework. Future studies may explore how techniques, such as robust statistics, bootstrap, and mixture models available in SEM, can be applied to meta-analysis.

In terms of graduate training in statistics, a single coherent framework can be introduced to students. This framework includes the general linear model, SEM, and meta-analysis. It helps student to appreciate the similarities and differences among the techniques under the same SEM framework. Graduate students may be more prepared to conduct both primary research and meta-analysis after their graduation.

1.4 Outline of the following chapters

Chapter 2 gives a brief overview on the key topics in SEM. These topics were selected in a way that they are relevant to the SEM-based meta-analysis. FIML estimation, definition variables, and phantom variables play a crucial role in the SEM-based meta-analysis. Chapter 3 provides a summary on how to calculate the effect sizes and their sampling variances and covariances for univariate and multivariate meta-analyses. We also introduce a general approach to derive the approximate sampling variances and covariances for *any* types of effect sizes using a delta method and SEM. Chapter 4 introduces univariate meta-analysis and how the meta-analytic models can be formulated as structural equation models. This chapter provides the foundation on understanding the SEM-based meta-analysis.

In Chapter 5, we extend the univariate meta-analysis to multivariate meta-analysis. We discuss the advantages of multivariate meta-analysis to the univariate meta-analysis. At the end of this chapter, we apply the multivariate meta-analysis to test mediation and moderation models on the effect sizes. Chapter 6 discusses issues of dependent effect sizes and several common strategies to

Table 1.1 Datasets used in this book.

Dataset	Type of meta-analysis	Key references	Topic
Mak09	Univariate	Cheung et al. (2012) and Mak et al. (2009)	(Log) odds ratio of atrial fibrillation between bisphosphonate and non-bisphosphonate users
Jaramillo05	Univariate	Jaramillo et al. (2005)	Correlation between organizational commitment and salesperson job performance
BCG	Multivariate	Colditz et al. (1994) and van Houwelingen et al. (2002)	(Log) odds ratio of BCG vaccine for preventing tuberculosis
wvs94a	Multivariate, mediation, and moderation	Cheung (2013) and World Values Study Group (1994)	Standardized mean differences between males and females on life satisfaction and life control
Bornmann07	Three-level	Bornmann et al. (2007), Cheung (2014b), and Marsh et al. (2009)	(Log) odds ratio of gender differences in peer reviews of grant proposals
Digman97	MASEM	Digman (1997), Cheung (2014a), and Cheung and Chan (2005a)	A higher-order confirmatory factor analytic model for the Big Five model
Becker94	MASEM	Becker and Schram (1994) and Cheung (2014a)	A regression model on SAT (Math) by using SAT (Verbal) and spatial ability as predictors
Hunter83	MASEM	Hunter (1983)	A path model for cognitive ability to supervisor ratings

handle the dependence when the degree of dependence is unknown. A three-level meta-analysis is proposed to handle the effect sizes nested within clusters. The relationship between a multivariate and a three-level meta-analyses is also discussed. Chapter 7 focuses on the MASEM. Several common methods for conducting MASEM are reviewed. The fixed- and random-effects two-stage structural equation modeling (TSSEM) approach are proposed and discussed in details. Issues related to the MASEM are discussed.

Chapter 8 addresses two advanced topics in the SEM-based meta-analysis. The first topic is the pros and cons of the restricted (or residual) maximum likelihood (REML) estimation and how it can be implemented in the SEM framework. The second topic is how to handle missing values in the moderators in a mixed-effects meta-analysis. Several common strategies for handling missing data are reviewed. Advantages and implementation of FIML to handle missing data are discussed. Chapter 9 gives an overview on how to implement the SEM-based meta-analysis in M*plus*, a popular SEM software. Most of the SEM-based meta-analysis except the TSSEM approach can be conducted in M*plus* by using a transformed variables approach. Appendix A gives a very brief introduction to the R statistical environment, the `OpenMx`, and the `metaSEM` packages.

1.4.1 Computer examples and data sets used in this book

Computer examples were provided to illustrate the techniques introduced in this book. The R statistical environment was mainly used as the platform of data analysis except Chapter 9 that used M*plus* as the statistical program. Several real data sets were used in the illustrations. All data sets are available in the `metaSEM` package. Table 1.1 summarizes these data sets. More details of the data sets will be given in the later chapters.

1.5 Concluding remarks and further readings

This book mainly covers the statistical models in the meta-analysis from an SEM approach. The SEM-based meta-analysis provides an alternative framework to conduct meta-analysis. It is useful to mention topics that will *not* be covered in this book. Conceptual issues, such as conceptualization, literature review, and coding study characteristics for moderator analysis in a meta-analysis, will not be covered. Readers may refer to, for example, Card (2012) and Cooper (2010) for details. Moreover, topics such as publication bias (Rothstein et al., 2005), graphical methods to display data (Anzures-Cabrera and Higgins, 2010), individual participant data (Whitehead, 2002), network meta-analysis (see Salanti and Schmid, 2012, for a special issue), correction for statistical artifacts (Schmidt and Hunter, 2015), and Bayesian meta-analysis (Whitehead, 2002) will not be covered in this book. These techniques have not been well explored in the SEM-based meta-analysis yet. Future research may investigate how these topics can be integrated into the SEM framework. Some matrix calculations are used in this book. Readers who are less familiar with them may refer to Fox (2009) or the online appendix of his book (Fox, 2008).

References

Aiken LS, West SG and Millsap RE 2008. Doctoral training in statistics, measurement, and methodology in psychology: replication and extension of Aiken, West, Sechrest, and Reno's (1990) survey of PhD programs in North America. *American Psychologist* **63**(1), 32–50.

Anzures-Cabrera J and Higgins JPT 2010. Graphical displays for meta-analysis: an overview with suggestions for practice. *Research Synthesis Methods* **1**(1), 66–80.

Bauer DJ 2003. Estimating multilevel linear models as structural equation models. *Journal of Educational and Behavioral Statistics* **28**(2), 135–167.

Becker BJ and Schram CM 1994. Examining explanatory models through research synthesis In *The handbook of research synthesis* (ed. Cooper H and Hedges LV). Russell Sage Foundation, New York, pp. 357–381.

Bentler PM 1986. Structural modeling and Psychometrika: an historical perspective on growth and achievements. *Psychometrika* **51**(1), 35–51.

Bentler PM 1990. Comparative fit indexes in structural models. *Psychological Bulletin* **107**(2), 238–246.

Bentler PM 2006. *EQS 6 structural equations program manual*. Multivariate Software, Encino, CA.

Bentler PM and Weeks D 1980. Linear structural equations with latent variables. *Psychometrika* **45**(3), 289–308.

Bollen K 2002. Latent variables in psychology and the social sciences. *Annual Review of Psychology* **53**, 605–634.

Bollen KA, Bauer DJ, Christ SL and Edwards MC 2010. Overview of structural equation models and recent extensions In *Statistics in the social sciences: current methodological developments* (ed. Kolenikov S, Steinley D and Thombs L). John Wiley & Sons, Inc., Hoboken, NJ, pp. 37–79.

Borenstein M, Hedges LV, Higgins JP and Rothstein HR 2010. A basic introduction to fixed-effect and random-effects models for meta-analysis. *Research Synthesis Methods* **1**(2), 97–111.

Bornmann L, Mutz R and Daniel HD 2007. Gender differences in grant peer review: a meta-analysis. *Journal of Informetrics* **1**(3), 226–238.

Card NA 2012. *Applied meta-analysis for social science research*. The Guilford Press, New York.

Chalmers I, Hedges LV and Cooper H 2002. A brief history of research synthesis. *Evaluation & the Health Professions* **25**(1), 12–37.

Cheung MWL 2013. Multivariate meta-analysis as structural equation models. *Structural Equation Modeling: A Multidisciplinary Journal* **20**(3), 429–454.

Cheung MWL 2014a. Fixed- and random-effects meta-analytic structural equation modeling: examples and analyses in R. *Behavior Research Methods* **46**(1), 29–40.

Cheung MWL 2014b. Modeling dependent effect sizes with three-level meta-analyses: a structural equation modeling approach. *Psychological Methods* **19**(2), 211–229.

Cheung MWL and Chan W 2005a. Classifying correlation matrices into relatively homogeneous subgroups: a cluster analytic approach. *Educational and Psychological Measurement* **65**(6), 954–979.

Cheung MWL and Chan W 2005b. Meta-analytic structural equation modeling: a two-stage approach. *Psychological Methods* **10**(1), 40–64.

Cheung MWL, Ho RCM, Lim Y and Mak A 2012. Conducting a meta-analysis: basics and good practices. *International Journal of Rheumatic Diseases* **15**(2), 129–135.

Cohen J 1968. Multiple regression as a general data-analytic system. *Psychological Bulletin* **70**(6), 426–443.

Cohen J and Cohen P 1975. *Applied multiple regression/correlation analysis for the behavioral sciences*. Erlbaum, Hillsdale, NJ.

Colditz GA, Brewer TF, Berkey CS, Wilson ME, Burdick E, Fineberg HV and Mosteller F 1994. Efficacy of BCG vaccine in the prevention of tuberculosis. Meta-analysis of the published literature. *JAMA: The Journal of the American Medical Association* **271**(9), 698–702.

Cooper HM 2010. *Research synthesis and meta-analysis: a step-by-step approach*, 4th edn. Sage Publications, Inc., Los Angeles, CA.

Cooper H and Hedges LV 2009. Research synthesis as a scientific process In *The handbook of research synthesis and meta-analysis* (ed. Cooper H, Hedges LV and Valentine JC), 2nd edn. Russell Sage Foundation, New York, pp. 3–16.

Curran PJ 2003. Have multilevel models been structural equation models all along? *Multivariate Behavioral Research* **38**(4), 529–569.

Curran PJ and Bauer DJ 2007. Building path diagrams for multilevel models. *Psychological Methods* **12**(3), 283–297.

Demidenko E 2013. *Mixed models: theory and applications with R*, 2nd edn. Wiley-Interscience, Hoboken, NJ.

Digman JM 1997. Higher-order factors of the Big Five. *Journal of Personality and Social Psychology* **73**(6), 1246–1256.

Fox J 2008. *Applied regression analysis and generalized linear models*, 2nd edn. Sage Publications, Inc., Thousand Oaks, CA

Fox J 2009. *A mathematical primer for social statistics*. Sage Publications, Inc., Los Angeles, CA.

Glass GV 1976. Primary, secondary, and meta-analysis of research. *Educational Researcher* **5**(10), 3–8.

Goldstein H 2011. *Multilevel statistical models*, 4th edn. John Wiley & Sons, Inc., Hoboken, NJ.

Hedges LV and Olkin I 1985. *Statistical methods for meta-analysis.* Academic Press, Orlando, FL.

Hox JJ 2010. *Multilevel analysis: techniques and applications*, 2nd edn. Routledge, New York.

Hunt M 1997. *How science takes stock: the story of meta-analysis*. Russell Sage Foundation, New York.

Hunter JE 1983. A causal analysis of cognitive ability, job knowledge, job performance, and supervisor ratings In *Performance measurement and theory* (ed. Landy F, Zedeck S and Cleveland J). Erlbaum Hillsdale, NJ, pp. 257–266.

Jaramillo F, Mulki JP and Marshall GW 2005. A meta-analysis of the relationship between organizational commitment and salesperson job performance: 25 years of research. *Journal of Business Research* **58**(6), 705–714.

Jöreskog KG 1969. A general approach to confirmatory maximum likelihood factor analysis. *Psychometrika* **34**(2), 183–202.

Joreskog KG 1970. A general method for analysis of covariance structures. *Biometrika* **57**(2), 239–251.

Jöreskog KG 1978. Structural analysis of covariance and correlation matrices. *Psychometrika* **43**(4), 443–477.

Jöreskog KG and Sörbom D 1996. *LISREL 8: a user's reference guide*. Scientific Software International, Inc., Chicago, IL.

Kaplan D 2009. *Structural equation modeling: foundations and extensions*, 2nd edn. SAGE Publications, Inc., Thousand Oaks, CA.

Lipsey MW and Wilson D 2000. *Practical meta-analysis*. Sage Publications, Inc., Thousand Oaks, CA.

MacCallum RC and Austin JT 2000. Applications of structural equation modeling in psychological research. *Annual Review of Psychology* **51**(1), 201–226.

Mak A, Cheung MWL, Ho RCM, Cheak AAC and Lau CS 2009. Bisphosphonates and atrial fibrillation: Bayesian meta-analyses of randomized controlled trials and observational studies. *BMC Musculoskeletal Disorders* **10**(1), 1–12.

Marsh HW, Bornmann L, Mutz R, Daniel HD and O'Mara A 2009. Gender effects in the peer reviews of grant proposals: a comprehensive meta-analysis comparing traditional and multilevel approaches. *Review of Educational Research* **79**(3), 1290–1326.

Matsueda RL 2012. Key advances in the history of structural equation modeling In *Handbook of structural equation modeling* (ed. Hoyle RH). Guilford Press New York, pp. 3–42.

Mehta PD and Neale MC 2005. People are variables too: multilevel structural equations modeling. *Psychological Methods* **10**(3), 259–284.

Mehta P and West S 2000. Putting the individual back into individual growth curves. *Psychological Methods* **5**(1), 23–43.

Muthén BO 1994. Multilevel covariance structure analysis. *Sociological Methods & Research* **22**(3), 376–398.

Muthén BO and Asparouhov T 2013. Item response modeling in Mplus: a multi-dimensional, multi-level, and multi-timepoint example In *Handbook of item response theory: models, statistical tools, and applications* (ed. Van der Linden WJ and Hambleton RK). Chapman & Hall/CRC Press, Boca Raton, FL, forthcoming.

Muthén BO and Muthén LK 2012. *Mplus user's guide*, 7th edn. Muthén & Muthén, Los Angeles, CA.

National Research Council 1992. *Combining information: statistical issues and opportunities for research*. National Academies Press, Washington, DC.

American Psychological Association 2007a. Karl G. Jöreskog: award for distinguished scientific applications of psychology. *American Psychologist* **62**(8), 768–769.

No authorship indicated 2007b. Peter M. Bentler: award for distinguished scientific applications of psychology. *American Psychologist* **62**(8), 769–782.

O'Rourke K 2007. An historical perspective on meta-analysis: dealing quantitatively with varying study results. *Journal of the Royal Society of Medicine* **100**(12), 579–582.

Pearson K 1904. Report on certain enteric fever inoculation statistics. *British Medical Journal* **2**(2288), 1243–1246.

Preacher K, Zyphur M and Zhang Z 2010. A general multilevel SEM framework for assessing multilevel mediation. *Psychological Methods* **15**(3), 209–233.

Raudenbush SW and Bryk AS 2002. *Hierarchical linear models: applications and data analysis methods*. Sage Publications, Inc., Thousand Oaks, CA.

Rothstein HR, Sutton AJ and Borenstein M 2005. *Publication bias in meta-analysis: prevention, assessment and adjustments*. John Wiley & Sons, Ltd, Chichester.

Rovine MJ and Molenaar PCM 2000. A structural modeling approach to a multilevel random coefficients model. *Multivariate Behavioral Research* **35**(1), 51–88.

Salanti G and Schmid CH 2012. Research synthesis methods special issue on network meta-analysis: introduction from the editors. *Research Synthesis Methods* **3**(2), 69–70.

Schmidt FL and Hunter JE 1977. Development of a general solution to the problem of validity generalization. *Journal of Applied Psychology* **62**(5), 529–540.

Schmidt FL and Hunter JE 2015. *Methods of meta-analysis: correcting error and bias in research findings*, 3rd edn. Sage Publications, Inc., Thousand Oaks, CA.

Skrondal A and Rabe-Hesketh S 2004. *Generalized latent variable modeling: multilevel, longitudinal, and structural equation models*. Chapman & Hall/CRC, Boca Raton, FL.

Song XY and Lee SY 2012. *Basic and advanced Bayesian structural equation modeling: with applications in the medical and behavioral sciences*. John Wiley & Sons, Inc., Hoboken, NJ.

Takane Y and Deleeuw J 1987. On the relationship between item response theory and factor-analysis of discretized variables. *Psychometrika* **52**(3), 393–408.

van Houwelingen HC, Arends LR and Stijnen T 2002. Advanced methods in meta-analysis: multivariate approach and meta-regression. *Statistics in Medicine* **21**(4), 589–624.

Viswesvaran C and Ones DS 1995. Theory testing: combining psychometric meta-analysis and structural equations modeling. *Personnel Psychology* **48**(4), 865–885.

Whitehead A 2002. *Meta-analysis of controlled clinical trials*. John Wiley & Sons, Ltd, Chichester.

World Values Study Group 1994. *World Values Survey, 19811984 and 19901993 [Computer file]*. Inter-University Consortium for Political and Social Research, Ann Arbor, MI.

Wu W, West SG and Taylor AB 2009. Evaluating model fit for growth curve models: integration of fit indices from SEM and MLM frameworks. *Psychological Methods* **14**(3), 183–201.

2

Brief review of structural equation modeling

This chapter reviews selected topics in structural equation modeling (SEM) that are relevant to the SEM-based meta-analysis. It provides a quick introduction to SEM for those who are less familiar with the techniques. This chapter begins by introducing three different model specifications—path diagrams, equations, and matrix specification. It then introduces popular structural equation models such as path analysis, confirmatory factor analytic (CFA) models, SEMs, latent growth models, and multiple-group analysis. How to obtain parameter estimates, standard errors (SEs), confidence intervals (CIs), test statistics, and various goodness-of-fit indices are introduced. Finally, we introduce phantom variables, definition variables, and full information maximum likelihood (FIML). These concepts are the keys to formulating meta-analytic models as structural equation models.

2.1 Introduction

SEM, also known as covariance structure analysis and correlation structure analysis, is a generic term for many related statistical techniques. Many popular multivariate techniques, such as correlation analysis, regression analysis, analysis of variance (ANOVA), multivariate analysis of variance (MANOVA), factor analysis, and item response theory, can be considered as special models of SEM. Generally speaking, SEM is a statistical technique to model the first and the second moments of the data when the data are multivariate normal. The first moment represent the mean structure, while the second moment represents the covariance matrix of the variables. If we are only interested in the covariance matrix among the variables, we may skip the mean structure.

Meta-Analysis: A Structural Equation Modeling Approach, First Edition. Mike W. -L. Cheung.

Companion Website: www.wiley.com/go/cheung/meta_analysis

SEM is widely used in psychology and the social sciences to test hypotheses involving observed and latent variables (e.g., Bentler, 1986; Bollen, 2002; MacCallum and Austin, 2000). Latent variables are hypothetical constructs that cannot be observed directly. They have to be represented by the observed variables known as indicators. By using the indicators to measure the latent variables, the amount of measurement errors can be quantified and taken into account when estimating the relationship among the latent variables.

There are several steps involved in fitting a structural equation model (see, e.g., Kline, 2011). A proposed model is specified based on the hypothesized relationship among the observed and latent variables. The proposed model is fitted against the data. When the solution for the optimization is convergent, parameter estimates, their SEs, and test statistics are available for inspection. Users may determine whether the proposed model fits the data well. If the proposed model does not fit the data, we may modify the model to see if the model fit can be improved. Interpretations on the overall model and the individual parameter estimates can be made.

2.2 Model specification

There are three equivalent approaches to specify a structural equation model. They are path diagrams, equations, and matrix specification (e.g., Mulaik, 2009). Let us illustrate these approaches by using a model on simple regression.

2.2.1 Equations

The first approach is to specify the models by equations. The model for the simple regression is

$$y = \beta_0 + \beta_1 x + e_y, \tag{2.1}$$

where x, y, e_y, β_0, and β_1 are the independent variable, dependent variable, the residual, the intercept, and the regression coefficient, respectively. As equations only allow us to specify the effects from one variable to another, we need to specify further constraints on the models. For example, we may need to indicate that x and e_y are uncorrelated, that is, $\mathsf{Cov}(x, e_y) = 0$. On the basis of the above model, we derive the expected means and the expected covariance matrix for the variables:

$$\begin{aligned} \mathsf{E}\left(\begin{bmatrix} y \\ x \end{bmatrix}\right) &= \begin{bmatrix} \beta_0 + \beta_1 \mu_x \\ \mu_x \end{bmatrix} \quad \text{and} \\ \mathsf{Cov}\left(\begin{bmatrix} y \\ x \end{bmatrix}\right) &= \begin{bmatrix} \beta_1^2 \sigma_x^2 + \sigma_{e_y}^2 & \\ \beta_1 \sigma_x^2 & \sigma_x^2 \end{bmatrix}, \end{aligned} \tag{2.2}$$

where μ_x, σ_x^2, and $\sigma_{e_y}^2$ are the population mean of x, the population variance of x, and the population variance of e_y, respectively. The expected means and the expected covariance matrix are used to compare against the observed means and covariance matrix in order to obtain parameter estimates.

2.2.2 Path diagram

One of the reasons for the popularity of SEM is its ability to specify models graphically. Path diagrams can be used to represent the mathematical models. Besides the conventional models, such as path analysis, CFA, and SEM, path diagrams have been extended to represent multilevel models (Curran and Bauer, 2007; Muthén and Muthén, 2012; Skrondal and Rabe-Hesketh, 2004) and meta-analysis (Cheung, 2008, 2013, 2014). Graphical models are convenient devices to represent the mathematical models.

There are slight variations on how graphical models are presented in SEM (Arbuckle, 2006; Bentler, 2006; Jöreskog and Sörbom, 1996; Muthén and Muthén, 2012; Neale et al., 2006). For example, some authors prefer to explicitly draw the means and the latent variables of the measurement errors while others do not. We follow the conventions in the `OpenMx` package (see Boker and McArdle, 2005) in this book.

Rectangles (or squares) and ellipses (or circles) are used to represent the observed and latent variables, respectively. Triangles represent a vector of constant one that is used to represent the intercepts. Single and double arrows represent prediction and covariance among the variables. Strictly speaking, a double arrow (variance) on the triangle is required to fulfill the tracing rules to calculate the model-implied means and covariance matrix (Boker and McArdle, 2005). Conventionally, this double arrow is not shown to simplify the figures.

Figure 2.1 shows two graphical model representations of the simple regression. The model in Figure 2.1a explicitly includes the error e_y and its variance $\sigma^2_{e_y}$. The main advantage of this representation is that it includes both latent and observed variables in the figure. Readers may easily map the figure to the equations and the matrix representation. The main disadvantage is that the latent variables of the residuals are required in the figure. Suppose that we are fitting a CFA model with 20 observed variables and 4 latent variables; we have to include 20 latent variables for the residuals. This may make the figure unnecessarily crowded.

Figure 2.1b shows an alternative representation for the same model. The main difference of this representation is that the latent variables for the residuals are not shown in the figure. The double arrows are drawn directly on the observed and the latent variables. When the double arrows are drawn on the independent variables, they represent the variances; when the double arrows are drawn on the dependent variables, they represent the error or residual variances. Even though the figure has been simplified, it essentially carries the same information as the figure with the explicit errors. The only drawback is that the e_y is not explicitly shown in the figure.

2.2.3 Matrix representation

Regardless of whether we specify the models in equations or path diagrams, most SEM packages convert these models to matrices for analysis. There are several matrix representations. The most traditional approach is the `LISREL` model (Jöreskog and Sörbom, 1996). The other popular model representations are the

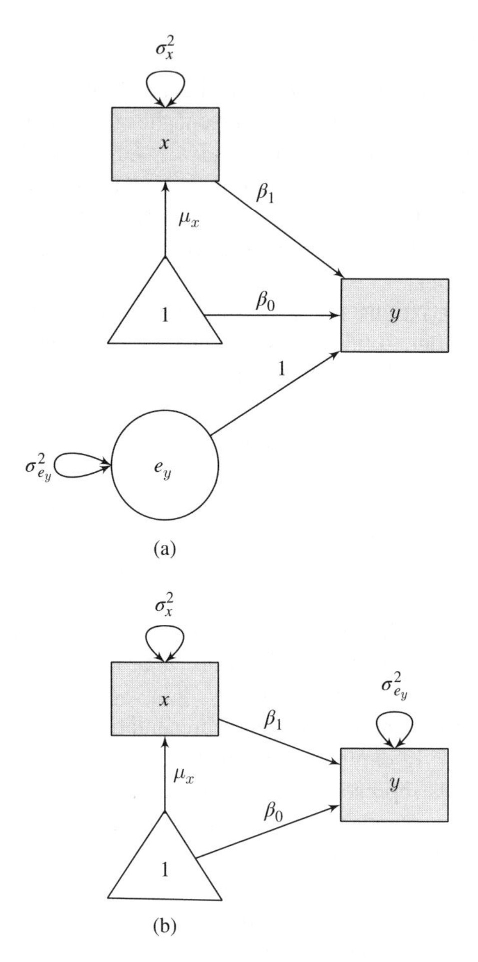

Figure 2.1 Two graphical model representations of a simple regression.

models used in EQS (Bentler, 2006), M*plus* (Muthén and Muthén, 2012), and the reticular action model (RAM) (McArdle, 2005; McArdle and McDonald, 1984). Although the specifications look different, models specified in one formulation can be translated to the other formulations. In this book, we mainly use the RAM formulation. Some common structural equation models are introduced in this chapter.

Suppose that there are p_o observed and p_l latent variables in the model and $p = p_o + p_l$ is the total number of variables; the RAM formulation involves four matrices: $\underset{p_o \times p}{\boldsymbol{F}}$, $\underset{p \times 1}{\boldsymbol{M}}$, $\underset{p \times p}{\boldsymbol{A}}$, and $\underset{p \times p}{\boldsymbol{S}}$. We may include the dimensions for the matrices for the ease of reference. Let $\underset{p \times 1}{\boldsymbol{v}}$ be a vector that includes all variables in the model. The $\boldsymbol{A}$ matrix links the variables by

$$\underset{p \times 1}{\boldsymbol{v}} = \underset{p \times p}{\boldsymbol{A}} \underset{p \times 1}{\boldsymbol{v}}, \tag{2.3}$$

where $\boldsymbol{A}$ denotes the asymmetric paths, such as the regression coefficients and the factor loadings, with a_{ij} in $\boldsymbol{A}$ representing the regression coefficient from v_j to v_i. The main purpose of $\boldsymbol{A}$ is to specify the single arrows in path diagrams.

$\boldsymbol{S}$ is a symmetric matrix representing the variances and covariances of $\boldsymbol{v}$. It is used to specify the double arrows in path diagrams. The diagonal elements represent the variances of the variables. If the elements in $\boldsymbol{v}$ are independent variables, the corresponding diagonals in $\boldsymbol{S}$ denote the variances; otherwise, the corresponding diagonals in $\boldsymbol{S}$ represent the residuals of the dependent variables. The off-diagonals in $\boldsymbol{S}$ represent the covariances of the variables. $\boldsymbol{M}$ represents the means or intercepts of the variables. $\boldsymbol{F}$ is a selection matrix consisting 1 and 0. It is used to select the observed variables.

Regarding the simple regression example, we stack the variables into a column vector $\boldsymbol{v} = \begin{bmatrix} y & x & e_y \end{bmatrix}^{\mathrm{T}}$, where $\boldsymbol{v}^{\mathrm{T}}$ is the transpose of $\boldsymbol{v} = \begin{bmatrix} y \\ x \\ e_y \end{bmatrix}$. Equation 2.4 shows the RAM formulation explicitly including e_y in the model that is equivalent to the model in Figure 2.1a.

$$
\begin{aligned}
\boldsymbol{A} &= \begin{matrix} y \\ x \\ e_y \end{matrix}\overset{\begin{matrix} y & x & e_y \end{matrix}}{\begin{bmatrix} 0 & \beta_1 & 1 \\ 0 & 0 & 0 \\ 0 & 0 & 0 \end{bmatrix}}, \quad \boldsymbol{S} = \begin{matrix} y \\ x \\ e_y \end{matrix}\overset{\begin{matrix} y & x & e_y \end{matrix}}{\begin{bmatrix} 0 & 0 & 0 \\ 0 & \sigma_x^2 & 0 \\ 0 & 0 & \sigma_{e_y}^2 \end{bmatrix}}, \\
\boldsymbol{F} &= \begin{matrix} y \\ x \\ e_y \end{matrix}\overset{\begin{matrix} y & x & e_y \end{matrix}}{\begin{bmatrix} 1 & 0 & 0 \\ 0 & 1 & 0 \\ 0 & 0 & 0 \end{bmatrix}}, \quad \text{and } \boldsymbol{M} = \begin{matrix} y \\ x \\ e_y \end{matrix}\begin{bmatrix} \beta_0 \\ \mu_x \\ 0 \end{bmatrix}.
\end{aligned} \tag{2.4}
$$

Equation 2.5 shows the RAM formulation with $\boldsymbol{v} = \begin{bmatrix} y & x \end{bmatrix}^{\mathrm{T}}$. The latent variable for the residual e_y is not shown in the model. It is equivalent to the model in Figure 2.1b. The dimensions of the matrices are smaller than those in Equation 2.4.

$$
\begin{aligned}
\boldsymbol{A} &= \begin{matrix} x \\ y \end{matrix}\overset{\begin{matrix} y & x \end{matrix}}{\begin{bmatrix} 0 & \beta_1 \\ 0 & 0 \end{bmatrix}}, \quad \boldsymbol{S} = \begin{matrix} y \\ x \end{matrix}\overset{\begin{matrix} y & x \end{matrix}}{\begin{bmatrix} \sigma_{e_y}^2 & 0 \\ 0 & \sigma_x^2 \end{bmatrix}}, \\
\boldsymbol{F} &= \begin{matrix} x \\ y \end{matrix}\overset{\begin{matrix} y & x \end{matrix}}{\begin{bmatrix} 1 & 0 \\ 0 & 1 \end{bmatrix}}, \quad \text{and } \boldsymbol{M} = \begin{matrix} y \\ x \end{matrix}\begin{bmatrix} \beta_0 \\ \mu_x \end{bmatrix},
\end{aligned} \tag{2.5}
$$

It can be shown that the model-implied means $\boldsymbol{\mu}(\boldsymbol{\theta})$ and covariance matrix $\boldsymbol{\Sigma}(\boldsymbol{\theta})$ are

$$
\begin{aligned}
\boldsymbol{\mu}(\boldsymbol{\theta}) &= \boldsymbol{F}(\boldsymbol{I} - \boldsymbol{A})^{-1}\boldsymbol{M} \quad \text{and} \\
\boldsymbol{\Sigma}(\boldsymbol{\theta}) &= \boldsymbol{F}(\boldsymbol{I} - \boldsymbol{A})^{-1}\boldsymbol{S}((\boldsymbol{I} - \boldsymbol{A})^{-1})^{\mathrm{T}}\boldsymbol{F}^{\mathrm{T}},
\end{aligned} \tag{2.6}
$$

where $\boldsymbol{X}^{-1}$ is the inverse of $\boldsymbol{X}$ with $\boldsymbol{X}\boldsymbol{X}^{-1} = \boldsymbol{X}^{-1}\boldsymbol{X} = \boldsymbol{I}$ and $\boldsymbol{I}$ is an identity matrix (McArdle, 2005; McArdle and McDonald, 1984). Applying an inverse to a matrix is similar to applying the division operator to a scalar. $\boldsymbol{\Sigma}(\boldsymbol{\theta})$ (or $\boldsymbol{\mu}(\boldsymbol{\theta})$) means that the population covariance matrix $\boldsymbol{\Sigma}$ (or the population mean vector $\boldsymbol{\mu}$) is a function of the unknown parameters $\boldsymbol{\theta}$. When the models are specified using the RAM formulation, the model-implied means and covariance matrix of an arbitrary model can be derived automatically. Given the model-implied means and covariance matrix, we may obtain the parameter estimates and the test statistics by comparing the sample moments to the model-implied moments (see Section 2.4.1).

2.3 Common structural equation models

Many multivariate statistics, such as ANOVA, MANOVA, multiple regression, confirmatory factor analysis, item response theory, and multilevel model, can be considered as special cases of SEM. We briefly review some of these models in this section.

2.3.1 Path analysis

Path analysis was developed by Wright (1921) to specify relationships among observed variables. Wright also developed the tracing rules to calculate the model-implied correlation elements based on the proposed structural model (e.g., Mulaik, 2009). This provides the foundation of SEM. One of the most popular applications of path analysis is mediation analysis (MacKinnon, 2008; Preacher and Hayes, 2004). The equations for the regression model are shown in Equation 2.7. As the means are not involved in estimating the indirect effect, we present the models without the means. That is, all variables are based on the centered scores:

$$\begin{aligned} y &= \beta m + \gamma x + e_y \quad \text{and} \\ m &= \alpha x + e_m. \end{aligned} \tag{2.7}$$

Figure 2.2 shows the mediation model, while Equation 2.8 lists the model in RAM formulation.

$$\begin{aligned} &\boldsymbol{A} = \begin{matrix} \\ y \\ m \\ x \end{matrix} \begin{matrix} y & m & x \\ \left[\begin{matrix} 0 \\ 0 \\ 0 \end{matrix}\right. & \begin{matrix} \beta \\ 0 \\ 0 \end{matrix} & \left.\begin{matrix} \gamma \\ \alpha \\ 0 \end{matrix}\right] \end{matrix}, \quad \boldsymbol{S} = \begin{matrix} \\ y \\ m \\ x \end{matrix} \begin{matrix} y & m & x \\ \left[\begin{matrix} \sigma^2_{e_y} \\ 0 \\ 0 \end{matrix}\right. & \begin{matrix} 0 \\ \sigma^2_{e_m} \\ 0 \end{matrix} & \left.\begin{matrix} 0 \\ 0 \\ \sigma^2_x \end{matrix}\right] \end{matrix}, \quad \text{and} \\ &\boldsymbol{F} = \begin{matrix} \\ y \\ m \\ x \end{matrix} \begin{matrix} y & m & x \\ \left[\begin{matrix} 1 \\ 0 \\ 0 \end{matrix}\right. & \begin{matrix} 0 \\ 1 \\ 0 \end{matrix} & \left.\begin{matrix} 0 \\ 0 \\ 1 \end{matrix}\right] \end{matrix}. \end{aligned} \tag{2.8}$$

The product term $\alpha\beta$ represents the indirect effect via the mediator while γ is the direct effect after controlling for the mediator. The total effect between x and y

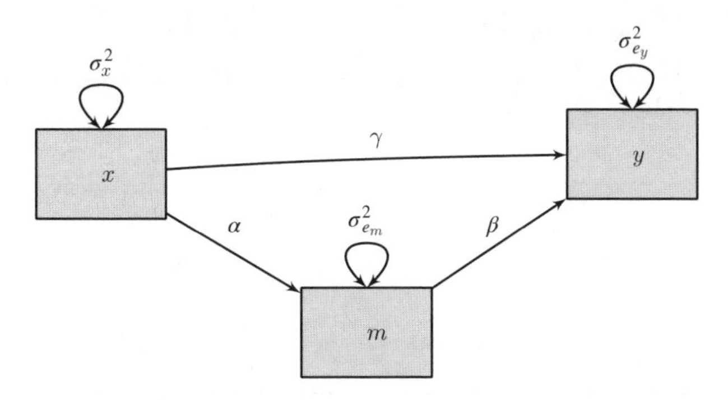

Figure 2.2 A path model with one mediator.

is $\alpha\beta + \gamma$. Methods on estimating SE or CI on the indirect effect can be found in MacKinnon (2008). As the indirect effect is based on the product of two random variables, its sampling distribution is not normally distributed unless the sample sizes are huge. Bootstrap CI and likelihood-based confidence interval (LBCI) are preferred to be used to construct the CIs to test the unstandardized and standardized indirect effects (e.g., Cheung, 2007a, 2009a).

2.3.2 Confirmatory factor analysis

A CFA model specifies how the observed variables are related to the latent variables (Brown and Prescott, 2006). It is usually used to study the psychometric properties of the measurements. Many research questions are related to the numbers of the latent variables and how the latent factors are related to the items. If the items are measuring similar constructs, their factor loadings on the same latent variable should all be reasonably high. Figure 2.3 shows a model with four items. Conventionally, three matrices are used to specify a CFA model in `LISREL`. The model-implied covariance matrix is

$$\boldsymbol{\Sigma}(\boldsymbol{\theta}) = \boldsymbol{\Lambda}\boldsymbol{\Phi}\boldsymbol{\Lambda}^{\mathrm{T}} + \boldsymbol{\Psi}, \tag{2.9}$$

where $\boldsymbol{\theta}$ includes the parameters from the $\boldsymbol{\Lambda}$, $\boldsymbol{\Phi}$, and $\boldsymbol{\Psi}$ matrices. $\boldsymbol{\Lambda}$ specifies the factor loadings that indicate how the latent variables are related to the items. $\boldsymbol{\Phi}$ specifies the factor covariance matrix among the latent factors. $\boldsymbol{\Psi}$ is the covariance matrix among the measurement errors. Because of the identification issue, either one factor variance or one factor loading per factor has to be fixed at a specific value, for example, 1. As the mean structure is rarely of interest unless in multiple-group analysis, it is usually skipped in the model specification.

To specify a CFA model in RAM formulation, we need to combine all the observed and latent variables together. Equation 2.10 shows the model for reference. It should be noted that the rows for f_1 and f_2 are missing in $\boldsymbol{F}$. As f_1 and f_2 are latent variables, there is no *real* data for them. They have to be filtered out

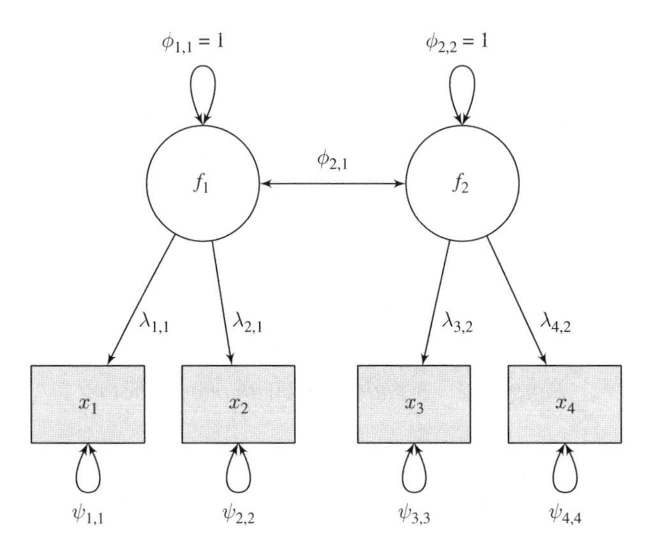

Figure 2.3 A confirmatory factor analytic model.

in the model-implied moments.

$$
\boldsymbol{A} = \begin{array}{c} \\ x_1 \\ x_2 \\ x_3 \\ x_4 \\ f_1 \\ f_2 \end{array}
\begin{array}{c}
\begin{array}{cccccc} x_1 & x_2 & x_3 & x_4 & f_1 & f_2 \end{array} \\
\begin{bmatrix}
0 & 0 & 0 & 0 & \lambda_{1,1} & 0 \\
0 & 0 & 0 & 0 & \lambda_{2,1} & 0 \\
0 & 0 & 0 & 0 & 0 & \lambda_{3,2} \\
0 & 0 & 0 & 0 & 0 & \lambda_{4,2} \\
0 & 0 & 0 & 0 & 0 & 0 \\
0 & 0 & 0 & 0 & 0 & 0
\end{bmatrix}
\end{array},
$$

$$
\boldsymbol{S} = \begin{array}{c} \\ x_1 \\ x_2 \\ x_3 \\ x_4 \\ f_1 \\ f_2 \end{array}
\begin{array}{c}
\begin{array}{cccccc} x_1 & x_2 & x_3 & x_4 & f_1 & f_2 \end{array} \\
\begin{bmatrix}
\psi_{1,1} & 0 & 0 & 0 & 0 & 0 \\
0 & \psi_{2,2} & 0 & 0 & 0 & 0 \\
0 & 0 & \psi_{3,3} & 0 & 0 & 0 \\
0 & 0 & 0 & \psi_{4,4} & 0 & 0 \\
0 & 0 & 0 & 0 & \phi_{1,1} = 1 & \phi_{2,1} \\
0 & 0 & 0 & 0 & \phi_{2,1} & \phi_{2,2} = 1
\end{bmatrix}
\end{array}, \quad \text{and} \tag{2.10}
$$

$$
\boldsymbol{F} = \begin{array}{c} \\ x_1 \\ x_2 \\ x_3 \\ x_4 \end{array}
\begin{array}{c}
\begin{array}{cccccc} x_1 & x_2 & x_3 & x_4 & f_1 & f_2 \end{array} \\
\begin{bmatrix}
1 & 0 & 0 & 0 & 0 & 0 \\
0 & 1 & 0 & 0 & 0 & 0 \\
0 & 0 & 1 & 0 & 0 & 0 \\
0 & 0 & 0 & 1 & 0 & 0
\end{bmatrix}
\end{array}.
$$

2.3.3 Structural equation model

When there are only associations among the latent variables, the models are called CFA or measurement models. If there are structural relationships imposed on the latent variables, they become structural equation models. Supposed that the association between the two latent variables in Figure 2.3 is changed to a direct path, it is a structural equation model. Figure 2.4 shows this model. As it is difficult to fix the variances of the latent dependent variables (cf. Steiger, 2002), the factor loadings of the latent dependent variables are usually fixed at 1 for identification purposes.

$$
\boldsymbol{A} = \begin{array}{c} \\ x_1 \\ x_2 \\ y_1 \\ y_2 \\ \xi_1 \\ \eta_1 \end{array}
\begin{array}{c}
\begin{array}{cccccc} x_1 & x_2 & y_1 & y_2 & \xi_1 & \eta_1 \end{array} \\
\begin{bmatrix}
0 & 0 & 0 & 0 & \lambda_{x1,1} = 1 & 0 \\
0 & 0 & 0 & 0 & \lambda_{x2,1} & 0 \\
0 & 0 & 0 & 0 & 0 & \lambda_{y1,1} = 1 \\
0 & 0 & 0 & 0 & 0 & \lambda_{y2,1} \\
0 & 0 & 0 & 0 & 0 & 0 \\
0 & 0 & 0 & 0 & \gamma_1 & 0
\end{bmatrix}
\end{array},
$$

$$
\boldsymbol{S} = \begin{array}{c} \\ x_1 \\ x_2 \\ y_1 \\ y_2 \\ \xi_1 \\ \eta_1 \end{array}
\begin{array}{c}
\begin{array}{cccccc} x_1 & x_2 & y_1 & y_2 & \xi_1 & \eta_1 \end{array} \\
\begin{bmatrix}
\theta_{\delta 1,1} & 0 & 0 & 0 & 0 & 0 \\
0 & \theta_{\delta 2,2} & 0 & 0 & 0 & 0 \\
0 & 0 & \theta_{\epsilon 1,1} & 0 & 0 & 0 \\
0 & 0 & 0 & \theta_{\epsilon 2,2} & 0 & 0 \\
0 & 0 & 0 & 0 & \phi_{1,1} & 0 \\
0 & 0 & 0 & 0 & 0 & \psi_{1,1}
\end{bmatrix}
\end{array}, \text{ and} \tag{2.11}
$$

$$
\boldsymbol{F} = \begin{array}{c} \\ x_1 \\ x_2 \\ y_1 \\ y_2 \end{array}
\begin{array}{c}
\begin{array}{cccccc} x_1 & x_2 & y_1 & y_2 & \xi_1 & \eta_1 \end{array} \\
\begin{bmatrix}
1 & 0 & 0 & 0 & 0 & 0 \\
0 & 1 & 0 & 0 & 0 & 0 \\
0 & 0 & 1 & 0 & 0 & 0 \\
0 & 0 & 0 & 1 & 0 & 0
\end{bmatrix}
\end{array}.
$$

One cautionary note is about the possibility of equivalent models (e.g., Raykov and Marcoulides, 2001, 2007). Equivalent models are models with the same model fit but with different substantive interpretations. They have the same model-implied mean vector and model-implied covariance matrix, chi-square statistic, degrees of freedom (dfs), and goodness-of-fit indices. Therefore, we cannot differentiate which one is better than the others. The main issue is that these models may have different substantive meanings and interpretations. For example, the CFA model in Figure 2.3 and the SEM in Figure 2.4 are equivalent models. We cannot tell which one is better from a statistical point of view. Researchers should specify models based on theories in order to support the directions of the structural models.

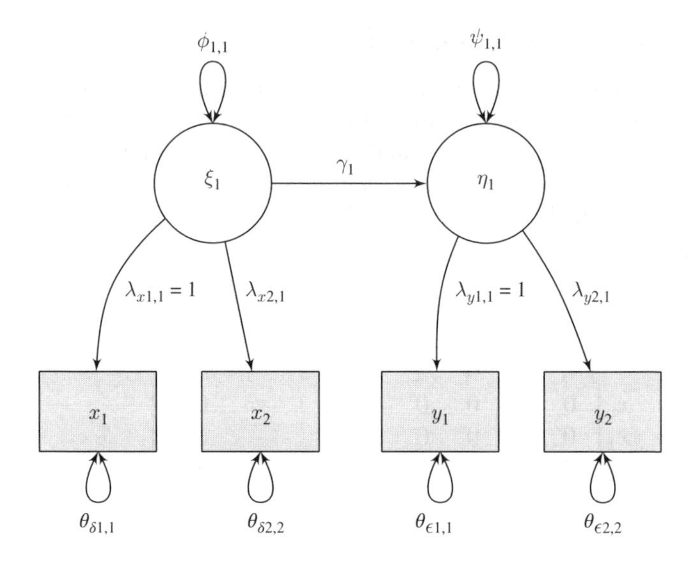

Figure 2.4 A structural equation model.

2.3.4 Latent growth model

Latent growth model is another popular applications of SEM (e.g., Bollen and Curran, 2006). It is used to model longitudinal data. Latent growth model can be formulated as a CFA with patterned factor loadings. Two latent factors are used to represent the intercept and the slope factors.

Figure 2.5 shows a latent growth model with a linear growth, while Equation 2.12 displays the correspondent RAM formulation. The mean structure is often skipped in path models, CFA, and even SEM because the mean structure does not carry useful information on the association among the variables. The mean structure $\boldsymbol{M}$ is crucial in latent growth model. The means of the intercept μ_I and the slope μ_S represent the average intercept and the average slope of the participants. They are known as the fixed effects in mixed-effects or multilevel models. The variances of the intercept $\mathsf{Var}(I)$ and the slope $\mathsf{Var}(S)$ represent the variation of the intercept and the slope of the participants (e.g., Cheung, 2007b). They are known as the random effects in mixed-effects models.

As shown in the factor loadings on the latent factor S, the time measured are equally spaced from 0, 1, 2, and 3. This approach may not be flexible enough to handle subjects measured at different time points. This limitation can be easily released by the use of definition variables. Arbitrary time can be *assigned* into the factor loadings S so that different subjects can be measured in different time points (Mehta and West, 2000). As each subject has his/her model with different parameter estimates, it is difficult to define a baseline model for all subjects. This may complicate the issue in calculating some goodness-of-fit indices, such as the CFI and Tucker–Lewis index (TLI), that depend on the baseline model (see Wu et al., 2009) (see Sections 2.5.2 and 2.5.3).

$$\mathbf{A} = \begin{array}{c} \\ y_1 \\ y_2 \\ y_3 \\ y_4 \\ I \\ S \end{array} \begin{array}{c} \begin{array}{cccccc} y_1 & y_2 & y_3 & y_4 & I & S \end{array} \\ \begin{bmatrix} 0 & 0 & 0 & 0 & 1 & 0 \\ 0 & 0 & 0 & 0 & 1 & 1 \\ 0 & 0 & 0 & 0 & 1 & 2 \\ 0 & 0 & 0 & 0 & 1 & 3 \\ 0 & 0 & 0 & 0 & 0 & 0 \\ 0 & 0 & 0 & 0 & 0 & 0 \end{bmatrix} \end{array},$$

$$\mathbf{S} = \begin{array}{c} \\ y_1 \\ y_2 \\ y_3 \\ y_4 \\ I \\ S \end{array} \begin{array}{c} \begin{array}{cccccc} y_1 & y_2 & y_3 & y_4 & I & S \end{array} \\ \begin{bmatrix} \mathrm{Var}(e_{y1}) & 0 & 0 & 0 & 0 & 0 \\ 0 & \mathrm{Var}(e_{y2}) & 0 & 0 & 0 & 0 \\ 0 & 0 & \mathrm{Var}(e_{y3}) & 0 & 0 & 0 \\ 0 & 0 & 0 & \mathrm{Var}(e_{y4}) & 0 & 0 \\ 0 & 0 & 0 & 0 & \mathrm{Var}(I) & \mathrm{Cov}(I,S) \\ 0 & 0 & 0 & 0 & \mathrm{Cov}(I,S) & \mathrm{Var}(S) \end{bmatrix} \end{array},$$

$$\mathbf{M} = \begin{array}{c} y_1 \\ y_2 \\ y_3 \\ y_4 \\ I \\ S \end{array} \begin{bmatrix} 0 \\ 0 \\ 0 \\ 0 \\ \mu_I \\ \mu_S \end{bmatrix}, \quad \text{and} \quad \mathbf{F} = \begin{array}{c} \\ y_1 \\ y_2 \\ y_3 \\ y_4 \end{array} \begin{array}{c} \begin{array}{cccccc} y_1 & y_2 & y_3 & y_4 & I & S \end{array} \\ \begin{bmatrix} 1 & 0 & 0 & 0 & 0 & 0 \\ 0 & 1 & 0 & 0 & 0 & 0 \\ 0 & 0 & 1 & 0 & 0 & 0 \\ 0 & 0 & 0 & 1 & 0 & 0 \end{bmatrix} \end{array}. \tag{2.12}$$

2.3.5 Multiple-group analysis

SEM can be easily extended to multiple-group (or multi-sample) analysis. Each group may have its own model and parameters. Equality and nonequality constraints on some of these parameters can be imposed. This approach enables researchers to test a variety of research hypotheses. For example, Figure 2.6 shows a multiple-group model on mediation. Suppose that Groups 1 and 2 represent males and females, respectively; we can test whether the indirect effect (and direct effect) are the same in males and females. This is known as a moderated mediation. The null hypothesis of equal indirect effect can be tested by imposing the nonlinear constraint $\alpha_{(1)}\beta_{(1)} = \alpha_{(2)}\beta_{(2)}$, where the subscripts in parentheses indicate the groups (e.g., Cheung, 2007a). Alternatively, we can estimate the CI on the difference $(\alpha_{(1)}\beta_{(1)} - \alpha_{(2)}\beta_{(2)})$. If the CI includes 0, it is not statistically significant with the predefined significance level.

Multiple-group analysis can be used to test the measurement invariance of the scale (e.g., Byrne and Watkins, 2003; Millsap, 2007; Vandenberg and Lance, 2000). The key issue is whether the scale is measuring the same construct

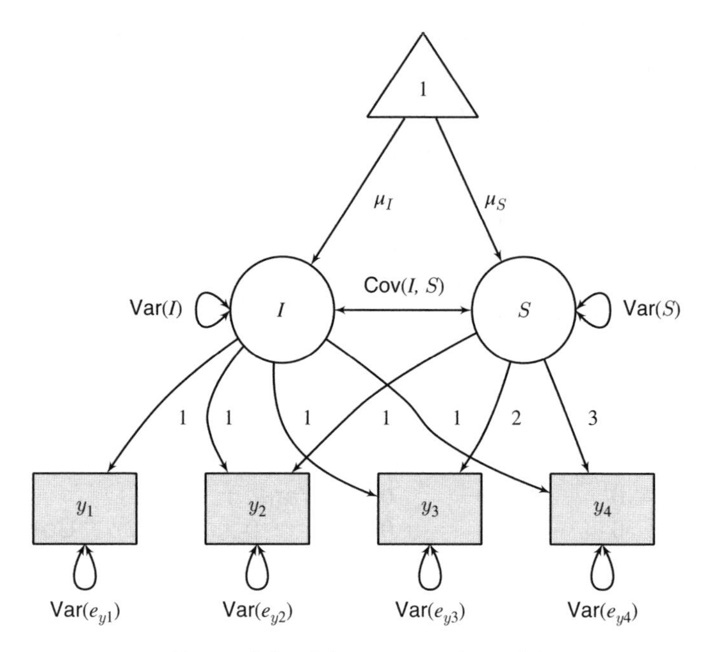

Figure 2.5 A latent growth model.

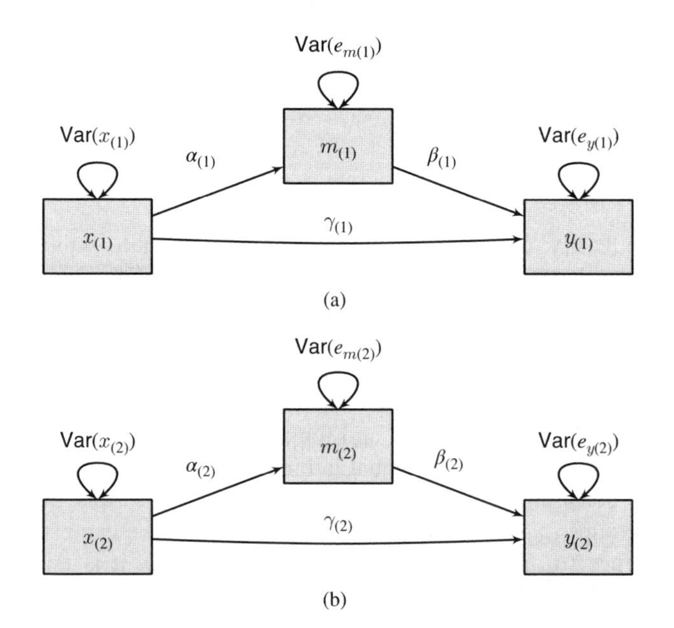

Figure 2.6 A moderated mediation model. (a) Group 1 and (b) Group 2.

to the same extent in different groups. If this is not true, results on simple comparisons on the means may be misleading. The invariance issue may be tested within a multiple-group CFA. Restrictions on some of the parameters are imposed gradually:

(i) configural invariance: the same pattern of fixed and free factor loadings across groups;

(ii) weak factorial invariance, also known as metric invariance: invariant factor loadings across groups;

(iii) strong factorial invariance, also known as scalar invariance: invariant factor loadings and intercepts across groups; and

(iv) strict factorial invariance: invariant factor loadings, intercepts, and factor variances across groups.

2.4 Estimation methods, test statistics, and goodness-of-fit indices

After specifying a model, we need to obtain the parameter estimates and the test statistics by comparing the model-implied moments with the sample moments. Summary statistics, such as the means and the covariance matrix, were usually used as input in the past when the computational power was not as powerful as that of today. Jöreskog (1967, 1969) showed that we could fit structural equation models based on the sample means and the covariance matrix only. The estimation is usually based on a *discrepancy function* (Browne, 1982). A discrepancy function returns a scalar value of the difference between the sample covariance matrix (and the means) and the model-implied covariance matrix (and the means). The discrepancy is zero if and only if the sample covariance matrix and the model-implied covariance matrix are exactly the same; otherwise, the discrepancy is positive. The parameters are estimated by minimizing the discrepancy function.

There are several discrepancy functions, such as the maximum likelihood estimation (MLE), the generalized least squares (GLS), and the weighted least squares (WLS) in SEM. It should be noted that the meanings of WLS and GLS are slightly different in the context of meta-analysis and SEM. In the literature of meta-analysis and regression analysis, WLS and GLS are used to handle the error structures with a diagonal and a block-diagonal variance–covariance matrix, respectively. In other words, WLS is a special case of GLS with uncorrelated residuals. In the context of SEM, GLS is a special case of WLS. The GLS estimation method is used with a normality assumption on the data, whereas the WLS estimation method can be used for data with arbitrary distributions.

2.4.1 Maximum likelihood estimation

MLE is probably one of the most popular estimation methods in statistics. Under some regularity conditions (e.g., Millar, 2011), maximum likelihood (ML) estimators have many desirable properties. For instance, they are consistent, asymptotically unbiased, asymptotically efficient, and asymptotically normally distributed. When the data are multivariate normal, the sample means and the covariance matrix are sufficient statistics. Analysis using the raw data and summary statistics are equivalent when there is no missing data. The parameter estimates of the covariance structure can be obtained by minimizing the ML discrepancy function $F_{\text{ML}}(\boldsymbol{\theta})$:

$$\begin{aligned} F_{\text{ML}}(\boldsymbol{\theta}) = {} & \log|\boldsymbol{\Sigma}(\boldsymbol{\theta})| + \text{tr}(\boldsymbol{S}\boldsymbol{\Sigma}(\boldsymbol{\theta})^{-1}) - \log|\boldsymbol{S}| - p \\ & - (\bar{\boldsymbol{y}} - \boldsymbol{\mu}(\boldsymbol{\theta}))^{\text{T}}\boldsymbol{\Sigma}(\boldsymbol{\theta})^{-1}(\bar{\boldsymbol{y}} - \boldsymbol{\mu}(\boldsymbol{\theta})), \end{aligned} \tag{2.13}$$

where $\text{tr}(\boldsymbol{X})$ is the trace of $\boldsymbol{X}$ that takes the sum of the diagonal elements of $\boldsymbol{X}$ and p is the number of variables in the model. When the mean structure is not involved, the fit function can be simplified to

$$F_{\text{ML}}(\boldsymbol{\theta}) = \log|\boldsymbol{\Sigma}(\boldsymbol{\theta})| + \text{tr}(\boldsymbol{S}\boldsymbol{\Sigma}(\boldsymbol{\theta})^{-1}) - \log|\boldsymbol{S}| - p. \tag{2.14}$$

2.4.2 Weighted least squares

When the MLE is used, it is assumed that the data are normally distributed. If this assumption is questionable, there are several alternatives. These include robust statistics, bootstrap statistics, and WLS estimation, which is also known as asymptotically distribution-free (ADF) estimation method (Browne, 1984). We illustrate the idea of WLS estimation here.

We only focus on the analysis of the covariance structure here by stacking the $p \times p$ sample covariance matrix $\boldsymbol{S}$ into a $\tilde{p} \times 1$ vector $\boldsymbol{s}$, where $\tilde{p} = p(p+1)/2$. We may use $\underset{\tilde{p}\times 1}{\boldsymbol{s}} = \text{vech}(\underset{p \times p}{\boldsymbol{S}})$ to represent this process where vech() is a function to half-vectorize a square matrix to a column vector by column majorization. For example, if $\boldsymbol{X} = \begin{bmatrix} 1 & 2 & 3 & 4 \\ 2 & 5 & 6 & 7 \\ 3 & 6 & 8 & 9 \\ 4 & 7 & 9 & 10 \end{bmatrix}$, $\text{vech}(\boldsymbol{X}) = \begin{bmatrix} 1 & 2 & 3 & 4 & 5 & 6 & 7 & 8 & 9 & 10 \end{bmatrix}^{\text{T}}$.

We may then define $F_{\text{WLS}}(\boldsymbol{\theta})$ as the criterion to be minimized to obtain the parameter estimates:

$$F_{\text{WLS}}(\boldsymbol{\theta}) = (\boldsymbol{s} - \boldsymbol{\sigma}(\boldsymbol{\theta}))^{\text{T}}\boldsymbol{V}_S^{-1}(\boldsymbol{s} - \boldsymbol{\sigma}(\boldsymbol{\theta})), \tag{2.15}$$

where $\boldsymbol{\sigma}(\boldsymbol{\theta}) = \text{vech}(\boldsymbol{\Sigma}(\boldsymbol{\theta}))$ and $\boldsymbol{V}_S$ is a $\tilde{p} \times \tilde{p}$ positive-definite matrix (Browne, 1984). Although $\boldsymbol{V}_S$ can be any consistent matrix, it is usually chosen to represent the sampling covariance matrix of $\boldsymbol{s}$ (Jöreskog et al., 1999).

Let us illustrate the above idea with the simple regression model in Equation 2.1. For the ease of illustration, we exclude the mean structure by focusing on the covariance structure only. The sample covariance matrix is $\boldsymbol{S} = \begin{bmatrix} \text{Var}(y) & \\ \text{Cov}(x,y) & \text{Var}(x) \end{bmatrix}$, while the model-implied covariance matrix is $\boldsymbol{\Sigma}(\boldsymbol{\theta}) = \begin{bmatrix} \beta_1^2\sigma_x^2 + \sigma_{e_y}^2 & \\ \beta_1\sigma_x^2 & \sigma_x^2 \end{bmatrix}$ with $\boldsymbol{\theta} = \begin{bmatrix} \beta_1 \\ \sigma_x^2 \\ \sigma_{e_y}^2 \end{bmatrix}$ is the vector of the parameters. We may vectorize these matrices by using $\boldsymbol{s} = \text{vech}(\boldsymbol{S}) = \begin{bmatrix} \mathsf{Var}(y) \\ \mathsf{Cov}(x,y) \\ \mathsf{Var}(x) \end{bmatrix}$ and $\boldsymbol{\sigma}(\boldsymbol{\theta}) = \text{vech}(\boldsymbol{\Sigma}(\boldsymbol{\theta})) = \begin{bmatrix} \beta_1^2\sigma_x^2 + \sigma_{e_y}^2 \\ \beta_1\sigma_x^2 \\ \sigma_x^2 \end{bmatrix}$. Finally, we also need to estimate the 3×3 asymptotic sampling covariance matrix $\boldsymbol{V}_S$ of $\boldsymbol{s}$. Section 3.3.2 discusses how to estimate it under the assumption of multivariate normality. We may obtain the parameter estimates $\hat{\boldsymbol{\theta}}$ by minimizing the differences between $\boldsymbol{s}$ and $\boldsymbol{\sigma}(\boldsymbol{\theta})$ weighted by $\boldsymbol{V}_S$ with Equation 2.15.

Most applications in SEM focus on the covariance structure while correlation structure analysis is less popular. One reason is that SEM was developed to analyze covariance structure. Sometimes, correlation structure analysis is also of theoretical and practical interests (Bentler, 2007), especially in the context of meta-analysis (Hunter and Hamilton, 2002).

If we are analyzing a correlation matrix, the diagonals are always 1. The diagonals do not carry any useful information. We may stack the $p \times p$ sample correlation matrix $\boldsymbol{R}$ into a $\vec{p} \times 1$ vector $\boldsymbol{r}$ where $\vec{p} = p(p-1)/2$, that is, $\underset{\vec{p}\times 1}{\boldsymbol{r}} = \text{vechs}(\underset{p\times p}{\boldsymbol{R}})$ where vechs() is a function to strict half-vectorize a square matrix to a column vector by column majorization. For example, $\text{vechs}(\boldsymbol{X}) = \begin{bmatrix} 2 & 3 & 4 & 6 & 7 & 9 \end{bmatrix}^{\text{T}}$ in our previous example.

Similarly, a WLS approach can also be applied to analyze correlation matrix (Bentler and Savalei, 2010; Fouladi, 2000). The criteria to be minimized for a correlation structure $\boldsymbol{P}(\boldsymbol{\gamma})$ is

$$F_{\text{WLS}}(\boldsymbol{\gamma}) = (\mathbf{r} - \boldsymbol{\rho}(\boldsymbol{\gamma}))^{\text{T}}\mathbf{V}_R^{-1}(\mathbf{r} - \boldsymbol{\rho}(\boldsymbol{\gamma})), \tag{2.16}$$

where $\boldsymbol{\rho}(\boldsymbol{\gamma}) = \text{vechs}(\mathbf{P}(\boldsymbol{\gamma}))$ and $\boldsymbol{V}_R$ is a $\vec{p} \times \vec{p}$ positive-definite matrix. It should be noted that the dimensions of $\boldsymbol{V}_R$ is smaller than that of $\boldsymbol{V}_S$.

Correlation structure analysis is also used to handle binary or ordinal categorical variables in SEM (Muthén, 1983, 1984). The idea is that the binary or ordinal variables are indicators of the latent continuous variables. We may use the appropriate link functions to link the latent continuous variables to the observed variables. The first step of the analysis is to estimate the polychoric correlation matrix among the latent continuous variables. WLS or diagonally weighted least squares (DWLS) may be used as the estimation method with the asymptotic covariance matrix of the elements of the polychoric correlation matrix as the weight matrix. Robust statistic is usually used to adjust for the SEs and the test statistics (e.g., Yuan and Bentler, 2007).

Special care has to be taken to ensure that the diagonals of $\boldsymbol{P}(\hat{\boldsymbol{\gamma}})$ are always ones; otherwise, the fitted model is not a correlation structure any more. More details regarding these issues are addressed in MASEM in Chapter 7.

2.4.3 Multiple-group analysis

Multiple-group SEM was first proposed by Jöreskog (1971) and Sörbom (1974). The fit function to be minimized is the sum of individual fit functions weighted by the sample sizes:

$$F_{MG}(\boldsymbol{\theta}) = \frac{\sum_{i=1}^{k}(n_i - 1)F_i(\boldsymbol{\theta})}{\sum_{i=1}^{k}(n_i - 1)}, \tag{2.17}$$

where $F_i(\boldsymbol{\theta})$ is the fit function in the ith group. We are usually interested in imposing some constraints in the model. For example, some of the factor loadings may be assumed equal in testing the measurement invariance of the data. The point that the models and the data assumptions can be different in different groups (Bentler et al., 1987) should be noted.

2.4.4 Likelihood ratio test and Wald test

Once we have fitted the proposed model, we need to evaluate the appropriateness of the model. There are two objectives in this step. The first one is to evaluate whether the proposed model, $H_0 : \boldsymbol{\Sigma} = \boldsymbol{\Sigma}(\boldsymbol{\theta})$ and $H_0 : \boldsymbol{\mu} = \boldsymbol{\mu}(\boldsymbol{\theta})$ if the mean structure is present, as a whole fits the data. Under some regularity conditions, the test statistic T based on the minimum of any of the above fit functions $F_{\min}(\boldsymbol{\theta})$,

$$T = (n - 1)F_{\min}(\boldsymbol{\theta}), \tag{2.18}$$

has an approximate chi-square distribution with the appropriate dfs if the proposed model is correct. If the test statistic is not significant, the proposed model is consistent with the data. We will discuss the use of various goodness-of-fit indices to evaluate the proposed models later.

The second objective is to test whether the individual parameter estimates $\hat{\boldsymbol{\theta}}$ are statistically significant. There are generally three types of test statistics—the likelihood ratio (LR) test, the Wald test, and the Lagrange multiplier test. These test statistics are asymptotically equal when the sample sizes are large (Buse, 1982; Engle, 1984). Under small samples, however, they can be different. As the LR test and the Wald test are more popular in SEM and meta-analysis, we mainly focus on these two tests here.

Suppose that $\underset{p\times 1}{\boldsymbol{\theta}}$ is a vector of p parameters of interests in the model. $\underset{p\times 1}{\hat{\boldsymbol{\theta}}}$ is the vector of the parameter estimates with $\underset{p\times p}{\hat{\mathbf{V}}}$ as the asymptotic sampling covariance matrix of the parameter estimates. $\boldsymbol{V}$ indicates the variability in estimating $\hat{\boldsymbol{\theta}}$. If we are interested in testing a null hypothesis on some parameters, say the ith parameter $\boldsymbol{\theta}_{[i]}$ to a specific value θ_0, we first obtain the SE of $\hat{\boldsymbol{\theta}}_{[i]}$ by taking the square root of the ith diagonal element in $\hat{\boldsymbol{V}}$, that is, $\text{SE}(\hat{\boldsymbol{\theta}}_{[i]}) = \sqrt{\hat{\boldsymbol{V}}_{[i,i]}}$. Strictly speaking, there is a "hat" on SE. As it is clear that the SE is estimated rather than known, the "hat" is often dropped in the formula. We test the null hypothesis $H_0 : \theta_{[i]} = \theta_0$ by using the test statistic z,

$$z = \frac{\hat{\theta}_{[i]} - \theta_0}{\mathrm{SE}(\hat{\theta}_{[i]})}, \tag{2.19}$$

which has an approximate standard normal distribution. The null hypothesis is rejected at $\alpha = 0.05$ when $|z| \geq z_{1-\alpha/2}$ with $z_{1-\alpha/2}$ is the $(1 - \alpha/2)$th percentile of the standard normal score. As most computer packages provide the SEs as a by-product after the estimation, the Wald statistic is easy to use to test the significance of the individual parameter estimates.

The above test is a special case of the Wald (W) test with only one parameter. If we want to test multiple or all parameters in the model, for example, $H_0 : \underset{p\times 1}{\boldsymbol{\theta}} = \underset{p\times 1}{\boldsymbol{\theta}_0}$, we calculate the W test by

$$W = \underset{1\times p}{(\hat{\boldsymbol{\theta}} - \boldsymbol{\theta}_0)^{\mathrm{T}}}\ \underset{p\times p}{\hat{\mathbf{V}}^{-1}}\ \underset{p\times 1}{(\hat{\boldsymbol{\theta}} - \boldsymbol{\theta}_0)}, \tag{2.20}$$

which has an approximate chi-square distribution with p dfs (e.g., Fox, 2008).

Another approach to test $H_0 : \boldsymbol{\theta} = \boldsymbol{\theta}_0$ is to apply the LR test. We first fit two models—one with the constraints $\boldsymbol{\theta}_0$ and the other without the constraints. We calculate an LR statistic, the difference of the log-likelihood (LL between these two models,

$$\mathrm{LR} = 2(\log L(\hat{\boldsymbol{\theta}}) - \log L(\boldsymbol{\theta}_0)), \tag{2.21}$$

has an approximate chi-square distribution with p dfs. The LR test can be used to test one or more parameters.

2.4.5 Confidence intervals on parameter estimates

Besides testing the significance of the parameter estimates with a Wald test or an LR test, we may construct CIs on the parameter estimates. A $100(1 - \alpha)\%$ CI on a parameter (θ) is a random interval, calculated from the sample, that contains θ with the prespecified probability in the long run, that is,

$$\Pr(\hat{\theta}_{\mathrm{L}} < \theta < \hat{\theta}_{\mathrm{U}}) = 1 - \alpha, \tag{2.22}$$

where α is the significance level and $\hat{\theta}_{\mathrm{L}}$ and $\hat{\theta}_{\mathrm{U}}$ are the estimates of the lower and upper limits on θ, respectively (e.g., Rice, 2007). It should be noted that both $\hat{\theta}_{\mathrm{L}}$ and $\hat{\theta}_{\mathrm{U}}$ are random variables. Therefore, they may vary from samples to samples. Moreover, the test statistic may be approximate rather than exact. The coverage probability may only approximately equal 95%.

One common method of constructing CIs is by inverting a test statistic with a known distribution (e.g., Casella and Berger, 2002). This inversion CI principle has been frequently used to construct CIs for the effect sizes (see Steiger and Fouladi, 1997). Both the Wald CI and the LBCI can be constructed by this principle. We first discuss how to construct these CIs. Then we compare their similarities and differences. This section is based on the work of Cheung (2009b).

2.4.5.1 Wald CIs

From Equation 2.19, we may compute the $100(1 - \alpha)\%$ Wald CI for θ with

$$\hat{\theta} \pm z_{1-\alpha/2}\mathrm{SE}(\hat{\theta}). \tag{2.23}$$

The Wald CIs are symmetric around $\hat{\theta}$. As the SEs are usually available in most statistical packages, the Wald CI is probably the most popular method to creating the CIs. If the parameter of interest is a function of other parameters, the SE may not be directly obtainable. In these cases, the delta method may be used to approximate the SE (e.g., Casella and Berger, 2002; Rice, 2007) (see Section 3.4.1).

2.4.5.2 Likelihood-based CIs

Besides inverting a z statistic to form a Wald CI, we may also construct an LBCI by inverting an LR statistic. When the null hypothesis ($H_0 : \theta = \theta_0$) is correct, the LR statistic in Equation 2.21 is asymptotically distributed as a chi-square variate with p dfs where p is the number of independent constraints imposed by the null hypothesis. The difference of two LR statistics on two estimates of the same parameter, where one is the ML estimate (treated as fixed) and the other is varied, is asymptotically distributed as a chi-square variate with 1 df.

To construct a $100(1 - \alpha)\%$ LBCI ($\hat{\theta}_L$ and $\hat{\theta}_U$) on a parameter, we move the parameter estimate (treated as varied) as far away as possible to the right from its ML estimate such that it is just statistically significant at the desired α significance level. That is, the $\hat{\theta}_U$ may be obtained by gradually increasing the estimate until

$$\chi^2_{1,1-\alpha} = \chi^2_U - \chi^2_{ML}, \tag{2.24}$$

where $\chi^2_{1,1-\alpha}$ is the critical value of the chi-square statistic with 1 df and with a α significance level and χ^2_U and χ^2_{ML} are the chi-square statistics of $\hat{\theta}_U$ and the ML estimate $\hat{\theta}$, respectively. The $\hat{\theta}_L$ may also be obtained by the same procedure except that the estimate for $\hat{\theta}_L$ is moving to the opposite direction.

If there is more than one parameter in the likelihood function, the above approach may be modified by using the profile likelihood method (Pawitan, 2001). The profile likelihood method reduces the likelihood function with multiple parameters to a likelihood function with a single parameter by treating other parameters as nuisance parameters and maximizing over them (e.g., Neale and Miller, 1997).

2.4.5.3 Relationship between the Wald test and the likelihood-based test

As both Wald test or its CI and the LR test or LBCI may be used to test hypotheses on the parameter estimates, we may compare their properties and determine which one is preferable. Let us illustrate the relationship between the Wald and the LR statistics when there is only one parameter. The Wald statistic is an approximation of the LR statistic by using a second-order Taylor's expansion of the LL function around the ML estimate (e.g., Pawitan, 2001, p.33),

$$\mathrm{LL}(\theta) \approx \mathrm{LL}(\hat{\theta}) + \frac{d(\mathrm{LL}(\hat{\theta}))}{d\theta}(\theta - \hat{\theta}) + \frac{1}{2}\frac{d^2(\mathrm{LL}(\hat{\theta}))}{d\theta^2}(\theta - \hat{\theta})^2, \tag{2.25}$$

where $\frac{d(\mathrm{LL}(\hat{\theta}))}{d\theta}$ and $\frac{d^2(\mathrm{LL}(\hat{\theta}))}{d\theta^2}$ are the first and second derivatives of $\mathrm{LL}(\theta)$ evaluated at $\hat{\theta}$, respectively. As the first derivative on the LL function evaluated at $\hat{\theta}$ is zero at the MLE, the above equation reduces to

$$\mathrm{LL}(\theta) \approx \mathrm{LL}(\hat{\theta}) - \frac{1}{2}I(\hat{\theta})(\theta - \hat{\theta})^2, \tag{2.26}$$

where $I(\hat{\theta}) = \frac{-d^2(\mathrm{LL}(\hat{\theta}))}{d\theta^2}$ is the observed Fisher information that indicates the curvature of the quadratic approximation of the LL function. The asymptotic sampling variance of the parameter estimate $\mathsf{Var}(\hat{\theta})$ can be obtained by

$$\mathsf{Var}(\hat{\theta}) = \frac{1}{I(\hat{\theta})}. \tag{2.27}$$

From the above equation, we can construct a Wald statistic for testing $H_0 : \theta = \theta_0$ versus $H_1 : \theta \neq \theta_0$,

$$W = \frac{(\hat{\theta} - \theta_0)^2}{\mathsf{Var}(\hat{\theta})}, \tag{2.28}$$

which has a chi-square distribution with 1 df.

Figure 2.7 shows the Wald test and the LR test when the $\mathrm{LL}(\theta)$ can be well approximated by the $-\frac{1}{2}I(\hat{\theta})(\theta - \hat{\theta})^2$. To test the null hypothesis $\theta = \theta_0$, the LR test compares the vertical differences between $\mathrm{LL}(\hat{\theta})$ and $\mathrm{LL}(\theta_0)$. If the difference is sufficiently large, it is statistically significant. In contrast, the W test compares the horizontal differences between $\hat{\theta}$ and θ_0. As the curvature of the quadratic approximation of the LL function also influences the appropriateness of the null hypothesis, we weigh the square difference by $I(\hat{\theta})$ (Equation 2.28). If the discrepancy is large, the null hypothesis is rejected.

The observed (or expected) Fisher information matrix, thus the $\mathrm{SE}(\hat{\theta})$, is usually available after obtaining the MLE. Therefore, the Wald test and the Wald CIs are usually available in nearly all statistical packages. As shown in Equation 2.26, the Wald statistic is based on the second-order quadratic approximation of the LL function. The Wald CI and the likelihood-based CI would be exactly the same in a few special cases only (see Buse, 1982).

In most cases, the Wald CI and the LBCI will differ. The appropriateness of using the Wald statistic to approximate the LL function depends on several factors, such as the model being analyzed, the sampling distribution of the parameter estimate, and the sample size. In reality, however, it is hard to tell whether the

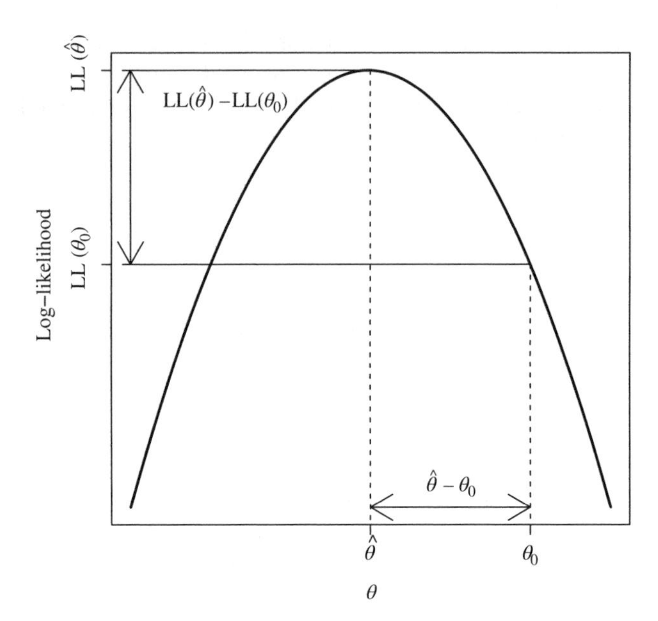

Figure 2.7 Relationship between a likelihood ratio and a Wald tests.

quadratic approximation is good or not. One method is to plot the true LL and its quadratic approximation graphically. Researchers may visually check whether or not the quadratic approximation is good (see Pawitan (2001) for some examples).

For example, Figure 2.8 shows two LLs of two data sets. The ML estimates for both data sets are $\hat{\theta} = 0.8$. For Figure 2.8a, the sample size is only 10, the quadratic approximation of the LL (and the SE) is not very good. The $\hat{\theta}_L$ of the Wald SE is

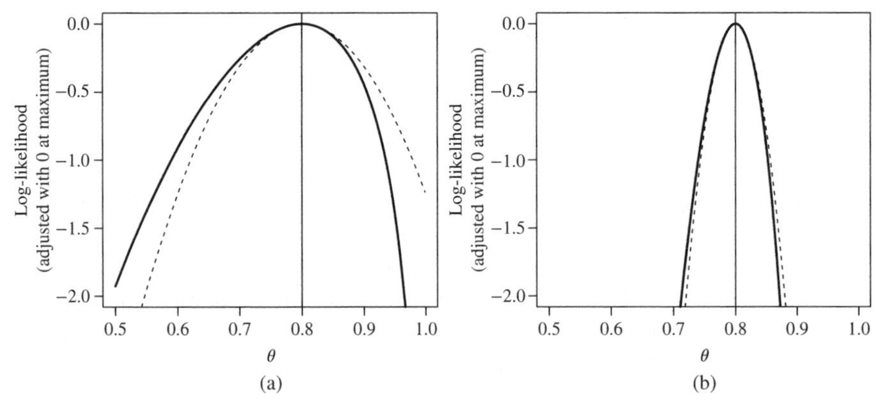

Figure 2.8 Binomial distribution with $n = 10$ and $x = 8$ (a) and $n = 100$ and $x = 80$ (b) and with solid lines (true log-likelihood) and dashed lines (quadratic approximation of log-likelihood).

too small, whereas the $\hat{\theta}_U$ is too big. The sample size for the data in Figure 2.8b is 100. The quadratic approximation of the LL is nearly the same as that of the true LL. CIs based on the Wald statistic and the LL are nearly identical. We may apply the Wald test for the data in Figure 2.8b. But we should be careful when applying the Wald test for the data in Figure 2.8a.

There are several criticisms about the use of Wald CIs in the literature. Wald CIs based on SEs assume implicitly that the LL function for the quantity of interest can be closely approximated by a quadratic function and, hence, is symmetric. Therefore, Wald CIs are always symmetric around the ML estimates. However, in many cases, the actual likelihood functions are asymmetric, in which cases the quadratic approximation underlying the Wald CIs may work poorly unless large samples are involved (Pawitan, 2001). For example, the sampling distributions of the parameters of the variances, correlation coefficients that are close to ± 1, and the product term of two random variables in indirect effects are usually asymmetric. In these cases, the symmetric Wald CIs are too optimistic in ruling out values of the parameter at one end and too pessimistic in ruling out values of the parameter at the other end (DiCiccio and Efron, 1996).

A related issue with the symmetric CIs is that the CIs may be out of the meaningful boundaries, for example, a negative lower limit for the variance. Although these CIs may be truncated to the meaningful bounds, say zero for variance and ± 1 for correlation, Steiger and Fouladi (1997) warned that the coverage probability for the truncated CI is maintained; however, the width of the CI may be suspicious as an indicator for the precision of the measurement because of the truncation of the nonsensible values.

Another problem with Wald CIs (and the significance tests based on Wald statistics) is that they are not invariant to monotonic transformations on the parameters (DiCiccio and Efron, 1996; Neale and Miller, 1997). For example, the coverage probabilities of the Wald CIs on a parameter, say a Pearson correlation, and its monotonic transformation, say a Fisher's z transformed score, need not be the same. This issue is particularly annoying in SEM because there are many equivalent models formed by different model parameterizations. For example, we may fix either a factor loading or the factor variance at some specific values, usually 1, in order to identify a latent variable. Many researchers have shown that the Wald test may indicate different conclusions (significance vs nonsignificance) depending on how the models are parameterized (Gonzalez and Griffin, 2001; Neale and Miller, 1997; Steiger, 2002).

LBCI offers many improvements over Wald CI (e.g., Meeker and Escobar, 1995; Neale and Miller, 1997; Pawitan, 2001). The LBCI uses the LL function directly instead of its quadratic approximation. They are asymmetric in capturing the sampling distribution of the parameter estimates; and they are invariant to monotonic transformations.

LBCIs have been suggested as alternatives to Wald CIs in areas where the Wald CIs are known to perform poorly. For example, Agresti (2002) recommended the use of LBCIs over Wald CIs in analyzing categorical data when the sample sizes are small to moderate. Similar suggestions have been offered in nonlinear regressions (e.g., Bates, 1988; Seber, 2003), random effects in meta-analyses (Hardy and Thompson, 1996; Viechtbauer, 2005), logistic regressions, and generalized linear models (Agresti, 2002).

Although the properties of the LBCIs seem appealing, there are still issues surrounding their use. First and foremost, researchers need to make distributional assumptions on the data in order to construct LBCI. The use of LBCIs is questionable when the specified LL function is inappropriate. For some models, the LBCIs may be out of the meaningful boundaries. For example, if we do not impose any restrictions on the variance component $\boldsymbol{T}^2$ on a multivariate meta-analysis, $\hat{\boldsymbol{T}}^2$ can be negative definite (see Section 5.3.2). Even $\hat{\boldsymbol{T}}^2$ is nonnegative definite, their LBCIs can be negative definite when there is no boundary imposed on $\hat{\boldsymbol{T}}^2$ in the analysis. As Casella and Berger (2002, p. 430), among others, have cautioned, there is no guarantee that the LBCIs will be optimal, although they will seldom be too bad. Despite this, they still "recommend constructing a confidence set based on inverting an LRT (likelihood ratio test), if possible."

2.4.6 Test statistics versus goodness-of-fit indices

As summarized by Steiger and Fouladi (1997), there are two general approaches in statistical analysis. They are the *reject-support* approach and the *accept-support* approach. Most statistical methods, such as t test, regression analysis, and ANOVA, are based on the reject-support approach. Researchers usually formulate research hypotheses with an intention to reject the null hypothesis, for example, $H_0 : \rho = 0$, between two variables. If the null hypothesis is rejected, the finding supports the researcher's belief. On the other hand, SEM researchers propose a CFA or an SEM model, say $H_0 : \boldsymbol{\Sigma} = \boldsymbol{\Sigma}(\theta)$. Most SEM users have the intention of *not* to reject the proposed models. Therefore, SEM applications fall into the accept-support approach.

2.4.6.1 LR test statistic

The LR test discussed earlier is the most common test statistic in SEM. It indicates the *badness-of-fit* of the proposed model. Under the appropriate assumptions, the LR statistic has an approximate chi-square distribution with the appropriate dfs. The empirical coverage of the LR statistic is close to the nominal values when some assumptions are correct. In order for the LR test to behave as a chi-square distribution, we need the following assumptions:

(i) the distribution assumptions are correct;

(ii) the proposed model is correctly specified; and

(iii) the sample sizes are large enough.

However, some of these assumptions may not be fulfilled in applied research. For example, Micceri (1989) examined 440 large-sample achievement and psychometric measures and found all data were statistically significant in testing the normality. The issue of nonnormality has been extensively studied in SEM. When the data are not normally distributed, the general findings are that the LR test will be inflated, whereas the SEs will be deflated (e.g., Curran et al., 1996). Satorra and Bentler (1988, 1994) proposed some scaled statistics to *correct* for this bias. This is generally known as the Satorra–Bentler scaled statistic (see Yuan and Bentler, 2007). The empirical Type I error of the correction statistic is roughly close to the prespecified value, for example, 0.05, even the data are not normally distributed.

Another assumption is that the proposed model is correctly specified. When there are latent variables, it means that the measurement model is literally correct. That is, the indicators are loaded in the correspondent latent factors and there are no minor or trial loadings. As most structural equation models are highly restrictive, they are likely wrong at the population level. Therefore, many SEM users consider their proposed models are only *approximation* of the reality. They do not expect that the proposed models are literally true at the population. If the LR statistic is used to assess the model fit in SEM, the test will be significant when there are trivial differences between the proposed model and the data when the sample sizes are large enough.

The third assumption is that the sample sizes are large enough. Reasonable sample sizes are required so that the LR statistic behaves as a chi-square distribution when the proposed model is correct. It should be noted that the empirical Type I error should be close to the nominal level, for example, $\alpha = 0.05$, when the proposed model is correct, no matter how large the sample sizes are. This rarely happens in reality because the proposed models are likely approximation of the phenomena being studied. When there are trial or minor model misspecification, the LR statistic tends to reject the proposed models when the sample sizes are getting bigger.

2.4.6.2 Goodness-of-fit indices

Many goodness-of-fit indices have been proposed to address some of these issues. We briefly introduce a few here. One popular type of fit indices is the incremental fit indices. This type of fit indices compares the model fit of a target model with the test statistic T_{T} and df_{T} against a baseline model with the test statistic T_{B} and df_{B}. A common baseline model is a model with all covariances fixed at 0 (but see Widaman and Thompson (2003) for suggestions on alternative baseline models). In other words, the incremental fit indices indicate how good the proposed model is when it is compared to the *worst* possible model (the baseline model).

For example, the TLI (Tucker and Lewis, 1973), also known as the non-normed fit index (NNFI), defined as

$$\mathrm{TLI} = \frac{\chi^2_{\mathrm{B}}/\mathrm{df}_{\mathrm{B}} - \chi^2_{\mathrm{T}}/\mathrm{df}_{\mathrm{T}}}{\chi^2_{\mathrm{B}}/\mathrm{df}_{\mathrm{B}} - 1}. \tag{2.29}$$

TLI measures the proportion reduction in the chi-square values when comparing the baseline to the hypothesized model after adjusting the complexity of the model. The usual range is from 0 to 1; however, it can exceed this range.

Another popular incremental fit index is the comparative fit index (CFI) (Bentler, 1990). It is similar to the TLI except that it is bounded within 0 and 1,

$$\mathrm{CFI} = 1 - \frac{\max((\chi^2_{\mathrm{T}} - \mathrm{df}_{\mathrm{T}}), 0)}{\max((\chi^2_{\mathrm{T}} - \mathrm{df}_{\mathrm{T}}), (\chi^2_{\mathrm{B}} - \mathrm{df}_{\mathrm{B}}), 0)}. \tag{2.30}$$

Some researchers, (e.g., Hu and Bentler, 1999), put CFI > 0.95 as an indication that the proposed model fits the data well.

When the model is correctly specified, the test statistic asymptotically follows a *central* chi-square distribution. However, the test statistic follows a *noncentral* chi-square distribution when the model is misspecified. The noncentrality parameter λ depends on the alternative model. It can be estimated by

$$\hat{\lambda} = \frac{(\chi^2_{\mathrm{T}} - \mathrm{df}_{\mathrm{T}})}{(n-1)}. \tag{2.31}$$

Steiger and Lind (1980) proposed to use a modified version of $\hat{\lambda}$ that is nonnegative and takes the model complexity (df_{T}) into account. They proposed the root mean square error of approximation (RMSEA) which is defined as

$$\mathrm{RMSEA} = \sqrt{\frac{\max((\chi^2_{\mathrm{T}} - \mathrm{df}_{\mathrm{T}})/(n-1), 0)}{\mathrm{df}_{\mathrm{T}}}}. \tag{2.32}$$

It assesses the misfit of the model per df. When there are more than 1 group, Steiger (1998) proposed the following modification to calculate the RMSEA,

$$\mathrm{RMSEA} = \sqrt{k}\sqrt{\frac{\max((\chi^2_{\mathrm{Total}} - \mathrm{df}_{\mathrm{Total}})/N_{\mathrm{Total}}, 0)}{\mathrm{df}_{\mathrm{Total}}}}, \tag{2.33}$$

where k is the number of groups and χ^2_{Total}, $\mathrm{df}_{\mathrm{Total}}$, and N_{Total} are the chi-square statistic, df, and sample size of all groups, respectively. CIs on RMSEA can be constructed by iterative procedures (Steiger and Fouladi, 1997). The RMSEA is bounded at 0. In theory, it does not have a maximum because the alternative model can be badly deviated from the proposed model.

Browne and Cudeck (1993) suggested that the proposed model can be considered as "close fit" when the population RMSEA is close to 0.05. If the population RMSEA is larger than 0.10, the proposed model does not fit the data. They further suggested using the 90% CI on the RMSEA to test the null hypotheses of "close fit." If the 90% CI of the RMSEA includes the null hypothesis of RMSEA = 0.05, the proposed model is a good fitted model.

Another index based on the residuals is the root mean square residual (RMR), which was introduced by Jöreskog and Sörbom (1981). The RMR is an index of the average discrepancy between the sample covariance matrix $\boldsymbol{S}$ and the model-implied covariance matrix $\hat{\boldsymbol{\Sigma}}$. The RMR is defined as

$$\mathrm{RMR} = \sqrt{\frac{\sum_{i=1}^{p}\sum_{j=1}^{i}(s_{ij} - \hat{\sigma}_{ij})^2}{\frac{p(p+1)}{2}}}, \tag{2.34}$$

where s_{ij} and $\hat{\sigma}_{ij}$ are the elements in $\boldsymbol{S}$ and $\hat{\boldsymbol{\Sigma}}(\hat{\boldsymbol{\theta}})$, respectively. If the proposed model is correct, the model-implied covariance matrix $\hat{\boldsymbol{\Sigma}}(\hat{\boldsymbol{\theta}})$ will be very close to the sample covariance matrix $\boldsymbol{S}$. As the variables are likely to be on different scales (variances), it is difficult to interpret the RMR. Bentler (1995) proposed to standardize the RMR by using s_{ii} and s_{jj}. The standardized root mean square residual (SRMR) for a covariance structure analysis is defined as

$$\mathrm{SRMR_S} = \sqrt{\frac{\sum_{i=1}^{p}\sum_{j=1}^{i}\left(\frac{s_{ij}-\hat{\sigma}_{ij}}{\sqrt{s_{ii}s_{jj}}}\right)^2}{\frac{p(p+1)}{2}}}. \tag{2.35}$$

When a correlation structure analysis is conducted, the diagonal elements of the sample correlation matrix and the model-implied correlation matrix are always fixed at 1. There is no need to standardize the residuals. Moreover, we need to exclude the diagonals from the calculations. The SRMR for a correlation structure analysis is modified as

$$\mathrm{SRMR_R} = \sqrt{\frac{\sum_{i=2}^{p}\sum_{j=1}^{i-1}(r_{ij} - \hat{\rho}_{ij})^2}{\frac{p(p-1)}{2}}}, \tag{2.36}$$

where r_{ij} and $\hat{\rho}_{ij}$ are the elements in $\boldsymbol{R}$ and $\hat{\boldsymbol{P}}$, respectively. The above definitions on SRMR may be extended to multiple-group analysis with k groups. The SRMR for the multiple-group analysis is

$$\mathrm{SRMR} = \frac{\sum_{i=1}^{k}(n_i - 1)\mathrm{SRMR}_i}{\sum_{j=1}^{k}(n_j - 1)}, \tag{2.37}$$

where SRMR_i is the SRMR in the ith group. Thus, it is simply a weighted mean of the individual SRMR. When SRMR is 0, the proposed model perfectly fits the data.

The theoretical maximum of SRMR is 1, which indicates that the proposed model fits the data extremely bad. Conventionally, SRMR < 0.05 indicates a reasonable fitted model.

If the model comparison involves non-nested models, we may use either the Akaike information criterion (AIC) (Akaike, 1987) or the Bayesian information criterion (BIC) (Schwarz, 1978) to compare among these models. The AIC is defined as

$$\text{AIC} = \chi^2_{\text{T}} - 2\text{df}_{\text{T}}, \tag{2.38}$$

whereas the BIC is defined as

$$\text{BIC} = \chi^2_{\text{T}} - \log(n)\text{df}_{\text{T}}. \tag{2.39}$$

They measure the parsimonious fit that considers both the model fit and the number of parameter estimated. Smaller value indicates that the model fits better in compromising between the model fit and the model complexity. We may choose the model with the smallest (better parsimonious fit) AIC or BIC. It should be noted that different programs may use slightly different formulas to calculate the AIC and BIC. Therefore, the values may not be comparable across programs.

Even though there are lots of goodness-of-fit indices, there is no universally accepted ones or cutoffs. Different authors may cite different sources to support the proposed models using different goodness-of-fit indices. Readers interested in this topic may refer to Barrett (2007) and the commentaries regarding the arguments for and against the goodness-of-fit indices.

2.5 Extensions on structural equation modeling

This section addresses three important extensions in SEM. They are the phantom variables, definition variables, and FIML. These concepts are crucial to formulating meta-analytic models as structural equation models.

2.5.1 Phantom variables

Phantom variables are "latent variables with no observed indicators" (Rindskopf, 1984, p. 38). Another feature of the phantom variables is that the (error) variances of the phantom variables are fixed at zero. They are also known as the nodes (Horn and McArdle, 1980) and the auxiliary variables (Raykov and Shrout, 2002). It is also similar to the dummy latent variables used by Chan (2007). Applications of the phantom variables, such as testing dependent correlations, squared multiple R, standardized regression coefficient, reliability estimates, and mediating effects can be found in Cheung (2007a, 2009b).

We illustrate what phantom variables are by a simple regression used in Cheung (2010). Suppose that we would like to conduct a regression analysis by regressing y on x. On the basis of the theory, the regression coefficient is nonnegative. There are several methods to ensure that the estimated regression coefficient is nonnegative.

Most SEM packages, for example, M*plus* (Muthén and Muthén, 2012), `OpenMx` (Boker et al., 2011), and `lavaan` (Rosseel, 2012), allow nonlinear constraints on the parameters. We may set a nonlinear constraint to impose that the estimated regression coefficient is nonnegative. An alternative method is to use a phantom variable to ensure that the regression coefficient is nonnegative. Without loss of generality, we exclude the intercept in the model specification. The model with a phantom variable is

$$\begin{aligned} y &= \beta P + e_y \quad \text{and} \\ P &= \beta x. \end{aligned} \tag{2.40}$$

The model is shown in Figure 2.9. As $\sigma^2_{e_P} = 0$, P has no impact on the model fit. The effect of x on y is β^2, which is always nonnegative.

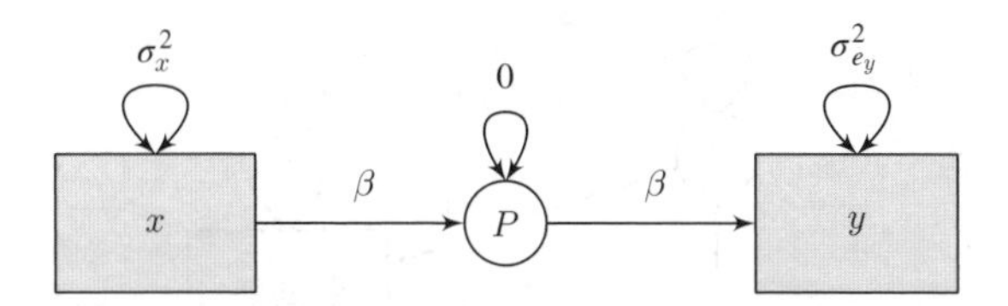

Figure 2.9 A regression model with a phantom variable.

2.5.2 Definition variables

Definition variables can be used to fix subject-specific values to any parameters in the model. These parameters can be, for example, path coefficients, factor loadings, means, and error variables. This means that the implied model may vary across subjects. Definition variables have been used to implement moderated regression (Neale, 2000) and multilevel SEM (Mehta and Neale, 2005). Applications of definition variables in meta-analysis (Cheung, 2008, 2010, 2013, 2014) will be discussed in details in later chapters.

Let us illustrate the idea of definition variables with a simple regression. Figure 2.10a shows a simple regression in the SEM literature. The effect from x on y is estimated via the parameter β_1. It should be noted that x is treated as a variable in the model. We have to estimate its variance σ^2_x and its mean μ_x. The model-implied means and covariance matrix for this model have been shown in Equation 2.2.

Figure 2.10b shows the same simple regression model in the regression literature. P is a phantom variable with a variance of 0. The mean of P is fixed at x_i via a definition variable. As there is a subscript i in x_i, the value of x_i varies across subjects. The path coefficient from P to y_i is β_1. The error variance $\sigma^2_{e_y}$ is the same across subjects after controlling for x_i. This is known as the assumption of homoscedasticity in regression analysis. By using the tracing rules, the model-implied conditional mean and variance of y_i are

$$\begin{aligned} \mathsf{E}(y_i|x_i) &= \beta_0 + \beta_1 x_i \quad \text{and} \\ \mathsf{Var}(y_i|x_i) &= \sigma^2_{e_y}. \end{aligned} \tag{2.41}$$

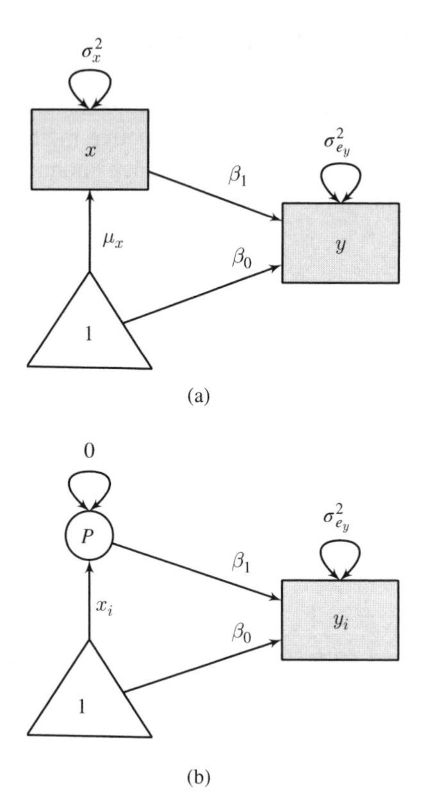

Figure 2.10 A regression model with and without a definition variable.

It is of importance to note that x_i is not a variable; there is no estimate on neither μ_x nor σ_x^2.

The models in Figure 2.10a and b are known as the *random-x* and the *fixed-x* regressions, respectively (e.g., Fox, 2008). Although the parameter estimates are the same in both panels, there are subtle differences between these two approaches. The most important one is that there is a subscript in Figure 2.10b indicating that it is a subject level model. Each subject has his/her own model. There is no a single model-implied mean and variance for all subjects. In our example, the mean of P is x_i that varies across subjects. On the other hand, there is no subscript in Figure 2.10a. There is a single model-implied mean and variance for all subjects. Summary statistics such as the means and covariance matrix may be used to fit the model using the discrepancy functions in Figure 2.10a, whereas FIML estimation must be used to analyze the model in Figure 2.10b. Moreover, there is no distribution assumption on x for the model in Figure 2.10b, whereas distribution assumption (usually normal distribution) is required on x for the model in Figure 2.10a.

2.5.3 Full information maximum likelihood estimation

By assuming that the data are multivariate normal, the −2∗log-likelihood (−2LL) of the ith subject for the proposed mean structure $\boldsymbol{\mu}_i(\boldsymbol{\theta})$ and the covariance structure $\boldsymbol{\Sigma}_i(\boldsymbol{\theta})$ is

$$-2\text{LL}_i(\boldsymbol{\theta}) = k_i \log(2\pi) + \log\left|\boldsymbol{\Sigma}_i(\boldsymbol{\theta})\right| + (y_i - \boldsymbol{\mu}_i(\boldsymbol{\theta}))^{\text{T}}\boldsymbol{\Sigma}_i(\boldsymbol{\theta})^{-1}(y_i - \boldsymbol{\mu}_i(\boldsymbol{\theta})), \quad (2.42)$$

where k_i is the number of non-missing observed variables in the ith subject; $|\boldsymbol{X}|$ is the determinant of $\boldsymbol{X}$, which is also known as the generalized variance; and $\log(x)$ is the natural logarithm of x. As there is a subscript i in the quantities in Equation 2.42, it indicates that raw data are required. As the observations are independent, the −2LL for all subjects is

$$-2\text{LL} = \sum\nolimits_{i=1}^{k} - 2\text{LL}_i(\boldsymbol{\theta}), \quad (2.43)$$

where k is the number of subjects. The parameter estimates $\hat{\theta}$ are obtained by minimizing the −2LL. The sampling covariance matrix of the parameter estimates $\mathsf{Cov}(\boldsymbol{\theta})$ can be obtained from the observed (or expected) Fisher information. This approach is usually known as the FIML in the SEM literature in contrast to the ML approach based on the summary statistics. FIML plays an important role in handling incomplete data in SEM and the SEM-based meta-analysis.

Another difference between the models in Figure 2.10 is how the model fit is calculated. When the summary statistics, the means, and the covariance matrix are used as the input for the model in Figure 2.10a, there are totally five pieces of information with five unknown parameters. Thus, the model is saturated with 0 df. Chi-square statistic and goodness-of-fit indices can be easily calculated. Regarding the model in Figure 2.10b, it is more complicated. As there is no single model for all subjects, FIML estimation has to be used (e.g., Mehta and Neale, 2005). For example, the `OpenMx` package reports the −2LL of the proposed model when raw data are analyzed. The df reported is the number of pieces of data (number of variables times number of subjects of the complete data) less the number of parameters. If there are 100 subjects without missing value in the simple regression example, the number of pieces of data is $100 = 100 \times 1$ as x is not treated as a variable. There are a total of three parameters in the model in Figure 2.8b. Thus, the df is 97.

It should be noted that the −2LL is *not* the LR test statistic in testing the proposed model. To calculate the LR on the proposed model, we need to calculate the difference in the −2LL between the proposed model and the saturated model. The saturated model is the model with all variables correlated and all the means are estimated. The parameter estimates are simply the sample covariance matrix and the sample mean. The proposed model is more restrictive than the saturated model. Suppose that the -2LL_T and -2LL_S are the −2LL of the proposed model and the saturated model, respectively, and df_T and df_S are the degrees of freedom for the proposed model and the saturated model, respectively. From Equation 2.21, we may calculate the test statistic LR_T as

$$LR_T = 2(LL_S - LL_T), \tag{2.44}$$

which has a chi-square distribution with $df = df_T - df_S$ under the null hypothesis that the proposed model is correct. If we want to calculate the incremental fit indices for the proposed model, we need to calculate the LR_B and df_B by applying Equation 2.44 on the baseline model.

2.6 Concluding remarks and further readings

This chapter briefly reviewed some key concepts in SEM. Several common structural equation models and how to specify these models were introduced. How to conduct statistical inferences in SEM was also reviewed. Because of the space constraint, the topics covered were very selective. Readers may refer to the textbooks, for example, Bollen (1989), Kline (2011), and Mulaik (2009) for the general concepts and more applications in SEM. The concepts of FIML (e.g., Enders, 2010), definition variables (Mehta and Neale, 2005), and phantom variables (Rindskopf, 1984) are crucial in understanding the SEM-based meta-analysis. The RAM formulation is also useful in fitting MASEM (McArdle, 2005). Readers who are new to SEM may be benefited by reading some of these readings before moving to the SEM-based meta-analysis.

References

Agresti A 2002. *Categorical data analysis*, *Wiley series in probability and statistics*, 2nd edn. John Wiley & Sons, Inc., New York.

Akaike H 1987. Factor analysis and AIC. *Psychometrika* **52**(3), 317–332.

Arbuckle JL 2006. *Amos 7.0 User's Guide*. SPSS, Chicago.

Barrett P 2007. Structural equation modelling: adjudging model fit. *Personality and Individual Differences* **42**(5), 815–824.

Bates DM 1988. *Nonlinear regression analysis and its applications*. John Wiley & Sons, Inc., New York.

Bentler PM 1986. Structural modeling and psychometrika: an historical perspective on growth and achievements. *Psychometrika* **51**(1), 35–51.

Bentler PM 1990. Comparative fit indexes in structural models. *Psychological Bulletin* **107**(2), 238–246.

Bentler PM 1995. *EQS structural equations program manual*. Multivariate Software, Encino, CA.

Bentler PM 2006. *EQS 6 structural equations program manual*. Multivariate Software, Encino, CA.

Bentler P 2007. Can scientifically useful hypotheses be tested with correlations? *American Psychologist* **62**(8), 772–782.

Bentler P, Lee S and Weng LJ 1987. Multiple population covariance structure analysis under arbitrary distribution theory. *Communications in Statistics - Theory and Methods* **16**(7), 1951–1964.

Bentler PM and Savalei V 2010. Analysis of correlation structures: current status and open problems In *Statistics in the social sciences* (ed. Kolenikov S, Steinley D and Thombs L). John Wiley & Sons, Inc., Hoboken, NJ, pp. 1–36.

Boker SM and McArdle JJ 2005. Path analysis and path diagrams In *Encyclopedia of statistics in behavioral science* (ed. Everitt BS and Howell DC), vol. 3. John Wiley & Sons, Ltd, Chichester, pp. 1529–1531.

Boker S, Neale M, Maes H, Wilde M, Spiegel M, Brick T, Spies J, Estabrook R, Kenny S, Bates T, Mehta P and Fox J 2011. OpenMx: an open source extended structural equation modeling framework. *Psychometrika* **76**(2), 306–317.

Bollen KA 1989. *Structural equations with latent variables*. John Wiley & Sons, Inc., New York.

Bollen K 2002. Latent variables in psychology and the social sciences. *Annual Review of Psychology* **53**, 605–634.

Bollen KA and Curran P 2006. *Latent curve models: a structural equation perspective*. Wiley-Interscience, Hoboken, NJ.

Brown H and Prescott R 2006. *Applied mixed models in medicine*. John Wiley & Sons, Inc.

Browne MW 1982. Covariance structures In *Topics in applied multivariate analysis* (ed. Hawkins DM). Cambridge University Press, Cambridge, pp. 72–141.

Browne MW 1984. Asymptotically distribution-free methods for the analysis of covariance structures. *British Journal of Mathematical and Statistical Psychology* **37**(1), 62–83.

Browne MW and Cudeck R 1993. Alternative ways of assessing model fit In *Testing structural equation models* (ed. Bollen KA and Long JS). Sage Publications, Inc., Newbury Park, CA, pp. 136–162.

Buse A 1982. The likelihood ratio, Wald, and Lagrange Multiplier tests: an expository note. *American Statistician* **36**(3), 153–157.

Byrne BM and Watkins D 2003. The issue of measurement invariance revisited. *Journal of Cross-Cultural Psychology* **34**(2), 155–175.

Casella G and Berger RL 2002. *Statistical inference*, 2nd edn. Brooks/Cole Publishing Company, Pacific Grove, CA.

Chan W 2007. Comparing indirect effects in SEM: a sequential model fitting method using covariance-equivalent specifications. *Structural Equation Modeling: A Multidisciplinary Journal* **14**(2), 326–346.

Cheung MWL 2007a. Comparison of approaches to constructing confidence intervals for mediating effects using structural equation models. *Structural Equation Modeling: A Multidisciplinary Journal* **14**(2), 227–246.

Cheung MWL 2007b. Comparison of methods of handling missing time-invariant covariates in latent growth models under the assumption of missing completely at random. *Organizational Research Methods* **10**(4), 609–634.

Cheung MWL 2008. A model for integrating fixed-, random-, and mixed-effects meta-analyses into structural equation modeling. *Psychological Methods* **13**(3), 182–202.

Cheung MWL 2009a. Comparison of methods for constructing confidence intervals of standardized indirect effects. *Behavior Research Methods* **41**(2), 425–438.

Cheung MWL 2009b. Constructing approximate confidence intervals for parameters with structural equation models. *Structural Equation Modeling: A Multidisciplinary Journal* **16**(2), 267–294.

Cheung MWL 2010. Fixed-effects meta-analyses as multiple-group structural equation models. *Structural Equation Modeling: A Multidisciplinary Journal* **17**(3), 481–509.

Cheung MWL 2013. Multivariate meta-analysis as structural equation models. *Structural Equation Modeling: A Multidisciplinary Journal* **20**(3), 429–454.

Cheung MWL 2014. Modeling dependent effect sizes with three-level meta-analyses: a structural equation modeling approach. *Psychological Methods* **19**(2), 211–229.

Curran PJ and Bauer DJ 2007. Building path diagrams for multilevel models. *Psychological Methods* **12**(3), 283–297.

Curran PJ, West SG and Finch JF 1996. The robustness of test statistics to nonnormality and specification error in confirmatory factor analysis. *Psychological Methods* **1**(1), 16–29.

DiCiccio TJ and Efron B 1996. Bootstrap confidence intervals. *Statistical Science* **11**(3), 189–212.

Enders CK 2010. *Applied missing data analysis*. Guilford Press, New York.

Engle RF 1984. Wald, likelihood ratio, and Lagrange multiplier tests in econometrics In *Handbook of econometrics* (ed. Griliches Z and Intriligator MD), Vol. 2. Elsevier, Amsterdam pp. 775–826.

Fouladi RT 2000. Performance of modified test statistics in covariance and correlation structure analysis under conditions of multivariate nonnormality. *Structural Equation Modeling: A Multidisciplinary Journal* **7**(3), 356–410.

Fox J 2008. *Applied regression analysis and generalized linear models*, 2nd edn. Sage Publications, Inc., Los Angeles

Gonzalez R and Griffin D 2001. Testing parameters in structural equation modeling: every "one" matters. *Psychological Methods* **6**(3), 258–269.

Hardy RJ and Thompson SG 1996. A likelihood approach to meta-analysis with random effects. *Statistics in Medicine* **15**(6), 619–629.

Horn JL and McArdle JJ 1980. Perspective on mathematical and statistical model building (MASMOB) in aging research In *Aging in the 1980s: Psychological Issues* (ed. Poon LW). American Psychological Association, Washington, DC, pp. 503–541.

Hu Lt and Bentler PM 1999. Cutoff criteria for fit indexes in covariance structure analysis: conventional criteria versus new alternatives. *Structural Equation Modeling: A Multidisciplinary Journal* **6**(1), 1–55.

Hunter JE and Hamilton MA 2002. The advantages of using standardized scores in causal analysis. *Human Communication Research* **28**(4), 552–561.

Jöreskog KG 1967. Some contributions to maximum likelihood factor analysis. *Psychometrika* **32**(4), 443–482.

Jöreskog KG 1969. A general approach to confirmatory maximum likelihood factor analysis. *Psychometrika* **34**(2), 183–202.

Jöreskog KG 1971. Simultaneous factor analysis in several populations. *Psychometrika* **36**(4), 409–426.

Jöreskog KG and Sörbom D 1981. *LISREL V: analysis of linear structural relationships by maximum likelihood and least squares methods*. International Educational Services, Chicago, IL.

Jöreskog KG and Sörbom D 1996. *LISREL 8: a user's reference guide*. Scientific Software International, Inc., Chicago, IL.

Jöreskog KG, Sörbom D, Du Toit S and Du Toit M 1999. *LISREL 8: new statistical features*. Scientific Software International, Inc., Chicago, IL.

Kline RB 2011. *Principles and practice of structural equation modeling*, 3rd edn. Guilford Press, New York.

MacCallum RC and Austin JT 2000. Applications of structural equation modeling in psychological research. *Annual Review of Psychology* **51**(1), 201–226.

MacKinnon D 2008. *Introduction to statistical mediation analysis*. Lawrence Erlbaum Associates, New York.

McArdle JJ 2005. The development of the RAM rules for latent variable structural equation modeling In *Contemporary psychometrics: a festschrift for Roderick P. McDonald* (ed. Maydeu-Olivares A and McArdle JJ). Lawrence Erlbaum Associates, Mahwah, NJ, pp. 225–273.

McArdle JJ and McDonald RP 1984. Some algebraic properties of the reticular action model for moment structures. *British Journal of Mathematical and Statistical Psychology* **37**(2), 234–251.

Meeker WQ and Escobar LA 1995. Teaching about approximate confidence regions based on maximum likelihood estimation. *American Statistician* **49**(1), 48–53.

Mehta PD and Neale MC 2005. People are variables too: multilevel structural equations modeling. *Psychological Methods* **10**(3), 259–284.

Mehta P and West S 2000. Putting the individual back into individual growth curves. *Psychological Methods* **5**(1), 23–43.

Micceri T 1989. The unicorn, the normal curve, and other improbable creatures. *Psychological Bulletin* **105**(1), 156–166.

Millar RB 2011. *Maximum likelihood estimation and inference: with examples in R, SAS, and ADMB*. John Wiley & Sons, Inc., Hoboken, NJ.

Millsap RE 2007. Invariance in measurement and prediction revisited. *Psychometrika* **72**(4), 461–473.

Mulaik SA 2009. *Linear causal modeling with structural equations*. CRC Press, Boca Raton, FL.

Muthén B 1983. Latent variable structural equation modeling with categorical data. *Journal of Econometrics* **22**(1-2), 43–65.

Muthén B 1984. A general structural equation model with dichotomous, ordered categorical, and continuous latent variable indicators. *Psychometrika* **49**(1), 115–132.

Muthén BO and Muthén LK 2012. *Mplus user's guide*, 7th edn. Muthén & Muthén, Los Angeles, CA.

Neale MC 2000. Individual fit, heterogeneity, and missing data in multigroup structural equation modeling In *Modeling longitudinal and multilevel data: practical issues, applied approaches and specific examples* (ed. Little TD, Schnabel KU and Baumert J). Erlbaum, Hillsdale, NJ, pp. 249–267.

Neale MC, Boker SM, Xie G and Maes HH 2006. Mx: statistical modeling, 7th edn. Technical report, Virginia Commonwealth University, Department of Psychiatry, Richmond, VA.

Neale MC and Miller MB 1997. The use of likelihood-based confidence intervals in genetic models. *Behavior Genetics* **27**(2), 113–120.

Pawitan Y 2001. *In all likelihood: statistical modelling and inference using likelihood*. Oxford University Press, Oxford, NY.

Preacher KJ and Hayes AF 2004. SPSS and SAS procedures for estimating indirect effects in simple mediation models. *Behavior Research Methods, Instruments, & Computers* **36**(4), 717–731.

Raykov T and Marcoulides GA 2001. Can there be infinitely many models equivalent to a given covariance structure model? *Structural Equation Modeling: A Multidisciplinary Journal* **8**(1), 142–149.

Raykov T and Marcoulides G 2007. Equivalent structural equation models: a challenge and responsibility. *Structural Equation Modeling: A Multidisciplinary Journal* **14**(4), 695–700.

Raykov T and Shrout PE 2002. Reliability of scales with general structure: point and interval estimation using a structural equation modeling approach. *Structural Equation Modeling: A Multidisciplinary Journal* **9**(2), 195–212.

Rice JA 2007. *Mathematical statistics and data analysis*, 3rd edn. Thomson/Brooks/Cole, Belmont, CA.

Rindskopf D 1984. Using phantom and imaginary latent-variables to parameterize constraints in linear structural models. *Psychometrika* **49**(1), 37–47.

Rosseel Y 2012. Lavaan: an R package for structural equation modeling. *Journal of Statistical Software* **48**(2), 1–36.

Satorra A and Bentler PM 1988. Scaling corrections for chi-square statistics in covariance structure analysis *Proceedings of the Business and Economic Section of the American Statistical Association*. American Statistical Association, Alexandria, VA, pp. 308–313.

Satorra A and Bentler PM 1994. Corrections to test statistics and standard errors in covariance structure analysis In *Latent variables analysis: applications for developmental research* (ed. von Eye A and Clogg CC). Sage Publications, Inc., Thousand Oaks, CA, pp. 399–419.

Schwarz G 1978. Estimating the dimension of a model. *Annals of Statistics* **6**(2), 461–464.

Seber GAF 2003. *Nonlinear regression*. Wiley-Interscience, Hoboken, NJ.

Skrondal A and Rabe-Hesketh S 2004. *Generalized latent variable modeling: multilevel, longitudinal, and structural equation models*. Chapman & Hall/CRC, Boca Raton, FL.

Sörbom D 1974. A general method for studying differences in factor means and factor structure between groups. *British Journal of Mathematical and Statistical Psychology* **27**(2), 229–239.

Steiger JH 1998. A note on multiple sample extensions of the RMSEA fit index. *Structural Equation Modeling: A Multidisciplinary Journal* **5**(4), 411–419.

Steiger JH 2002. When constraints interact: a caution about reference variables, identification constraints, and scale dependencies in structural equation modeling. *Psychological Methods* **7**(2), 210–227.

Steiger JH and Fouladi RT 1997. Noncentrality interval estimation and the evaluation of statistical models In *What if there were no significance tests?* (ed. Harlow LL, Mulaik SA and Steiger JH). Lawrence Erlbaum Associates, Inc., Mahwah, NJ, pp. 221–257.

Steiger JH and Lind JC 1980. Statistically based tests for the number of common factors. *Paper presented at the annual meeting of the Psychometric Society*, Iowa City, IA.

Tucker LR and Lewis C 1973. A reliability coefficient for maximum likelihood factor analysis. *Psychometrika* **38**(1), 1–10.

Vandenberg RJ and Lance CE 2000. A review and synthesis of the measurement invariance literature: suggestions, practices, and recommendations for organizational research. *Organizational Research Methods* **3**(1), 4–70.

Viechtbauer W 2005. Bias and efficiency of meta-analytic variance estimators in the random-effects model. *Journal of Educational and Behavioral Statistics* **30**(3), 261–293.

Widaman KF and Thompson JS 2003. On specifying the null model for incremental fit indices in structural equation modeling. *Psychological Methods* **8**(1), 16–37.

Wright S 1921. Correlation and causation. *Journal of Agricultural Research* **20**(7), 557–585.

Wu W, West SG and Taylor AB 2009. Evaluating model fit for growth curve models: integration of fit indices from SEM and MLM frameworks. *Psychological Methods* **14**(3), 183–201.

Yuan KH and Bentler PM 2007. Robust procedures in structural equation modeling In *Handbook of latent variable and related models* (ed. Lee SY). Elsevier/North-Holland Amsterdam, Boston, MA, pp. 367–397.

3

Computing effect sizes for meta-analysis

This chapter covers how to estimate common effect sizes and their sampling variances in a meta-analysis. We begin by introducing the formulas to compute effect sizes and their sampling variances for a univariate meta-analysis. Formulas to calculate effect sizes for a multivariate meta-analysis are then introduced. This chapter then introduces a general approach to calculate the sampling variances for the univariate effect sizes and the sampling covariance matrices for the multivariate effect sizes. The delta method is used to the approximate sampling variances of the effect sizes by considering the effect sizes as functions of the summary statistics. We show how structural equation modeling (SEM) can be used as a computational device to simplify the procedures on estimating the approximate sampling variances and covariances. Examples are used to illustrate the procedures in the R statistical environment.

3.1 Introduction

The main difference between a narrative review and a meta-analysis is that effect sizes are explicitly calculated and synthesized in a meta-analysis. An effect size summarizes the result of a study. A meta-analysis cannot be conducted without the effect sizes and their sample variances. This chapter introduces how to calculate some common effect sizes and their sample variances. These effect sizes serve as the ingredients for the statistical modeling that will be introduced in later chapters.

There are many definitions of an effect size. One common definition is that an effect size is a quantitative measure of the magnitude of some phenomenon that is used for the purpose of addressing a question of interest (see Kelley and Preacher

Meta-Analysis: A Structural Equation Modeling Approach, First Edition. Mike W. -L. Cheung.

Companion Website: www.wiley.com/go/cheung/meta_analysis

(2012) for a review). The effect size has to be scaled properly. It is sometimes appropriate to use the raw scores, whereas some forms of standardization are required in many settings.

There are several properties for the effect sizes in a meta-analysis. First, the effect size should be directional. As the effect can be either positive or negative, the effect size should indicate the direction of the effect size. For example, the percentage of variance explained R^2 is a popular index to summarize the effect of prediction in regression analysis. It is rarely used in the meta-analysis because R^2 does not indicate the directions of the predictors. R^2 is always nonnegative regardless of the signs of the regression coefficients. Moreover, it combines the effects of all predictors, while different studies may include different numbers of predictors. It is difficult to synthesize R^2 based on different numbers of predictors. Similarly, the percentage of variance explained in analysis of variance (ANOVA), such as η^2 and ω^2, is usually not appropriate to be used as effect sizes in a meta-analysis.

The effect size should also be *relatively* independent of the sample size. In other words, the measured effect size should not get larger (or smaller) simply because a large sample size is used. If the effect sizes are biased at the small samples, some corrections may be applied to adjust for the bias. In this book, we let f_i be a population or *true* effect size, and y_i be its observed effect size in the ith study. The effect size can be a mean difference, a correlation coefficient, or a (log) odds ratio (OR). The theory behind the meta-analysis is general enough to be applied to any type of effect size. Besides extracting the effect sizes, we also need to calculate the sampling variance v_i of y_i. v_i quantifies the precision of y_i. As most meta-analytic procedures need to weigh the studies by their precision, it is of importance to obtain v_i in a meta-analysis. It should also be noted that it is usually the v_i, not the sample size, is used in the meta-analysis (cf. Schmidt and Hunter, 2015). Although v_i is primarily determined by the sample size, v_i may also be affected by other factors for some effect sizes. Once we have calculated v_i, the sample sizes are rarely involved in the meta-analysis.

When the sample sizes in the primary studies are *reasonably* large, it is assumed that y_i is normally distributed around the *true* effect size f_i in the ith study with a known sampling variance v_i, that is, $y_i \sim \mathcal{N}(f_i, v_i)$. This is known as the conditional sampling variance. Another way to express it is

$$y_i = f_i + e_i, \tag{3.1}$$

with $\mathsf{Var}(e_i) = v_i$. We assume that y_i is unbiased in estimating the *true* effect size f_i or $\mathsf{E}(y_i) = f_i$. As we are going to assume that y_i is normally distributed in this book, it is necessary to discuss when this assumption is appropriate. The appropriateness of the normality assumption depends on several factors, for example,

(i) the sample size in the primary studies;

(ii) the distribution of the raw data;

(iii) the type of effect size; and

(iv) the population value of the *true* effect size.

Sample size is the most obvious factor to determining the distributions of the effect sizes. Effect sizes are usually calculated from the summary statistics, such as the means, variances, and covariances. Because of the central limit theorem, the effect sizes will be normally distributed when the sample sizes are getting larger and larger. If the sample sizes in the primary studies are sufficiently large, all effect sizes will be normally distributed. However, how large is the sample sizes depends on other factors.

The second factor is the distribution of the raw data. When the raw data are normally distributed, the effect size will approach normal distribution much faster than the cases in which the raw data are not normally distributed, given the same conditions. The third factor is the type of effect size. For example, the raw mean difference (RMD) approaches a normal distribution faster than that based on the standardized mean difference (SMD), given the same data. In order to make some effect sizes approach normal distribution faster, some transformations may be applied. For example, we may apply a Fisher's z transformation on the correlation coefficient and a logarithm transformation on the OR. This is known as the variance stabilizing transformation. The last factor is the population value of the *true* effect size. For example, a correlation coefficient tends to be skewed when the population correlation is away from 0. Fisher's z transformation can be applied to normalize its sampling distribution.

3.2 Effect sizes for univariate meta-analysis

In some research settings, a single effect size is sufficient to summarize the effect of the study. Most published studies on meta-analysis were based on one single effect size. This section reviews some common effect sizes used in the literature. There are generally three types of effect sizes. They are the mean differences, correlation, and binary data. When different studies report different types of effect sizes, it is sometimes possible to convert the effect sizes from one type to another (e.g., Borenstein, 2009; Borenstein et al., 2009).

3.2.1 Mean differences

Many outcome measures are continuous. When there are a treatment and a control groups, the mean difference may be used to quantify the treatment effect. In educational settings, for example, researchers may want to compare the teaching effectiveness of class size by comparing small class size (treatment) versus normal class size (control). The dependent variable can be the academic performance at the end of the intervention. Both the RMD and SMD can be used as the effect sizes, depending on whether the meanings of the scales are clearly defined and comparable across studies.

3.2.1.1 Raw mean difference

When the scales are comparable across studies, the raw (or unstandardized) mean difference may be used as the effect size. The key feature of the RMD is that its meaning is the same as that of the raw data. For example, if the dependent variable is the grade point average (GPA) in educational research or blood pressure in medical research, the RMD refers to the differences on GPA or blood pressure, respectively. Bond Jr. et al. (2003) argued that the RMD is preferable because results based on the SMD may be distorted when the meanings of the scales are comparable across studies. As it is understood that we can calculate the effect size for each study, we do not include the subscript i in y_i and v_i in the following formulas to simplify the notation. The population RMD is defined as

$$f_{\mathrm{RMD}} = \mu_{\mathrm{T}} - \mu_{\mathrm{C}}, \tag{3.2}$$

where μ_{T} and μ_{C} are the population means in the treatment and control groups, respectively. The so-called *treatment* and *control* are just for the ease of reference. It is possible that there is no intervention or control in the studies. For example, we may use the same formula to calculate the gender difference on some objective test such as the Graduate Record Examinations (GRE).

The sample RMD (y_{RMD}) is estimated by

$$y_{\mathrm{RMD}} = \bar{Y}_{\mathrm{T}} - \bar{Y}_{\mathrm{C}}, \tag{3.3}$$

where $\bar{Y}_{\mathrm{T}}$ and $\bar{Y}_{\mathrm{C}}$ are the sample means in the treatment and control groups, respectively. Positive values indicate that the treatment effect is higher than that of the control group. If we assume that the population variances are the same (the assumption of homogeneity of variances), we may pool the variances together to increase the precision of the estimated variance. The pooled variance is

$$s_{\mathrm{p}}^2 = \frac{(n_{\mathrm{T}} - 1)s_{\mathrm{T}}^2 + (n_{\mathrm{C}} - 1)s_{\mathrm{C}}^2}{n_{\mathrm{T}} + n_{\mathrm{C}} - 2}, \tag{3.4}$$

where s_{T}^2 and s_{C}^2 are the sample variances in the treatment and control groups, respectively, and n_{T} and n_{C} are the sample sizes in the treatment and control groups, respectively. By applying the central limit theorem, we may estimate the sampling variances of $\bar{Y}_{\mathrm{T}}$ and $\bar{Y}_{\mathrm{C}}$ by

$$\begin{aligned} \mathsf{Var}(\bar{Y}_{\mathrm{T}}) &= \frac{s_{\mathrm{p}}^2}{n_{\mathrm{T}}} \quad \text{and} \\ \mathsf{Var}(\bar{Y}_{\mathrm{C}}) &= \frac{s_{\mathrm{p}}^2}{n_{\mathrm{C}}}. \end{aligned} \tag{3.5}$$

As Y_{T} and Y_{C} are independent, the sampling variance of y_{RMD} is simply the sum of their sampling variances, that is,

$$\begin{aligned} v_{\mathrm{RMD}} &= \mathsf{Var}(\bar{Y}_{\mathrm{T}}) + \mathsf{Var}(\bar{Y}_{\mathrm{C}}) \\ &= \left(\frac{1}{n_{\mathrm{T}}} + \frac{1}{n_{\mathrm{C}}}\right) s_{\mathrm{p}}^2. \end{aligned} \tag{3.6}$$

If we do not assume the homogeneity of variances, the sampling variances of $\bar{Y}_1$ and $\bar{Y}_2$ are

$$\widetilde{\mathsf{Var}}(\bar{Y}_\mathrm{T}) = \frac{s_\mathrm{T}^2}{n_\mathrm{T}} \quad \text{and} \qquad \widetilde{\mathsf{Var}}(\bar{Y}_\mathrm{C}) = \frac{s_\mathrm{C}^2}{n_\mathrm{C}}. \tag{3.7}$$

Then, the sampling variance of y_RMD is

$$\tilde{v}_\mathrm{RMD} = \frac{s_\mathrm{T}^2}{n_\mathrm{T}} + \frac{s_\mathrm{C}^2}{n_\mathrm{C}}. \tag{3.8}$$

It appears that $\tilde{v}_\mathrm{RMD}$ is more appealing because it does not require the assumption of homogeneity of variances. However, $\tilde{v}_\mathrm{RMD}$ is less accurate than v_RMD when the assumption of homogeneity of variances is appropriate. A larger sample size may be required in order to treat $\tilde{v}_\mathrm{RMD}$ as known in a meta-analysis. Therefore, most published meta-analyses are based on the homogeneity of variances in calculating the sampling variances.

3.2.1.2 Standardized mean difference

If the scale is less clear or the measures are different across studies, an SMD may be used. In educational settings, for example, different schools may use their own internal grading systems. It is difficult to directly compare the performance across schools. An SMD may be used as a scale-free measure. As the mean difference is divided by a standard deviation, SMD does not carry any unit. Strictly speaking, the unit of SMD is measured in terms of standard deviation in the primary studies. If the raw scores are meaningful, for example, GRE scores in educational settings, the effect sizes calculated based on the RMD and SMD can be totally different when the standard deviations are different in the primary studies (Bond Jr. et al., 2003). Researchers should determine whether the scales are comparable across studies before deciding which one to use as the effect size.

The population SMD is defined as

$$f_\mathrm{SMD} = \frac{\mu_\mathrm{T} - \mu_\mathrm{C}}{\sigma}, \tag{3.9}$$

where σ is the population variance in both groups by assuming the homogeneity of variances. One of the most popular estimators was proposed by Cohen (1992), which is generally known as the Cohen's *d*,

$$y_\mathrm{d} = \frac{\bar{Y}_\mathrm{T} - \bar{Y}_\mathrm{C}}{s_\mathrm{p}}. \tag{3.10}$$

y_d is interpreted as the mean difference between these two groups in terms of a standard unit (s_p). The approximate sampling variance of y_d is

$$v_d = \frac{n_T + n_C}{n_T n_C} + \frac{y_d^2}{2(n_T + n_C)}. \tag{3.11}$$

Hedges (1981) showed that the estimator in Equation 3.10 is slightly overestimating the absolute value of the population parameter in Equation 3.9 in small samples. Hedges (1981) further proposed a modification to minimize the bias, which is generally known as the Hedges' g. The correction factor $c(\text{df})$ is calculated as

$$c(\text{df}) = 1 - \frac{3}{4\text{df} - 1}, \tag{3.12}$$

where $\text{df} = n_T + n_C - 2$ for two independent groups. The Hedges' g is calculated as

$$\begin{aligned} y_g &= c(\text{df}) y_d \quad \text{and} \\ v_g &= [c(\text{df})]^2 v_d. \end{aligned} \tag{3.13}$$

As the correction factor $c(\text{df})$ is always less than 1, y_g is usually smaller than y_d. In practice, the differences are usually small unless the sample sizes are very small.

3.2.1.3 Repeated measures

The above formulas are not appropriate when the data are not independent. For example, data based on pre-post scores or matched data (e.g., husbands and wives) are likely correlated. The effect size and its sampling variance should take the correlation into account. Let μ_{pre} and μ_{post} be the population means of the pretest and posttest scores. If the scale is comparable across studies, for example, age difference between husbands and wives, we may define the population difference score as

$$f_{\text{Diff}} = \mu_{\text{post}} - \mu_{\text{pre}}. \tag{3.14}$$

The sample estimate based on the sample means of the pretest ($\bar{y}_{\text{pre}}$) and posttest ($\bar{y}_{\text{post}}$) scores is

$$y_{\text{Diff}} = \bar{y}_{\text{post}} - \bar{y}_{\text{pre}}. \tag{3.15}$$

The variance of the difference score s^2_{Diff} of the subjects (not the sampling variance of y_{Diff}) can be calculated from the correspondent summary statistics,

$$s^2_{\text{Diff}} = s^2_{\text{post}} + s^2_{\text{pre}} - 2 s_{\text{post}} s_{\text{pre}} r_{\text{pre,post}}, \tag{3.16}$$

where s^2_{post}, s^2_{pre}, and $r_{\text{pre,post}}$ are the sample variance of y_{post}, the sample variance of y_{pre}, and the correlation between them, respectively. The sampling variance of y_{Diff} is estimated by the central limit theorem,

$$v_{\text{Diff}} = \frac{s^2_{\text{Diff}}}{n}, \tag{3.17}$$

where n is the number of pairs of the data.

When the scale is not clear or its meaning is different across studies, we may standardize the mean difference with an appropriate standardizer so that the calculated effect sizes are comparable across studies. One obvious choice is the standard deviation of the change score (CS) (σ_{Diff}, the population parameter of s_{Diff} in Equation 3.16). We may define the population mean difference on the CS (Gibbons et al., 1993) as

$$f_{\text{CS}} = \frac{\mu_{\text{post}} - \mu_{\text{pre}}}{\sigma_{\text{Diff}}}. \tag{3.18}$$

By using the standard deviation of the difference score in Equation 3.16 as the standardizer, the sample effect size is calculated as

$$y_{\text{CS}} = \frac{\bar{y}_{\text{post}} - \bar{y}_{\text{pre}}}{s_{\text{Diff}}}. \tag{3.19}$$

Positive values on y_{CS} are interpreted as the improvement ($\bar{y}_{\text{post}} - \bar{y}_{\text{pre}}$) in terms of standard deviation on the difference score (s_{Diff}).

y_{CS} is standardized by the standard deviation of the CS, whereas y_{d} and y_{g} are standardized by the standard deviation of the groups; it is of importance to note that y_{CS} is not comparable to those of y_{d} and y_{g} calculated from independent groups. When the correlation between y_{pre} and y_{post} is larger than 0.5, which is the typical case in repeated measures (RM), y_{CS} is larger than that calculated from the independent groups. In contrast, y_{CS} is smaller than that calculated from the independent groups when the correlation is smaller than 0.5.

Studies may involve both between study and within study designs. Researchers may want to combine the effect sizes based on independent groups and RM. Researchers can compare whether the design predicts the effects by the use of a mixed-effects meta-analysis (see Morris and DeShon (2002) for a thorough discussion). We can define the population effect size for a RM that is comparable to the independent groups as

$$f_{\text{RM}} = \frac{\mu_{\text{post}} - \mu_{\text{pre}}}{\sigma_{\text{com}}}, \tag{3.20}$$

where $\sigma_{\text{com}} = \sigma_{\text{pre}} = \sigma_{\text{post}}$ is the common population standard deviation in the pretest and posttest scores. As the standardizers have similar meanings for the independent groups and the repeated measures, the calculated effect sizes are comparable.

From Equation 3.16, the common variance for the pretest and posttest scores s^2_{com} can be computed as

$$s^2_{\text{com}} = \frac{s^2_{\text{Diff}}}{2(1 - r_{\text{pre,post}})}. \tag{3.21}$$

The sample estimate similar to the Cohen's d for the repeated measures and its sampling variance are

$$\begin{aligned} y_{\text{RM(d)}} &= \frac{\bar{y}_{\text{post}} - \bar{y}_{\text{pre}}}{s_{\text{Diff}}} \sqrt{2(1 - r_{\text{pre,post}})} \quad \text{and} \\ v_{\text{RM(d)}} &= \left(\frac{1}{n} + \frac{y^2_{\text{RM(d)}}}{2n} \right) 2(1 - r_{\text{pre,post}}). \end{aligned} \tag{3.22}$$

Similar to the SMD, we may obtain a less biased effect size that is similar to the Hedges' g by

$$\begin{aligned} y_{\text{RM(g)}} &= c(\text{df}) y_{\text{RM(d)}} \quad \text{and} \\ v_{\text{Rm(g)}} &= [c(\text{df})]^2 v_{\text{RM(d)}}, \end{aligned} \tag{3.23}$$

where $c(\text{df}) = 1 - \frac{3}{4n-5}$ in repeated measures.

3.2.2 Correlation coefficient and its Fisher's z transformation

Correlation coefficient is probably one of the most popular effect sizes in applied research. As it is standardized, its value can be compared and pooled across studies. Let r be the sample correlation coefficient with n sample size. Some approaches, for example, Schmidt and Hunter (2015), the pooled correlation coefficient is obtained by weighting the sample correlations by their correspondent sample sizes. Another approach is to weight the sample correlations by their correspondent sampling variances. The approximate sampling variance of y_r can be obtained by

$$\begin{aligned} y_{\text{r}} &= r \quad \text{and} \\ v_{\text{r}} &= \frac{(1 - r^2)^2}{n}. \end{aligned} \tag{3.24}$$

As the sampling distribution of the correlation coefficient is skewed unless the population correlation is close to zero or the sample size is sufficiently large, some researchers (e.g., Hedges and Olkin, 1985) prefer to use the Fisher's z transformed score, which is defined as

$$\begin{aligned} y_{\text{z}} &= 0.5 \log \left(\frac{1 + r}{1 - r} \right) \quad \text{and} \\ v_{\text{z}} &= \frac{1}{n - 3}. \end{aligned} \tag{3.25}$$

There are two advantages of this transformation. First, y_{z} is approximately normally distributed regardless of the population value of r. Therefore, y_{z} usually approaches a normal distribution faster than that for y_{r} does. Second, v_{z} does not depend on r; the bias or sampling error in estimating r does not accumulate in estimating v_{z}. Readers may refer to Field (2001, 2005) and Hafdahl and Williams (2009) for the simulation results comparing these two approaches.

3.2.3 Binary variables

In many research settings, the outcome of interest is binary, for example, pass or failure in educational research and survived or dead in medical research. We first discuss the effect size when there is only one binary variable of interest. We then discuss the case with two binary variables.

3.2.3.1 Proportion

The prevalence or the base rate probability of a disease or a psychological disorder plays an important role in medical research and clinical psychology. Researchers are interested in synthesizing and comparing the prevalence in different studies, countries, or even regions. Let X be distributed as a binary variable with π as the population probability. Suppose that we observe x counts of success from n trials; the maximum likelihood (ML) estimate is $p = x/n$. When n is sufficiently large, the sampling variance of p is $p(1-p)/n$. As p approaches normal distribution very slowly unless π is around 0.5, it is rarely used in meta-analysis. We usually normalize the distribution by the log-odds transformation. The log-odd of the population proportion is

$$f_{\text{log(odds)}} = \log\left(\frac{\pi}{1-\pi}\right), \tag{3.26}$$

where log() is the natural logarithm. The sample effect size and its sampling variance are

$$\begin{aligned} y_{\text{log(odds)}} &= \log\left(\frac{p}{1-p}\right) \quad \text{and} \\ v_{\text{log(odds)}} &= \frac{1}{np(1-p)}. \end{aligned} \tag{3.27}$$

When $y_{\text{log(odds)}}$ is zero, it indicates that the probability is 0.5. When $y_{\text{log(odds)}}$ is positive, it indicates that the probability is higher than 0.5. As we are usually interested in the probability p rather than the log-odds, we backtransform the log-odds into p after the analysis for ease of interpretations.

3.2.3.2 Odds ratio

When there are two binary variables, we are usually interested in the association between them. Suppose that the two binary variables represent the treatment group versus the control group and the outcome variable (yes versus no). Let π_T and π_C be the population probability of yes in the treatment and control groups, respectively. There are several measures to compare these two probabilities (see Fleiss and Berlin (2009) for a comparison). OR is usually recommended in meta-analysis because it is meaningful for both experimental and observational studies (Keith et al., 1998). The OR is 1 when the probabilities are the same in both groups. When the treatment group has a higher probability, the OR is larger than 1. To normalize the sampling distribution, we take the logarithm on the OR. The population OR is defined as

$$f_{\log(\text{OR})} = \log\left(\frac{\pi_T/(1-\pi_T)}{\pi_C/(1-\pi_C)}\right). \tag{3.28}$$

Let p_T and n_T be the probability of yes and the sample size in the treatment group, respectively, and p_C and n_C be the probability of yes and the sample size in the control group for the sample data, respectively. The sample log of the OR and its sampling variance can be estimated by

$$\begin{aligned}
y_{\log(\text{OR})} &= \log\left(\frac{p_T/(1-p_T)}{p_C/(1-p_C)}\right) \quad \text{and} \\
v_{\log(\text{OR})} &= \mathsf{Var}\left(\log\left(\frac{p_T/(1-p_T)}{p_C/(1-p_C)}\right)\right) \\
&= \mathsf{Var}\left(\log\left(\frac{p_T}{(1-p_T)}\right) - \log\left(\frac{p_C}{(1-p_C)}\right)\right) \\
&\quad \text{as} \quad \log\left(\frac{a}{b}\right) = \log(a) - \log(b) \\
&= \frac{1}{n_T p_T(1-p_T)} + \frac{1}{n_C p_C(1-p_C)} \\
&\quad \text{as} \quad p_T \quad \text{and} \quad p_C \quad \text{are independent.}
\end{aligned} \tag{3.29}$$

3.3 Effect sizes for multivariate meta-analysis

There are usually more than one dependent variables or treatment conditions in the primary studies. One single effect size may not be sufficient to summarize the results for a study. Researchers conducting a meta-analysis may need to synthesize multiple effect sizes rather than a univariate effect size. This section extends the calculations of univariate effect sizes to multiple effect sizes in a multivariate meta-analysis. We first introduce how to calculate the multiple effect sizes for the mean differences. Then we discuss how to calculate correlation matrices and the effect sizes for binary variables. It should be noted that this section focuses on the cases with more than one effect sizes per study rather than the effect sizes for a multivariate statistics. As the effect sizes for a multivariate statistic, for example, the Wilks's lambda λ in multivariate analysis of variance (MANOVA) (e.g., Kline, 2013), are usually nondirectional, they are not suitable for a meta-analysis.

3.3.1 Mean differences

There are two typical scenarios for the needs of using multiple effect sizes for the mean differences (Gleser and Olkin, 1994, 2009). They are the *multiple treatment studies* and the *multiple-endpoint studies*. The multiple treatment studies compare more than one treatment groups against the same control group. As the same control group is used in the comparison, the calculated effect sizes are nonindependent. The multiple-endpoint studies report more than one effect sizes in each primary studies. Again, these multiple effect sizes are not independent.

3.3.1.1 Multiple treatment studies

For most experimental or intervention studies, there are usually more than one experimental groups with different levels of manipulations, whereas there is only one control group in order to minimize the cost and the number of participants. This is reflected by the fact that analysis of variance (ANOVA) is usually more popular than the independent sample t test in experimental design. Suppose that there are a total of k treatment groups with one control group; let $\mu_{\mathrm{T}j}$ and μ_{C} be the population means of the jth treatment group and the control group, respectively, and $\bar{y}_{\mathrm{T}j}$ and $\bar{y}_{\mathrm{C}}$ be the sample means of the jth treatment group and the control group, respectively. We further assume that the variances are homogeneous with a common variance σ^2_{p} for all groups. With a slight abuse of notation (cf. Equation 3.4), we may calculate the pooled variance as

$$s^2_{\mathrm{p}} = \frac{\sum_{i=1}^{k}(n_{\mathrm{T}i} - 1)s^2_{\mathrm{T}i} + (n_{\mathrm{C}} - 1)s^2_{\mathrm{C}}}{\sum_{i=1}^{k}(n_{\mathrm{T}i} - 1) + (n_{\mathrm{C}} - 1)}, \tag{3.30}$$

where $s^2_{\mathrm{T}i}$ and $n_{\mathrm{T}i}$ are the sample variance and sample size in the ith group, respectively.

As there are more than one effect size, we use the notations $\boldsymbol{\mu}_{\mathrm{MT(d)}}$ and $\boldsymbol{y}_{\mathrm{MT(d)}}$ to represent the $k \times 1$ vectors of the population and sample Cohen's d for the multiple treatment (MT) studies and $\boldsymbol{V}_{\mathrm{MT(d)}}$ to represent its $k \times k$ sampling covariance matrix. The jth treatment and its sampling variance can be estimated similarly to those in Equations 3.10 and 3.11,

$$\begin{aligned} y_{\mathrm{MT(d)}j} &= \frac{\bar{y}_{\mathrm{T}j} - \bar{y}_{\mathrm{C}}}{s_{\mathrm{p}}} \quad \text{and} \\ V_{\mathrm{MT(d)}jj} &= \frac{1}{n_{\mathrm{T}j}} + \frac{1}{n_{\mathrm{C}}} + \frac{y^2_{\mathrm{MT(d)}j}}{2n_{\mathrm{Total}}}, \end{aligned} \tag{3.31}$$

where $n_{\mathrm{Total}} = \sum_{i=i}^{k} n_{\mathrm{T}j} + n_{\mathrm{C}}$ is the total sample size in the study. Gleser and Olkin (1994, 2009) showed that the approximate sampling covariance between the ith and jth treatment groups can be estimated by

$$V_{\mathrm{MT(d)}ij} = \frac{1}{n_{\mathrm{C}}} + \frac{y_{\mathrm{MT(d)}i}y_{\mathrm{MT(d)}j}}{2n_{\mathrm{Total}}}. \tag{3.32}$$

3.3.1.2 Multiple-endpoint studies

Another type of multiple effect sizes is known as the multiple-endpoint (ME) studies. Researchers may calculate the SMDs between two independent groups. For example, when comparing the gender differences on academic performance, researchers may report gender differences on both mathematical achievement and language achievement. We may calculate the SMDs on mathematical achievement and language achievement independently. However, this practice does not take into account the mathematical achievement and language achievement that are usually positively correlated and so does the calculated SMDs.

Suppose that there are p dependent variables in the treatment and control groups. By assuming that the covariance matrices are homogeneous across these two groups with a common population covariance matrix Σ_p, we may calculate a pooled covariance matrix

$$\boldsymbol{S}_\text{p} = \frac{(n_\text{T} - 1)\boldsymbol{S}_\text{T} + (n_\text{C} - 1)\boldsymbol{S}_\text{C}}{n_\text{T} + n_\text{C} - 2}, \tag{3.33}$$

where $\boldsymbol{S}_\text{T}$ and $\boldsymbol{S}_\text{C}$ are the sample covariance matrices in the treatment and control groups, respectively, and n_T and n_C are the sample sizes in the treatment and control groups, respectively. It is of importance to note that s^2 and its population counterpart σ^2 and $\boldsymbol{S}$ and its population counterpart $\boldsymbol{\Sigma}$ are usually used to represent the variance and the covariance matrix, respectively. Therefore, we need to use $\sqrt{\boldsymbol{S}_{ii}}$ when referring to the standard deviation in the ith variable.

$\boldsymbol{y}_{\text{ME(d)}}$ and $\boldsymbol{V}_{\text{ME(d)}}$ are used to denote the $p \times 1$ vector of the Cohen's d for the ME studies and its $p \times p$ sampling covariance matrix. The jth endpoint (dependent variable) can be estimated similar to that in Equations 3.10 and 3.11 by

$$\begin{aligned} y_{\text{ME(d)}j} &= \frac{\bar{y}_{\text{T}j} - \bar{y}_{\text{C}j}}{\sqrt{\boldsymbol{S}_{\text{p}jj}}} \quad \text{and} \\ V_{\text{ME(d)}jj} &= \frac{1}{n_\text{T}} + \frac{1}{n_\text{C}} + \frac{y^2_{\text{ME(d)}j}}{2(n_\text{T} + n_\text{C})}. \end{aligned} \tag{3.34}$$

Gleser and Olkin (1994, 2009) showed that large sample approximation of the sampling covariance between the ith and jth endpoints (effect sizes) is

$$V_{\text{ME(d)}ij} = \left(\frac{1}{n_\text{T}} + \frac{1}{n_\text{C}}\right) r_{ij} + \frac{y_{\text{ME(d)}i} y_{\text{ME(d)}j}}{2(n_\text{T} + n_\text{C})} r^2_{ij}, \tag{3.35}$$

where r_{ij} is the correlation between the ith and jth variables. Sometimes, the correlation between the variables may not be available in the primary studies. We need to estimate them from other sources of information.

3.3.2 Correlation matrix and its Fisher's z transformation

When there are more than two variables, a correlation or a covariance matrix is used to indicate their linear association. Many of the multivariate statistics, such as regression analysis, path analysis, exploratory and confirmatory factor analyses, and even SEM can be conducted based on the correlation or covariance matrices under the normality assumption. Correlation matrices are also the data for a meta-analytic structural equation modeling (MASEM) introduced in Chapter 7.

The sampling variances of the correlation coefficients can be estimated by Equation 3.24. As the correlation coefficients within the same study are likely correlated, we also need to estimate their sampling covariances. Several authors (e.g., Nel, 1985, Olkin and Siotani, 1976) have shown that the approximate sampling covariance between two correlation coefficients r_{ij} and r_{kl} can be estimated by

$$\begin{aligned}\mathsf{Cov}(r_{ij}, r_{kl}) = (0.5\rho_{ij}\rho_{kl}(\rho_{ik}^2 + \rho_{il}^2 + \rho_{jk}^2 + \rho_{jl}^2) + \rho_{ik}\rho_{jl} + \rho_{il}\rho_{jk} - \\ (\rho_{ij}\rho_{ik}\rho_{il} + \rho_{ij}\rho_{jk}\rho_{jl} + \rho_{ik}\rho_{jk}\rho_{kl} + \rho_{il}\rho_{jl}\rho_{kl}))/n.\end{aligned} \tag{3.36}$$

We may also convert the correlation vector into a vector of Fisher's z scores and their sampling variances with Equation 3.25. The sampling covariance between z_{ij} (the Fisher's z score of r_{ij}) and z_{kl} (the Fisher's z score of r_{kl}) can be estimated by (e.g., Hafdahl, 2007; Steiger, 1980)

$$\mathsf{Cov}(z_{ij}, z_{kl}) = \frac{n\mathsf{Cov}(r_{ij}, r_{kl})}{(n-3)(1-r_{ij}^2)(1-r_{kl}^2)}. \tag{3.37}$$

3.3.3 Odds ratio

Similar to continuous outcome variables, multiple treatment studies are also popular in studies with binary outcome variables. For example, there may be several intervention groups in testing the effectiveness of different types of therapies, whereas there is only one control group.

Suppose that there are a total of k intervention groups with one control group; $\boldsymbol{y}_{\log(\text{OR})}$ and $\boldsymbol{V}_{\log(\text{OR})}$ represent the $k \times 1$ vector of effect sizes and its $k \times k$ sampling covariance matrix. The jth treatment and its sampling variance can be estimated by Equation 3.29,

$$\begin{aligned} y_{\log(\text{OR})j} &= \log\left(\frac{p_{\text{T}j}/(1-p_{\text{T}j})}{p_{\text{C}}/(1-p_{\text{C}})}\right) \quad \text{and} \\ \boldsymbol{V}_{\log(\text{OR})jj} &= \frac{1}{n_{\text{T}j}p_{\text{T}j}(1-p_{\text{T}j})} + \frac{1}{n_{\text{C}}p_{\text{C}}(1-p_{\text{C}})}. \end{aligned} \tag{3.38}$$

As the control group is used as the reference in all treatment groups, Gleser and Olkin (2009) (see also Bagos, 2012) showed that the sampling covariance between the ith and jth treatment groups is simply

$$\boldsymbol{V}_{\log(\text{OR})ij} = \frac{1}{n_{\text{C}}p_{\text{C}}(1-p_{\text{C}})}. \tag{3.39}$$

3.4 General approach to estimating the sampling variances and covariances

The above sections reviewed some popular effect sizes and their sampling variances or covariances in a meta-analysis. Gleser and Olkin (1994, 2009) provided formulas to estimating the sampling covariance matrix for many common multiple effect sizes. There are cases in which the standard formulas are not sufficient for the researchers to calculate the sampling variance covariance matrix of the multiple effect sizes. For example, there may be studies involving both MT and ME studies. That is, there are multiple dependent variables per study and more than one treatment groups. Although it is still possible to extend the above formulas to this case, the derivations are complicated and prompted to errors.

On the other hand, there may be more than one definition of effect sizes for a particular research setting. Researchers may want to explore how sensitive the conclusions are when different definitions of effect sizes are used. This is an important part of the judgment calls in meta-analysis (e.g., Aguinis et al., 2011; Aytug et al., 2012). In calculating the SMD, for example, we may test the robustness of the results by considering with or without the assumption of homogeneity of variances in calculating the effect sizes in the between-group studies.

There are at least three different standardizers in calculating the effect sizes for repeated measures—(i) the standard deviation of the change score; (ii) the common standard deviation assuming the homogeneity of the pre- and posttest scores; and (iii) the standard deviation of the pretest score. This situation becomes even more complicated for multiple effect sizes. The calculated effect sizes and their sampling variances may be different in the multiple treatment studies if we do not assume the homogeneity of variances.

This section introduces a general approach to derive the required sampling variances and covariances. As most effect sizes are functions of the basic summary statistics, such as the means and covariance matrices, we use the delta method to estimate their sampling variances or covariance matrices. Moreover, we also introduce the use of SEM to do the numerical calculations. Researchers do not need to analytically derive the formulas. It should be noted that both the delta method and the applications of SEM are based on large samples. The calculated sampling variances or covariance matrices are approximately correct when the sample sizes are reasonably large. Simulation studies may be used to verify the accuracy of these estimates in typical sample sizes.

3.4.1 Delta method

We need both the effect size y_i and its sampling variance v_i (or $\mathbf{y}_i$ and $\mathbf{V}_i$ for multiple effect sizes) to conduct a meta-analysis. It is usually easy to compute y_i, whereas it is more challenging to estimate its sampling variance. The delta method can be used to derive the sampling variance of arbitrary functions (e.g., Bishop et al., 1975, Section 14.6). We illustrate the key ideas with the examples of log-odd on the probability (Equation 3.27) and the Fisher's z-transformed score on the correlation (Equation 3.25).

Suppose that there is a random variable x with the mean μ and variance σ^2. We would like to calculate a function $f(x)$ on it. We may approximate the function with the first-order Taylor series,

$$f(x) \approx f(\mu) + \frac{df(\mu)}{dx}(x - \mu), \tag{3.40}$$

where $\frac{df(\mu)}{dx}$ is the derivative of $f(x)$ evaluated at μ. As we rarely know the population mean, we substitute it by the sample mean.

Taking the expectations on both sides, we have

$$\begin{aligned} \mathsf{E}(f(x)) &\approx f(\mu) \quad \text{and} \\ \mathsf{Var}(f(x)) &\approx \left(\frac{df(\mu)}{dx}\right)^2 \mathsf{Var}(x-\mu) \\ &\approx \left(\frac{df(\mu)}{dx}\right)^2 \sigma^2. \end{aligned} \tag{3.41}$$

Suppose that an effect size y is a function of a parameter x with its sampling variance v_x; we may estimate the effect size and its sampling variance v_y by

$$\begin{aligned} y &= f(x) \quad \text{and} \\ v_y &= \left(\frac{df(x)}{dx}\right)^2 v_x. \end{aligned} \tag{3.42}$$

Let us illustrate the idea with the example on sampling variance of log-odds from Millar (2011). We derive the sampling variance of $y_{\log(\text{odds})}$, which is a function of p, shown in Equation 3.27 using the delta method. We first calculate $\frac{df(p)}{dp}$ by

$$\begin{aligned} \frac{df(p)}{dp} &= \frac{\mathrm{d}\log(p/(1-p))}{dp} \\ &= \frac{1-p}{p}\frac{\mathrm{d}(p/(1-p))}{dp} \quad \text{as} \quad \frac{\mathrm{d}\log(u)}{dx} = \frac{1}{u}\frac{du}{dx} \\ &= \frac{1-p}{p}\frac{(1-p)+p}{(1-p)^2} \quad \text{as} \quad \frac{d}{dx}\left(\frac{u}{v}\right) = \frac{v\frac{du}{dx} - u\frac{dv}{dx}}{v^2} \\ &= \frac{1}{p(1-p)}. \end{aligned} \tag{3.43}$$

The sampling variance is then calculated by

$$\begin{aligned} v_{\log(\text{odds})} &= \left(\frac{df(p)}{dp}\right)^2 \mathsf{Var}(p) \\ &= \left(\frac{1}{p(1-p)}\right)^2 \frac{p(1-p)}{n} \\ &= \frac{1}{np(1-p)}. \end{aligned} \tag{3.44}$$

As a second example, we derive the sampling variance of the Fisher's z-transformed score, which is a function of r. We first calculate $\frac{df(r)}{dr}$,

$$\begin{aligned}
\frac{df(r)}{dr} &= \frac{d(0.5\log((1+r)/(1-r)))}{dr} \\
&= 0.5\frac{1-r}{1+r}\frac{d((1+r)/(1-r))}{dr} \quad \text{as} \quad \frac{d\log(u)}{dx} = \frac{1}{u}\frac{du}{dx} \\
&= 0.5\frac{1-r}{1+r}\frac{(1-r)+(1+r)}{(1-r)^2} \quad \text{as} \quad \frac{d}{dx}\left(\frac{u}{v}\right) = \frac{v\frac{du}{dx} - u\frac{dv}{dx}}{v^2} \\
&= \frac{1}{(1+r)(1-r)}.
\end{aligned} \tag{3.45}$$

We calculate the approximate sampling variance of the Fisher's z-transformed score by

$$\begin{aligned}
v_z &= \left(\frac{df(r)}{dr}\right)^2 \mathsf{Var}(r) \\
&= \left(\frac{1}{(1+r)(1-r)}\right)^2 \frac{(1-r^2)^2}{n} \\
&= \left(\frac{1}{(1+r)(1-r)}\right)^2 \frac{((1+r)(1-r))^2}{n} \quad \text{as} \quad x^2 - y^2 = (x+y)(x-y) \\
&= \frac{1}{n}.
\end{aligned} \tag{3.46}$$

This result is the same as that in Equation 3.25 when the sample sizes are large. Using $(n-3)$ is preferred to n when the sample sizes are not large.

The delta method can be generalized to functions of more than one variable and to more than one functions e.g., Millar, (2011). We focus on the general case of more than one functions, which is also known as the multivariate delta method. Suppose that $\underset{g\times 1}{\boldsymbol{x}}$ is a vector of g random variables that has the mean vector $\underset{g\times 1}{\boldsymbol{\mu}}$ and variance–covariance matrix $\underset{g\times g}{\boldsymbol{V}_{\boldsymbol{x}}}$. We are interested in calculating p functions from these g random variables, that is, $\underset{p\times 1}{\boldsymbol{y}} = \boldsymbol{f}(\boldsymbol{x})$. For example, $\boldsymbol{x}$ includes the means and covariance matrices, while $\boldsymbol{y} = \boldsymbol{f}(\boldsymbol{x})$ is a vector of effect sizes for ME studies. More importantly, we also want to estimate the sampling covariance matrix of the multiple effect sizes $\underset{p\times p}{\boldsymbol{V}_{\boldsymbol{y}}}$.

Similar to the applications of univariate delta method, the multivariate delta method is based on the first-order Taylor series,

$$\underset{p\times 1}{\boldsymbol{f}(\boldsymbol{x})} \approx \underset{p\times 1}{\boldsymbol{f}(\boldsymbol{\mu})} + \underset{p\times g}{\frac{\partial \boldsymbol{f}(\boldsymbol{\mu})}{\partial \boldsymbol{x}}} \underset{g\times 1}{(\boldsymbol{x} - \boldsymbol{\mu})}, \tag{3.47}$$

where $\frac{\partial f(\mu)}{\partial x}$ is the partial derivatives evaluated at $\boldsymbol{\mu}$. As we rarely know the population parameters, we substitute it by the sample values. Suppose that a vector of effect size $\boldsymbol{y}$ is a function of some parameter estimates $\boldsymbol{x}$ with its sampling covariance matrix $V_{\boldsymbol{x}}$; we may estimate the effect size and its sampling covariance matrix by

$$\begin{aligned} \underset{p\times 1}{\boldsymbol{y}} &= \underset{p\times 1}{\boldsymbol{f}(\boldsymbol{x})} \quad \text{and} \\ \boldsymbol{V}_y &= \underset{p\times g}{\frac{\partial f(\boldsymbol{x})}{\partial \boldsymbol{x}}} \underset{g\times g}{\boldsymbol{V}_{\boldsymbol{x}}} \underset{g\times p}{\left(\frac{\partial f(\boldsymbol{x})}{\partial \boldsymbol{x}}\right)^{\mathrm{T}}}. \end{aligned} \tag{3.48}$$

Let us illustrate the ideas with the RMD that is defined as $f(\boldsymbol{x}) = \bar{Y}_{\mathrm{T}} - \bar{Y}_{\mathrm{C}}$. As there is only one function, the $\underset{p\times g}{\frac{\partial f(\mu)}{\partial x}}$ is

$$\begin{aligned} \frac{\partial f(\boldsymbol{x})}{\partial \boldsymbol{x}} &= \begin{bmatrix} \frac{\partial(\bar{Y}_{\mathrm{T}}-\bar{Y}_{\mathrm{C}})}{\partial \bar{Y}_{\mathrm{T}}} & \frac{\partial(\bar{Y}_{\mathrm{T}}-\bar{Y}_{\mathrm{C}})}{\partial \bar{Y}_{\mathrm{C}}} \end{bmatrix} \\ &= \begin{bmatrix} 1 & -1 \end{bmatrix}. \end{aligned} \tag{3.49}$$

As the two groups are independent, the sampling covariance is zero. From Equation 3.5, the sampling covariance matrix of $\bar{Y}_{\mathrm{T}}$ and $\bar{Y}_{\mathrm{C}}$ with the assumption of homogeneity of variances is

$$\boldsymbol{V}_{\boldsymbol{x}} = \begin{bmatrix} \frac{s_{\mathrm{p}}^2}{n_{\mathrm{T}}} & \\ 0 & \frac{s_{\mathrm{p}}^2}{n_{\mathrm{C}}} \end{bmatrix}. \tag{3.50}$$

Then the sampling variance of y_{RMD} in Equation 3.6 can be estimated by

$$\begin{aligned} v_{\mathrm{RMD}} &= \begin{bmatrix} 1 & -1 \end{bmatrix} \begin{bmatrix} \frac{s_{\mathrm{p}}^2}{n_{\mathrm{T}}} & \\ 0 & \frac{s_{\mathrm{p}}^2}{n_{\mathrm{C}}} \end{bmatrix} \begin{bmatrix} 1 \\ -1 \end{bmatrix} \\ &= \frac{n_{\mathrm{T}} + n_{\mathrm{C}}}{n_{\mathrm{T}} n_{\mathrm{C}}} s_{\mathrm{p}}^2. \end{aligned} \tag{3.51}$$

When the variances are involved in the calculations of the effect sizes, such as the SMD, we may need to obtain the approximate sampling variances of the variances. When the data are normally distributed, the sampling variance of the variance s^2 can be calculated by (e.g., Tamhane and Dunlop, 2000)

$$\mathsf{Var}(s^2) = \frac{2s^4}{n-1}. \tag{3.52}$$

3.4.2 Computation with structural equation modeling

The delta method is very useful to deriving the sampling variance and covariance matrix of the effect sizes. However, we still need to calculate the $\frac{\partial f(x)}{\partial x}$ and the sampling covariance matrix of the parameter estimates $\boldsymbol{V}_{\boldsymbol{x}}$. The calculations are usually not trivial and subjected to errors. Many authors have demonstrated how SEM can be used to conduct many multivariate statistics, such as ANOVA, MANOVA, regression analysis, and reliability analysis (e.g., Cheung, 2009, Cheung and Chan, 2004; Preacher, 2006; Raykov, 2001). After fitting the proposed model, the SEM packages provide the parameter estimates and their sampling covariance matrix as the standard outputs. More importantly, some SEM packages, for example, `LISREL`, M*plus*, and `lavaan`, can be directly used to compute functions of parameter estimates and their sampling covariance matrix using the delta method. Researchers only need to specify the functions without the need to derive the $\frac{\partial f(x)}{\partial x}$ and $\boldsymbol{V}_{\boldsymbol{x}}$. Another advantage of the SEM approach is that the constraints, for example, assumptions of homogeneity of variance or covariance matrices, can be tested with the use of a likelihood ratio (LR) statistic. We are going to illustrate the idea with some examples.

3.4.2.1 Repeated measures

Figure 3.1 shows a structural equation model for repeated measures. The latent variables η_{pre} and η_{post} represent the standardized variables for x_{pre} and x_{post} with σ_{pre} and σ_{post} as the standard deviations. μ_{pre} and μ_{post} represent the population means for x_{pre} and x_{post}, respectively. $\rho_{\text{pre,post}}$ is the correlation between the pre- and posttest scores.

Several definitions of effect sizes can be calculated. If the scale is meaningful, we may calculate an effect size on the mean difference,

$$f_{\text{Diff}} = \mu_{\text{post}} - \mu_{\text{pre}}. \tag{3.53}$$

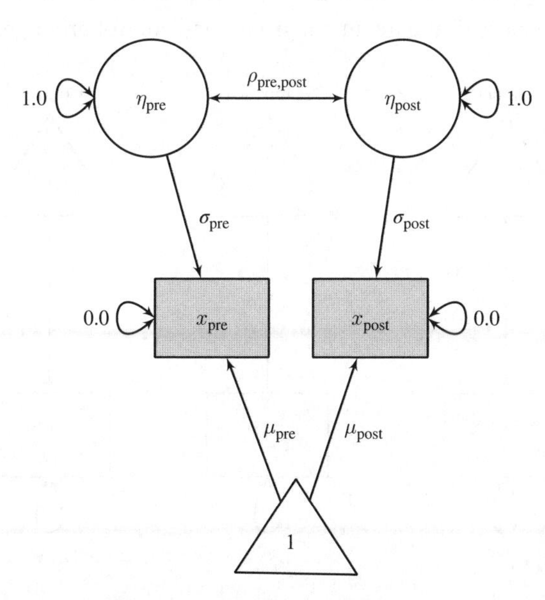

Figure 3.1 A structural equation model for repeated measures.

When the scale is arbitrary, we may calculate several standardized versions of the mean differences. The first effect size is to use the σ_{pre} as the standardizer. This effect size is useful when the standard deviations on the pre- and posttest scores are different. The function is

$$f_{\text{Pre}} = \frac{\mu_{\text{post}} - \mu_{\text{pre}}}{\sigma_{\text{pre}}}. \tag{3.54}$$

The second effect size is based on the homogeneity of variances at the pre- and posttest scores. As σ_{pre} and σ_{post} are parameters, we may impose an equality constraint $\sigma_{\text{com}} = \sigma_{\text{pre}} = \sigma_{\text{post}}$ in the SEM packages. Therefore, the function with the constraint is

$$\begin{aligned} f_{\text{RM}} &= \frac{\mu_{\text{post}} - \mu_{\text{pre}}}{\sigma_{\text{com}}} \quad \text{with the constraint} \\ \sigma_{\text{com}} &= \sigma_{\text{pre}} = \sigma_{\text{post}}. \end{aligned} \tag{3.55}$$

The third effect size is to use the standard deviation of the change score as the standardizer.

$$f_{\text{CS}} = \frac{\mu_{\text{post}} - \mu_{\text{pre}}}{\sqrt{\sigma^2_{\text{pre}} + \sigma^2_{\text{post}} - 2\sigma_{\text{pre}}\sigma_{\text{post}}\rho_{\text{pre,post}}}}. \tag{3.56}$$

3.4.2.2 Multiple treatment studies

Figure 3.2 shows a structural equation model for a multiple treatment study with two treatment groups and one control group. $\mu_1^{(T2)}$, $\mu_1^{(T1)}$, and $\mu_1^{(C)}$ are the means of the treatment groups and control group, respectively whereas $\sigma_1^{2(T2)}$, $\sigma_1^{2(T1)}$, and $\sigma_1^{2(C)}$ are the variances of the treatment groups and control group, respectively.

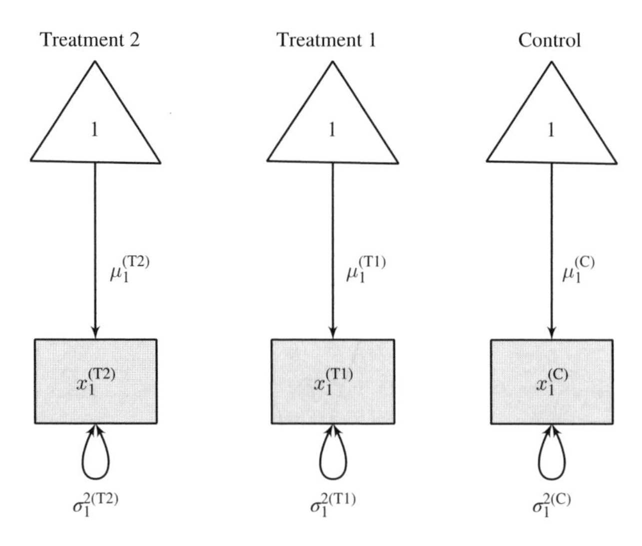

Figure 3.2 A structural equation model for multiple treatment study.

If the scales are comparable, we may calculate the two RMDs by

$$\begin{bmatrix} f_{\text{RMD1}} \\ f_{\text{RMD2}} \end{bmatrix} = \begin{bmatrix} \mu_1^{(\text{T1})} - \mu_1^{(\text{C})} \\ \mu_1^{(\text{T2})} - \mu_1^{(\text{C})} \end{bmatrix}. \tag{3.57}$$

If the scales are arbitrary, we may standardize the mean differences by the standard deviations. If we do not assume the homogeneity of variances, we may use the $\sigma_1^{(\text{C})}$ as the standardizer,

$$\begin{bmatrix} f_{\text{SMD1}} \\ f_{\text{SMD2}} \end{bmatrix} = \begin{bmatrix} \frac{\mu_1^{(\text{T1})} - \mu_1^{(\text{C})}}{\sqrt{\sigma_1^{2(\text{C})}}} \\ \frac{\mu_1^{(\text{T2})} - \mu_1^{(\text{C})}}{\sqrt{\sigma_1^{2(\text{C})}}} \end{bmatrix}. \tag{3.58}$$

If we assume the homogeneity of variances, we may estimate the effect sizes by imposing the equality constraints on the variances,

$$\begin{bmatrix} f_{\text{SMD1}} \\ f_{\text{SMD2}} \end{bmatrix} = \begin{bmatrix} \frac{\mu_1^{(\text{T1})} - \mu_1^{(\text{C})}}{\sqrt{\sigma_1^2}} \\ \frac{\mu_1^{(\text{T2})} - \mu_1^{(\text{C})}}{\sqrt{\sigma_1^2}} \end{bmatrix} \text{ with the constraint} \tag{3.59}$$
$$\sigma_1^2 = \sigma_1^{2(\text{C})} = \sigma_1^{2(\text{T1})} = \sigma_1^{2(\text{T2})}.$$

3.4.2.3 Multiple-endpoint studies

Figure 3.3 shows a structural equation model for the ME study with two outcome variables (x_1 and x_2). With this model parameterization, σ_1 and σ_2 are the standard

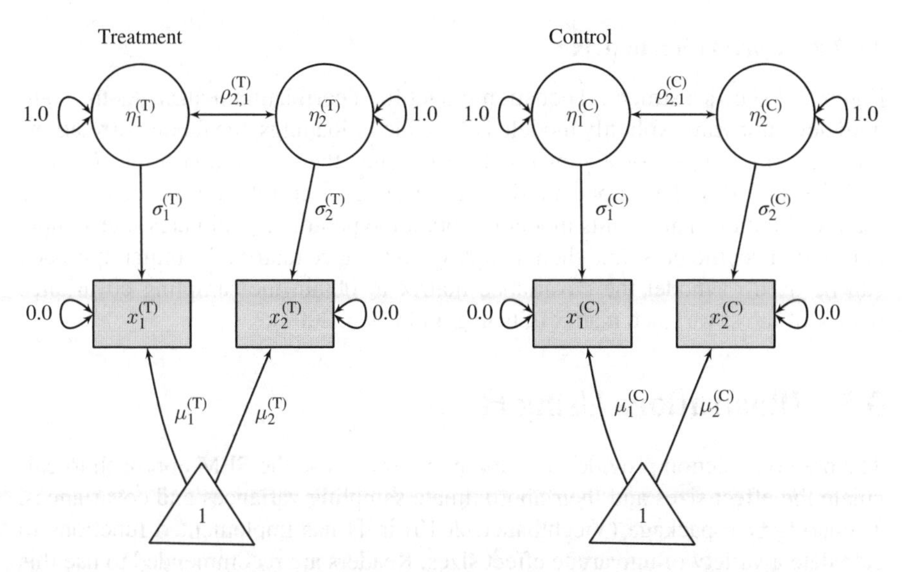

Figure 3.3 A structural equation model for multiple-endpoint study.

deviations of x_1 and x_2, respectively, whereas $\rho_{2,1}$ is the correlation between these two outcome variables. The main feature of the model is that the standard deviations (not the variances) are available for calculations.

If the scales are comparable across studies, we may define the RMDs as

$$\begin{bmatrix} f_{\text{RMD1}} \\ f_{\text{RMD2}} \end{bmatrix} = \begin{bmatrix} \mu_1^{(T)} - \mu_1^{(C)} \\ \mu_2^{(T)} - \mu_2^{(C)} \end{bmatrix}. \tag{3.60}$$

When the scales are not comparable across studies, we may calculate the SMDs to represent the effects. If we do not assume the homogeneity of covariance matrices, we may use the standard deviations in the control group as the standardizers,

$$\begin{bmatrix} f_{\text{SMD1}} \\ f_{\text{SMD2}} \end{bmatrix} = \begin{bmatrix} \frac{\mu_1^{(T)} - \mu_1^{(C)}}{\sigma_1^{(C)}} \\ \frac{\mu_2^{(T)} - \mu_2^{(C)}}{\sigma_2^{(C)}} \end{bmatrix}. \tag{3.61}$$

If we also assume the homogeneity of covariance matrices, we may estimate the effect sizes by imposing the equality constraints on the covariance matrices,

$$\begin{aligned} \begin{bmatrix} f_{\text{SMD1}} \\ f_{\text{SMD2}} \end{bmatrix} &= \begin{bmatrix} \frac{\mu_1^{(T)} - \mu_1^{(C)}}{\sigma_1} \\ \frac{\mu_2^{(T)} - \mu_2^{(C)}}{\sigma_2} \end{bmatrix} \quad \text{with the constraints} \\ \sigma_1 &= \sigma_1^{(C)} = \sigma_1^{(T)} \\ \sigma_2 &= \sigma_2^{(C)} = \sigma_2^{(T)} \\ \rho_{2,1} &= \rho_{2,1}^{(C)} = \rho_{2,1}^{(T)}. \end{aligned} \tag{3.62}$$

3.4.2.4 Correlation matrix

Figure 3.4 shows a structural equation model for a correlation matrix. As the standard deviations are explicitly modeled as the factor loadings, the factor correlations represent the sample correlation coefficients when there is no constraint (Cheung and Chan, 2004). The input matrix can be either a correlation or a covariance matrix. We may analyze this model to obtain the parameter estimates (the sample correlation coefficients) and their sampling covariance matrix. A similar approach can be used to model the covariance matrix to obtain the sampling covariance matrix of the covariance matrix (Cheung and Chan, 2009).

3.5 Illustrations Using R

The previous section provides an outline on how to use the SEM approach to calculate the effect sizes and their approximate sampling variances and covariances. The `metafor` package (Viechtbauer, 2010) in R has implemented functions to calculate a variety of univariate effect sizes. Readers are recommended to use this package to calculate the univariate effect sizes. When the required effect sizes are

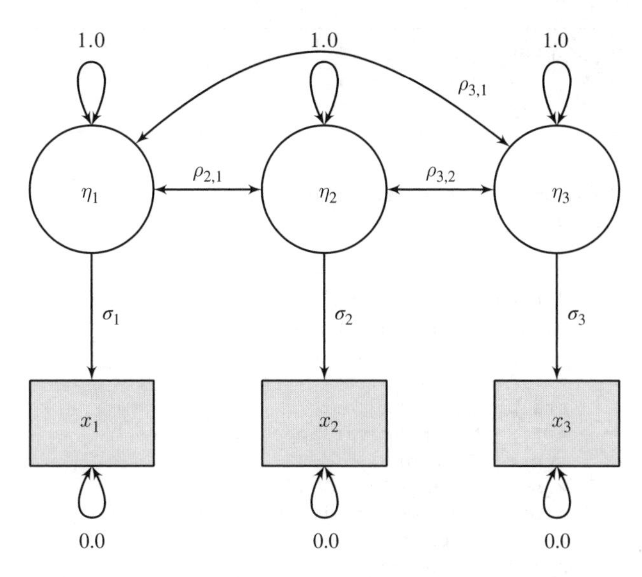

Figure 3.4 A structural equation model for a correlation matrix.

not available in the metafor package, we may use the SEM approach to approximate the sampling variances or covariance matrices. This section illustrates the SEM approach with the lavaan (Rosseel, 2012) and the metaSEM (Cheung, 2014) packages. The statistics reported in the illustrations were captured by using the Sweave function in R (Leisch, 2002). The numbers of decimal places may be slightly different for those reported in the selected output and in the text.

3.5.1 Repeated measures

Suppose the sample means for the pre- and posttest scores are 10 and 13, respectively; the sample covariance matrix between them and the sample size are $\begin{bmatrix} 10 & \\ 8 & 12 \end{bmatrix}$ and 50, respectively. The following R code fits the model showed in Figure 3.1. All the relevant parameter estimates are labeled so that they can be used to compute the effect sizes. When the labels are the same for the parameters, these parameters are constrained equally. Diff, SMD.cs, and SMD.pre are calculated based on the parameters defined earlier. After fitting the model, we may obtain the calculated effect sizes, their standard errors (SEs), and the approximate 95% Wald confidence intervals (CIs) with the parameterEstimates() function.

```
R> ## Library for the SEM
R> library("lavaan")
R> ## Sample covariance matrix on pre- and post-test scores
R> lower <- '10
             8 12'
R> ( Cov <- getCov(lower, diag=TRUE, names=c("x_pre","x_post")) )
```

```
       x_pre x_post
x_pre     10      8
x_post     8     12
```

```
R> ## Sample means for the pre- and post-test scores
R> Mean <- c(10, 13)
R> ## Sample size
R> N <- 50
R> model1 <- '# Label the sds with sd_pre and sd_post
              eta_pre =~ sd_pre*x_pre
              eta_post =~ sd_post*x_post
              # r: correlation betwen pre- and post-test
              eta_pre ~~ r*eta_post
              # Fix the error variances at 0
              x_pre ~~ 0*x_pre
              x_post ~~ 0*x_post
              # Label the means with m_pre and m_post
              x_pre ~ m_pre*1
              x_post ~ m_post*1
              # Calculate the effect sizes
              # Diff: change score
              Diff := m_post - m_pre
              # SMD.cs sd on change score as the standardizer
              SMD.cs := (m_post - m_pre)/sqrt(sd_pre^2+sd_post^2
                         -2*sd_pre*sd_post*r)
              # SMD.pre: sd_pre as the standardizer
              SMD.pre := (m_post - m_pre)/sd_pre'
R> ## Fit the model
R> fit1 <- cfa(model1, sample.cov=Cov, sample.mean=Mean,
               sample.nobs=N, std.lv=TRUE,
               sample.cov.rescale=FALSE)
R> ## Display the summary
R> ## summary(fit1)
R>
R> ## Display the selected output
R> parameterEstimates(fit1)[c(12,13,14), -c(1,2,3)]
```

```
    label   est    se     z pvalue ci.lower ci.upper
1    Diff 3.000 0.346 8.660      0    2.321    3.679
2  SMD.cs 1.225 0.187 6.547      0    0.858    1.591
3 SMD.pre 0.949 0.145 6.547      0    0.665    1.233
```

The above output shows that the y_{Diff} and $\text{SE}_{\text{Diff}} = \sqrt{v_{\text{Diff}}}$ are 3.0000 and 0.3464, respectively. The SMDs (and their SEs) using the change score and the pretest score standard deviations as the standardizers are 1.2247 (SE = 0.1871) and 0.9487 (SE = 0.1449), respectively.

We can also calculate the y_{RM} using a common standard deviation on the pretest and posttest scores. In the following model, we impose the equality constraint on the standard deviations of the pretest and posttest scores by using the same label sd for the standard deviations.

```
R> model2 <- '# Label the common sd
              eta_pre =~ sd*x_pre
              eta_post =~ sd*x_post
              # r: correlation between pre- and post-test
              eta_pre ~~ r*eta_post
              # Fix the error variances at 0
              x_pre ~~ 0*x_pre
              x_post ~~ 0*x_post
              # Label the means with m_pre and m_post
              x_pre ~ m_pre*1
              x_post ~ m_post*1
              # Calculate the effect sizes
              # Common sd
              SMD.common := (m_post-m_pre)/sd'
R> ## Fit the model
R> fit2 <- cfa(model2, sample.cov=Cov, sample.mean=Mean,
               sample.nobs=N, std.lv=TRUE,
               sample.cov.rescale=FALSE)
R> ## Display the selected output
R> parameterEstimates(fit2)[12, -c(1,2,3)]
```

```
       label   est    se     z pvalue ci.lower ci.upper
1 SMD.common 0.905 0.131 6.904      0    0.648    1.161
```

The above output shows that the effect size for the repeated measure y_{RM} (and its SE_{RM}) by assuming the same standard deviation on the pretest and posttest scores is 0.9045 (0.1310).

3.5.2 Multiple treatment studies

Suppose that the sample variances for the control and the two treatment groups are 10, 11, and 12, respectively; the sample means for them are 5, 7, and 9, respectively. The sample sizes for these groups are 50, 52, and 53, respectively. We fit the model with MT studies in Figure 3.2. As there are three groups, we need to create a list of variances as input (Var in the following example). We impose the homogeneity of variances by using the same label s2 for the variances in these groups, whereas the means are labeled as m1, m2, and m3 in these groups. When the labels are different, the parameter estimates can be different. Then, we define the effect sizes for the MT studies.

As there are two effect sizes, some extra steps are required to obtain the sampling covariance matrix of multiple effect sizes with the multivariate delta method.

```
R> ## Group 1 (control group): variance
R> var1 <- matrix(10, dimnames=list("x","x"))
```

```
R> ## Group 2 (treatment 1): variance
R> var2 <- matrix(11, dimnames=list("x","x"))
R> ## Group 3 (treatment 2): variance
R> var3 <- matrix(12, dimnames=list("x","x"))
R> ## Convert variances into a list
R> Var <- list(var1, var2, var3)
R> ## Means for the groups
R> Mean <- list(5, 7, 9)
R> ## Sample sizes for the groups
R> N <- c(50, 52, 53)
R> ## Assuming homogeneity of variances by using the same label "s2"
R> model3 <- 'x ~~ c("s2", "s2", "s2")*x
             x ~ c("m1", "m2", "m3")*1
             # SMD for treatment 1
             MT1 := (m2-m1)/sqrt(s2)
             # SMD for treatment 2
             MT2 := (m3-m1)/sqrt(s2)'
R> fit3 <- sem(model3, sample.cov=Var, sample.mean=Mean,
              sample.nobs=N, sample.cov.rescale=FALSE)
R> ## Obtain the free parameters in the model
R> ( x <- fit3@Fit@x )
```

```
[1] 11.02  5.00  7.00  9.00
```

```
R> ## Obtain the sampling covariance matrix of the parameter
R> ## estimate
R> ( VCOV <- vcov(fit3) )
```

```
   s2    m1    m2    m3
s2 1.567
m1 0.000 0.220
m2 0.000 0.000 0.212
m3 0.000 0.000 0.000 0.208
```

```
R> ## Compute the multiple effect sizes
R> ( MT <- fit3@Model@def.function(x=x) )
```

```
   MT1    MT2
0.6025 1.2050
```

```
R> ## Compute the jacobian for the 'defined parameters'
R> JAC <- lavaan:::lavJacobianD(func=fit3@Model@def.function, x=x)
R> ## Compute the sampling covariance matrix using delta method
R> MT.VCOV <- JAC %*% VCOV %*% t(JAC)
R> ## Add the variable names for ease of reference
R> dimnames(MT.VCOV) <- list(names(MT), names(MT))
R> MT.VCOV
```

```
        MT1     MT2
MT1 0.04040 0.02234
MT2 0.02234 0.04355
```

The computed multiple effect sizes and its sampling covariance matrix are $\begin{bmatrix} 0.6025 \\ 1.2050 \end{bmatrix}$ and $\begin{bmatrix} 0.0404 & \\ 0.0223 & 0.0436 \end{bmatrix}$, respectively.

3.5.3 Multiple-endpoint studies

Suppose that there are two groups (control vs treatment groups) with two dependent variables (x1 and x2). The sample covariance matrices for the control and the treatment groups are $\begin{bmatrix} 11 & \\ 5 & 10 \end{bmatrix}$ and $\begin{bmatrix} 12 & \\ 6 & 11 \end{bmatrix}$, respectively. The sample means for the control and the treatment groups are $\begin{bmatrix} 10 & 11 \end{bmatrix}^T$ and $\begin{bmatrix} 12 & 13 \end{bmatrix}^T$, respectively. The sample size for both groups is 50. We can implement the model for ME studies in Figure 3.3 in SEM. We impose the homogeneity of covariance matrices in the control and treatment groups by using the same labels on the variances (sd1 and sd2) and correlation (r). The two means in the control group are labeled with m1_1 and m2_1 and the two means in the treatment group are labeled with m1_2 and m2_2.

```
R> lower <- '11
             5, 10'
R> ## Convert a lower triangle data into a covariance matrix
R> Cov1 <- getCov(lower, diag=TRUE, names=c("x1", "x2"))
R> lower <- '12
             6, 11'
R> ## Convert a lower triangle data into a covariance matrix
R> Cov2 <- getCov(lower, diag=TRUE, names=c("x1", "x2"))
R> ## Convert covariance matrices into a list
R> Cov <- list(Cov1, Cov2)
R> ## Means for the two groups
R> Mean <- list(c(10,11), c(12,13))
R> ## Sample sizes for the groups
R> N <- c(50, 50)
R> ## Assuming homogeneity of covariance matrices by
R> ## using the same labels: "sd1", "sd2", and "r"
R> model4 <- 'eta1 =~ c("sd1", "sd1")*x1
              eta2 =~ c("sd2", "sd2")*x2
              eta1 ~~ c("r", "r")*eta2
              x1 ~ c("m1_1", "m1_2")*1
              x2 ~ c("m2_1", "m2_2")*1
              x1 ~~ 0*x1
              x2 ~~ 0*x2
              # Multiple endpoint effect size 1
              ME1 := (m1_2 - m1_1)/sd1
              # Multiple endpoint effect size 2
```

```
              ME2 := (m2_2 - m2_1)/sd2'
R> fit4 <- sem(model4, sample.cov=Cov, sample.mean=Mean,
               sample.nobs=N, std.lv=TRUE,
               sample.cov.rescale=FALSE)
R> ## Obtain the free parameters in the model
R> ( x <- fit4@Fit@x )
```

```
[1]  3.3912  3.2404  0.5005 10.0000 11.0000 12.0000 13.0000
```

```
R> ## Obtain the sampling covariance matrix of the parameter
R> ## estimates
R> ( VCOV <- vcov(fit4) )
```

```
     sd1   sd2   r     m1_1  m2_1  m1_2  m2_2
sd1  0.058
sd2  0.014 0.053
r    0.006 0.006 0.006
m1_1 0.000 0.000 0.000 0.230
m2_1 0.000 0.000 0.000 0.110 0.210
m1_2 0.000 0.000 0.000 0.000 0.000 0.230
m2_2 0.000 0.000 0.000 0.000 0.000 0.110 0.210
```

```
R> ## Compute the multivariate effect sizes
R> ( ME <- fit4@Model@def.function(x=x) )
```

```
   ME1    ME2
0.5898 0.6172
```

```
R> ## Compute the jacobian for 'defined parameters'
R> JAC <- lavaan:::lavJacobianD(func=fit4@Model@def.function, x=x)
R> ## Compute the sampling covariance matrix using delta method
R> ME.VCOV <- JAC %*% VCOV %*% t(JAC)
R> ## Add the variable names for ease of reference
R> dimnames(ME.VCOV) <- list(names(ME), names(ME))
R> ME.VCOV
```

```
         ME1     ME2
ME1 0.04174 0.02048
ME2 0.02048 0.04190
```

The computed multiple effect sizes and its sampling covariance matrix are $\begin{bmatrix}0.5898\\0.6172\end{bmatrix}$ and $\begin{bmatrix}0.0417 & \\ 0.0205 & 0.0419\end{bmatrix}$, respectively.

3.5.4 Multiple treatment with multiple-endpoint studies

The power of the SEM approach is that it is easy to define and compute new effect sizes and their sampling covariance matrix. Suppose that there are two treatment groups and one control group with two dependent variables per group; this study involves both MT (two treatment groups vs one control group) and ME (two dependent effect sizes). Suppose that the sample covariance matrices for the control and the two treatment groups are $\begin{bmatrix} 11 & \\ 5 & 10 \end{bmatrix}$, $\begin{bmatrix} 12 & \\ 6 & 11 \end{bmatrix}$, and $\begin{bmatrix} 13 & \\ 7 & 12 \end{bmatrix}$, respectively; the sample means for the control and the two treatment groups are $\begin{bmatrix} 10 & 11 \end{bmatrix}^{\mathrm{T}}$, $\begin{bmatrix} 12 & 13 \end{bmatrix}^{\mathrm{T}}$, and $\begin{bmatrix} 13 & 14 \end{bmatrix}^{\mathrm{T}}$, respectively. The sample size for all groups is 50. Although it is still possible to derive it analytically, it is tedious and subject to human errors. We can easily compute the effect sizes with their sampling covariance matrix with the SEM approach. By using the control group as the reference, the SMDs for treatment 1 are defined by `ES1_1` and `ES2_1`, while the SMDs for treatment 2 are defined by `ES1_2` and `ES2_2`.

```
R> ## Covariance matrix of the control group
R> lower <- '11
             5, 10'
R> ## Convert a lower triangle data into a covariance matrix
R> Cov1 <- getCov(lower, diag=TRUE, names=c("x1", "x2"))
R> ## Covariance matrix of the treatment group 1
R> lower <- '12
             6, 11'
R> Cov2 <- getCov(lower, diag=TRUE, names=c("x1", "x2"))
R> ## Covariance matrix of the treatment group 2
R> lower <- '13
             7, 12'
R> Cov3 <- getCov(lower, diag=TRUE, names=c("x1", "x2"))
R> ## Convert covariance matrices into a list
R> Cov <- list(Cov1, Cov2, Cov3)
R> ## Means for the three groups
R> ## 10 and 11 are the means for variables 1 and 2
R> Mean <- list(c(10,11), c(12,13), c(13,14))
R> ## Sample sizes for the groups
R> N <- c(50, 50, 50)
R> ## Assuming homogeneity of covariance matrices
R> model5 <- 'eta1 =~ c("sd1", "sd1", "sd1")*x1
              eta2 =~ c("sd2", "sd2", "sd2")*x2
              eta1 ~~ c("r", "r", "r")*eta2
              ## The subscripts 0, 1 and 2 represent the means
              ##  of the control and two  treatment groups
              x1 ~ c("m1_0", "m1_1", "m1_2")*1
              x2 ~ c("m2_0", "m2_1", "m2_2")*1
              ## The measurement errors are fixed at 0
              x1 ~~ 0*x1
              x2 ~~ 0*x2
```

```
                ## Multiple endpoint effect size 1 for
                ## treatment group 1
                ES1_1 := (m1_1 - m1_0)/sd1
                ## Multiple endpoint effect size 2 for treatment
                ## group 1
                ES2_1 := (m2_1 - m2_0)/sd2
                ## Multiple endpoint effect size 1 for
                ## treatment group 2
                ES1_2 := (m1_2 - m1_0)/sd1
                ## Multiple endpoint effect size 2 for treatment
                ## group 2
                ES2_2 := (m2_2 - m2_0)/sd2'
R> fit5 <- sem(model5, sample.cov=Cov, sample.mean=Mean,
                sample.nobs=N, std.lv=TRUE,
                sample.cov.rescale=FALSE)
R> ## Obtain the free parameters in the model
R> ( x <- fit5@Fit@x )
```

```
[1]  3.4641  3.3166  0.5222 10.0000 11.0000 12.0000 13.0000 13.0000
[9] 14.0000
```

```
R> ## Obtain the sampling covariance matrix of the parameter
R> ## estimates
R> ( VCOV <- vcov(fit5) )
```

```
      sd1   sd2   r     m1_0  m2_0  m1_1  m2_1  m1_2  m2_2
sd1   0.040
sd2   0.010 0.037
r     0.004 0.004 0.004
m1_0  0.000 0.000 0.000 0.240
m2_0  0.000 0.000 0.000 0.120 0.220
m1_1  0.000 0.000 0.000 0.000 0.000 0.240
m2_1  0.000 0.000 0.000 0.000 0.000 0.120 0.220
m1_2  0.000 0.000 0.000 0.000 0.000 0.000 0.000 0.240
m2_2  0.000 0.000 0.000 0.000 0.000 0.000 0.000 0.120 0.220
```

```
R> ## Compute the multivariate effect sizes
R> ( ES <- fit5@Model@def.function(x=x) )
```

```
 ES1_1  ES2_1  ES1_2  ES2_2
0.5774 0.6030 0.8660 0.9045
```

```
R> ## Compute the jacobian for 'defined parameters'
R> JAC <- lavaan:::lavJacobianD(func=fit5@Model@def.function, x=x)
R> ## Compute the sampling covariance matrix using delta method
R> ES.VCOV <- JAC %*% VCOV %*% t(JAC)
```

```
R> ## Add the variable names for ease of reference
R> dimnames(ES.VCOV) <- list(names(ES), names(ES))
R> ES.VCOV
```

```
        ES1_1   ES2_1   ES1_2   ES2_2
ES1_1 0.04111 0.02121 0.02167 0.01092
ES2_1 0.02121 0.04121 0.01092 0.02182
ES1_2 0.02167 0.01092 0.04250 0.02160
ES2_2 0.01092 0.02182 0.02160 0.04273
```

The computed SMDs for treatment group 1 are 0.5774 and 0.6030, while the SMDs for treatment group 2 are 0.8660 and 0.9045. The asymptotic sampling covariance matrix for the effect sizes is shown in `ES.VCOV`.

3.5.5 Correlation matrix

The sampling covariance matrix of the correlation matrix plays an crucial role in MASEM discussed in Chapter 7. The `metaSEM` package has an `asyCov()` function to calculate the sampling covariance matrix of the correlation matrix. If the `cor.analysis=FALSE` argument is specified, `asyCov()` calculates the sampling covariance matrix of the covariance matrix. The following examples illustrate the procedures.

```
R> library("metaSEM")
R> ## Sample correlation matrix
R> ( C1 <- matrix(c(1,0.5,0.4,0.5,1,0.2,0.4,0.2,1), ncol=3,
                  dimnames=list(c("x1","x2","x3"),
                                c("x1","x2","x3"))) )
```

```
    x1  x2  x3
x1 1.0 0.5 0.4
x2 0.5 1.0 0.2
x3 0.4 0.2 1.0
```

```
R> ## Standard deviations
R> SD <- diag(c(1.2, 1.3, 1.4))
R> ## Convert the correlation matrix to a covariance matrix
R> C2 <- SD %*% C1 %*% SD
R> dimnames(C2) <- list(c("x1","x2","x3"),
                        c("x1","x2","x3"))
R> C2
```

```
      x1    x2    x3
x1 1.440 0.780 0.672
x2 0.780 1.690 0.364
x3 0.672 0.364 1.960
```

```
R> ## Calculate the sampling covariance matrix of
R> ## the correlation matrix with n=50
R> asyCov(C2, n=50)
```

```
         x2x1     x3x1     x3x2
x2x1 0.011480 0.001286 0.005235
x3x1 0.001286 0.014400 0.007714
x3x2 0.005235 0.007714 0.018808
```

```
R> ## Calculate the sampling covariance matrix of
R> ## the covariance matrix with n=50
R> asyCov(C2, n=50, cor.analysis=FALSE)
```

```
        x1x1     x2x1    x3x1     x2x2    x3x2     x3x3
x1x1 0.08464 0.045845 0.03950 0.024833 0.02139 0.018432
x2x1 0.04584 0.062082 0.02139 0.053804 0.02897 0.009984
x3x1 0.03950 0.021394 0.06682 0.011589 0.03619 0.053760
x2x2 0.02483 0.053804 0.01159 0.116575 0.02511 0.005408
x3x2 0.02139 0.028971 0.03619 0.025109 0.07030 0.029120
x3x3 0.01843 0.009984 0.05376 0.005408 0.02912 0.156800
```

3.6 Concluding remarks and further readings

This chapter introduced several common effect sizes in univariate and multivariate meta-analyses. Readers may refer to, for example, Borenstein (2009), Fleiss and Berlin (2009), and Gleser and Olkin (2009) for more details. The delta method and the SEM approach were introduced as general methods to approximate the sampling variances and covariances for both univariate and multiple effect sizes. Researchers can calculate different types of effect sizes to address different research questions. Readers should be reminded that the results of the delta method are only approximately correct when the samples are reasonably large.

References

Aguinis H, Dalton DR, Bosco FA, Pierce CA and Dalton CM 2011. Meta-analytic choices and judgment calls: implications for theory building and testing, obtained effect sizes, and scholarly impact. *Journal of Management* **37**(1), 5–38.

Aytug ZG, Rothstein HR, Zhou W and Kern MC 2012. Revealed or concealed? Transparency of procedures, decisions, and judgment calls in meta-analyses. *Organizational Research Methods* **15**(1), 103–133.

Bagos PG 2012. On the covariance of two correlated log-odds ratios. *Statistics in Medicine* **31**(14), 1418–1431.

Bishop YM, Fienberg SE and Holland PW 1975. *Discrete multivariate analysis: theory and practice*. MIT Press, Cambridge, MA.

Bond CF Jr., Wiitala WL and Dan F 2003. Meta-analysis of raw mean differences. *Psychological Methods* **8**(4), 406–418.

Borenstein M 2009. Effect sizes for continuous data. In *The handbook of research synthesis and meta-analysis* (ed. Cooper H, Hedges LV and Valentine JC), 2nd edn. Russell Sage Foundation, New York, pp. 221–235.

Borenstein M, Hedges LV, Higgins JP and Rothstein HR 2009. *Introduction to meta-analysis*. John Wiley & Sons, Ltd, Chichester, West Sussex, Hoboken, NJ.

Cheung MWL 2009. Constructing approximate confidence intervals for parameters with structural equation models. *Structural Equation Modeling: A Multidisciplinary Journal* **16**(2), 267–294.

Cheung MWL 2014. metaSEM: An R package for meta-analysis using structural equation modeling. Frontiers in Psychology 5(1521).

Cheung MWL and Chan W 2004. Testing dependent correlation coefficients via structural equation modeling. *Organizational Research Methods* **7**(2), 206–223.

Cheung MWL and Chan W 2009. A two-stage approach to synthesizing covariance matrices in meta-analytic structural equation modeling. *Structural Equation Modeling: A Multidisciplinary Journal* **16**(1), 28–53.

Cohen J 1992. A power primer. *Psychological Bulletin* **112**(1), 155–159.

Field A 2001. Meta-analysis of correlation coefficients: a Monte Carlo comparison of fixed- and random-effects methods. *Psychological Methods* **6**(2), 161–180.

Field A 2005. Is the meta-analysis of correlation coefficients accurate when population correlations vary? *Psychological Methods* **10**(4), 444–467.

Fleiss JL and Berlin JA 2009. Effect sizes for dichotomous data In *The handbook of research synthesis and meta-analysis* (ed. Cooper HM, Hedges LV and Valentine JC), 2nd edn. Russell Sage Foundation, New York, pp. 237–253.

Gibbons RD, Hedeker DR and Davis JM 1993. Estimation of effect size from a series of experiments involving paired comparisons. *Journal of Educational and Behavioral Statistics* **18**(3), 271–279.

Gleser LJ and Olkin I 1994. Stochastically dependent effect sizes. In *The handbook of research synthesis* (ed. Cooper H and Hedges LV). Russell Sage Foundation, New York, pp. 339–355.

Gleser LJ and Olkin I 2009. Stochastically dependent effect sizes. In *The handbook of research synthesis and meta-analysis* (ed. Cooper H, Hedges LV and Valentine JC), 2nd edn. Russell Sage Foundation, New York, pp. 357–376.

Hafdahl AR 2007. Combining correlation matrices: simulation analysis of improved fixed-effects methods. *Journal of Educational and Behavioral Statistics* **32**(2), 180–205.

Hafdahl AR and Williams MA 2009. Meta-analysis of correlations revisited: attempted replication and extension of Field's (2001) simulation studies. *Psychological Methods* **14**(1), 24–42.

Hedges LV 1981. Distribution theory for glass's estimator of effect size and related estimators. *Journal of Educational and Behavioral Statistics* **6**(2), 107–128.

Hedges LV and Olkin I 1985. *Statistical methods for meta-analysis*. Academic Press, Orlando, FL.

Keith C, Rindskopf D and Shadish WR 1998. Using odds ratios as effect sizes for meta-analysis of dichotomous data: a primer on methods and issues. *Psychological Methods* **3**(3), 339–353.

Kelley K and Preacher KJ 2012. On effect size. *Psychological Methods* **17**(2), 137–152.

Kline RB 2013. *Beyond significance testing: statistics reform in the behavioral sciences*, 2nd edn. American Psychological Association, Washington, DC.

Leisch F 2002. Sweave: dynamic generation of statistical reports using literate data analysis In *Compstat 2002 - proceedings in computational statistics* (ed. Hrdle W and Rönz B). Physica Verlag, Heidelberg, pp. 575580. ISBN: 3-7908-1517-9.

Millar RB 2011. *Maximum likelihood estimation and inference: with examples in R, SAS, and ADMB*. John Wiley & Sons, Inc., Hoboken, NJ.

Morris SB and DeShon RP 2002. Combining effect size estimates in meta-analysis with repeated measures and independent-groups designs. *Psychological Methods* **7**(1), 105–125.

Nel D 1985. A matrix derivation of the asymptotic covariance matrix of sample correlation coefficients. *Linear Algebra and its Applications* **67**, 137–145.

Olkin I and Siotani M 1976. Asymptotic distribution of functions of a correlation matrix In *Essays in probability and statistics* (ed. Ideka S) Shinko Tsusho, Tokyo, pp. 235–251.

Preacher KJ 2006. Testing complex correlational hypotheses with structural equation models. *Structural Equation Modeling: A Multidisciplinary Journal* **13**(4), 520–543.

Raykov T 2001. Testing multivariable covariance structure and means hypotheses via structural equation modeling. *Structural Equation Modeling: A Multidisciplinary Journal* **8**(2), 2–24.

Rosseel Y 2012. Lavaan: an R package for structural equation modeling. *Journal of Statistical Software* **48**(2), 1–36.

Schmidt FL and Hunter JE 2015. *Methods of meta-analysis: correcting error and bias in research findings* 3rd edn. Sage Publications, Inc., Thousand Oaks, CA.

Steiger JH 1980. Tests for comparing elements of a correlation matrix. *Psychological Bulletin* **87**(2), 245–251.

Tamhane AC and Dunlop DD 2000. *Statistics and data analysis: from elementary to intermediate*. Prentice Hall, Upper Saddle River, NJ.

Viechtbauer W 2010. Conducting meta-analyses in R with the metafor package. *Journal of Statistical Software* **36**(3), 1–48.

4

Univariate meta-analysis

This chapter begins by introducing the basic ideas of the fixed-effects model. The extension to the random-effects model is then introduced. Conceptual and statistical differences between the fixed-effects and the random-effects models are discussed. By including study characteristics as moderators, we extend the random-effects model to the mixed-effects model. Key concepts in a meta-analysis are introduced, such as testing the homogeneity of effect sizes, estimating heterogeneity variance, quantifying the degree of heterogeneity in the random-effects model, and quantifying the explained variance in the mixed-effects model. These models are then formulated under the structural equation modeling (SEM) framework. This SEM-based meta-analysis provides the foundation for more advanced analyses such as the multivariate and the three-level meta-analyses introduced in later chapters. Graphical models are proposed to represent the meta-analytic models. Several applications are used to illustrate the procedures in the R statistical environment.

4.1 Introduction

We begin this chapter by considering a model with only one variable y that is normally distributed with a mean of μ_y and a variance of σ_y^2, that is, $y \sim \mathcal{N}(\mu_y, \sigma_y^2)$. As we are going to show later, this simple model has many similarities to those of a fixed-effects meta-analysis in terms of mathematical and graphical models. It is well known that the unbiased estimators for μ_y and σ_y^2 based on a random sample of n data points are

$$\begin{aligned} \bar{y} &= \frac{\sum_{i=1}^{n} y_i}{n} \quad \text{and} \\ s^2_{y(\text{unbiased})} &= \frac{\sum_{i=1}^{n} (y_i - \bar{y})^2}{n-1}. \end{aligned} \tag{4.1}$$

Meta-Analysis: A Structural Equation Modeling Approach, First Edition. Mike W. -L. Cheung.

Companion Website: www.wiley.com/go/cheung/meta_analysis

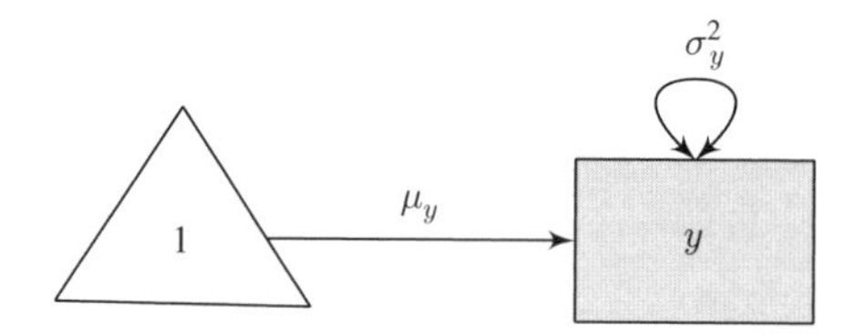

Figure 4.1 A model with one variable y.

Many statistical techniques such as SEM are meant to model the mean structure and the covariance structure of the data. If we fit the model in SEM with only one observed variable, the model-implied mean and the model-implied variance are

$$\begin{aligned} \mu_i(\boldsymbol{\theta}) &= \mu_y \quad \text{and} \\ \Sigma_i(\boldsymbol{\theta}) &= \sigma^2_y. \end{aligned} \tag{4.2}$$

Figure 4.1 shows the graphical model that includes μ_y and σ^2_y as the parameters. The triangle labeled with 1 represents the constant 1. It is used to represent either the means of the independent variables or the intercepts of the dependent variables. Simplifying Equation 2.42 for one observed variable here, the -2 $*$log-likelihood (-2LL_i) of the ith data point is

$$-2\text{LL}_i(\boldsymbol{\theta}) = \log(2\pi) + \log(\Sigma_i(\boldsymbol{\theta})) + \frac{(y_i - \mu_i(\boldsymbol{\theta}))^2}{\Sigma_i(\boldsymbol{\theta})}, \tag{4.3}$$

where $\log(x)$ is the natural logarithm of x.

As the data are independent, the -2LL of all data is $-2\text{LL}(\boldsymbol{\theta}) = -\sum_{i=1}^{n} 2\text{LL}_i(\boldsymbol{\theta})$. When maximum likelihood (ML) estimation, the default estimation method in SEM packages, is used, the estimators for μ_y and σ^2_y are

$$\begin{aligned} \bar{y} &= \frac{\sum_{i=1}^{n} y_i}{n} \quad \text{and} \\ s^2_{y(\text{ML})} &= \frac{\sum_{i=1}^{n} (y_i - \bar{y})^2}{n}. \end{aligned} \tag{4.4}$$

The ML estimate of variance in Equation 4.4 is slightly negatively biased in small samples. The choice between the $s^2_{y(\text{ML})}$ and $s^2_{y(\text{unbiased})}$ is similar to the issues in choosing between the ML and the restricted or (residual) maximum likelihood (REML) estimation. We will discuss this issue in more details in Section 8.1. As the ML estimation method is generally used in this book, we assume that the sample size (the number of studies in the meta-analysis) is not too small or that the degree of the bias on the variance is not a major concern in the research questions.

This example shows how SEM is usually used to analyze the data. The research questions are translated into either equations or graphical models representing the interrelationship of the variables. The proposed model is converted into the mean structure and the covariance structure. This model is fitted in the SEM software.

The models are then interpreted. In this chapter, we show how the SEM-based meta-analysis can be used to conduct univariate meta-analysis.

4.2 Fixed-effects model

We now consider the fixed-effects meta-analysis with one effect size y_i. y_i represents a generic effect size in the ith study. It can be an (log) odds ratio, a raw mean difference, a standardized mean difference, a correlation coefficient, a Fisher's z transformed score of the correlation coefficient, or some other effect sizes. When the sample sizes are reasonably large, it is assumed that y_i has a known sampling variance v_i, that is, $y_i \sim \mathcal{N}(f_i, v_i)$, where f_i is the *true* population effect size in the ith study (see the details in Chapter 3 on how to estimate v_i for various effect sizes). The *true* effect size f_i indicates what the population effect size is when there is no sampling error.

Under the fixed-effects model, the population effect sizes are usually assumed to be equal, that is, $\beta_F = f_1 = f_2 = \cdots = f_k$ is the common effect (readers may refer to Bonett (2008, 2009) for an alternative fixed-effects model without the assumption of a common effect). The univariate fixed-effects model for the ith study is

$$y_i = \beta_F + e_i, \tag{4.5}$$

where $\mathsf{Var}(e_i) = v_i$ is the known sampling variance. Under this model, the differences in the observed effect sizes are only due to the sampling error. The population effect sizes for all studies are the same once the sampling error has been taken into account.

4.2.1 Estimation and hypotheses testing

By treating v_i as known, we estimate β_F by minimizing the criterion F_{WLS} using the weighted least squares (WLS) estimation method,

$$F_{\text{WLS}} = \sum_{i=1}^{k} w_i (y_i - \beta_F)^2, \tag{4.6}$$

where $w_i = 1/v_i$. To solve the above equation, we take the derivative of F_{WLS} with respect to β_F,

$$\begin{aligned} \frac{dF_{\text{WLS}}}{d\beta_F} &= \frac{d\sum_{i=1}^{k}(w_i y_i^2 - 2w_i y_i \beta_F + w_i \beta_F^2)}{d\beta_F} \\ &= -2\sum_{i=1}^{k} w_i y_i + 2\sum_{i=1}^{k} w_i \beta_F. \end{aligned} \tag{4.7}$$

We set $\frac{dF_{\text{WLS}}}{d\beta_F} = 0$ in order to find the inflation point. The solution of the above equation is simply the weighted mean of y_i achieved by using w_i (the reciprocal of the precision) as the weight:

$$\hat{\beta}_{\mathrm{F}} = \frac{\sum_{i=1}^{k} w_i y_i}{\sum_{i=1}^{k} w_i}. \tag{4.8}$$

The idea behind the above formula is to put more weight on the observed effect sizes with higher precision (smaller v_i). To show that the solution is a local minimum (because we are minimizing F_{WLS}), we take the second derivative with respect to β_{F},

$$\frac{d^2 F_{\mathrm{WLS}}}{d\beta_{\mathrm{F}}^2} = 2\sum_{i=1}^{k} w_i. \tag{4.9}$$

As w_i is always positive, $2\sum_{i=1}^{k} w_i > 0$. Therefore, the solution is a local minimum. After estimating $\hat{\beta}_{\mathrm{F}}$, we also need to quantify the precision of the estimate by estimating the standard error (SE) or the confidence interval (CI). The estimated SE or CI can be used to test the null hypothesis and to draw statistical inferences.

Before introducing how to estimate the sampling variance on $\hat{\beta}_{\mathrm{F}}$, it is instructive to first introduce the following formula. Suppose X is a weighted mean of $x_1, x_2, \dots, x_k$, where x_1 to x_k are independent variables;

$$X = \frac{a_1 x_1}{A} + \frac{a_2 x_2}{A} + \cdots + \frac{a_k x_k}{A}, \tag{4.10}$$

where a_i is a constant and $A = \sum_{i=1}^{k} a_i$. We calculate the variance of X as

$$\begin{aligned}
\mathsf{Var}(X) &= \mathsf{Var}\left(\frac{a_1 x_1}{A} + \frac{a_2 x_2}{A} + \cdots + \frac{a_k x_k}{A}\right) \\
&= \frac{1}{A^2}\mathsf{Var}(a_1 x_1 + a_2 x_2 + \cdots + a_k x_k) \\
&= \frac{1}{A^2}\sum_{i=1}^{k} a_i^2 \mathsf{Var}(x_i) \quad \text{because } x_1 \text{ to } x_k \text{ are independent.}
\end{aligned} \tag{4.11}$$

As $\hat{\beta}_{\mathrm{F}}$ is a weighted mean of y_i, we apply the above formula to obtain the sampling variance of $\hat{\beta}_{\mathrm{F}}$ by treating $w_i/(\sum_{i=1}^{k} w_i)$ as constant,

$$\begin{aligned}
\mathsf{Var}(\hat{\beta}_{\mathrm{F}}) &= \mathsf{Var}\left(\frac{\sum_{i=1}^{k} w_i y_i}{\sum_{i=1}^{k} w_i}\right) \\
&= \frac{1}{(\sum_{i=1}^{k} w_i)^2}\sum_{i=1}^{k} \mathsf{Var}(w_i y_i) \\
&= \frac{1}{(\sum_{i=1}^{k} w_i)^2}\sum_{i=1}^{k} w_i^2 \frac{1}{w_i} \quad \text{as} \quad \mathsf{Var}(y_i) = \frac{1}{w_i} \\
&= \frac{\sum_{i=1}^{k} w_i}{(\sum_{i=1}^{k} w_i)^2} \\
&= \frac{1}{\sum_{i=1}^{k} w_i}.
\end{aligned} \tag{4.12}$$

Therefore, the $\text{SE}_{\hat{\beta}_\text{F}}$ of $\hat{\beta}_\text{F}$ is simply $\sqrt{\text{Var}(\hat{\beta}_\text{F})}$. $\hat{\beta}_\text{F}$ is an unbiased estimate of the population effect size if y_i is unbiased and the population effect sizes are homogeneous. Moreover, $\hat{\beta}_\text{F}$ has the smallest sampling variance among all possible weighted estimators in the class of unbiased estimators when the sampling variances are truly known (Hedges, 2007). We may test whether $H_0 : \beta_\text{F} = \beta_0$ by using

$$z = \frac{\hat{\beta}_\text{F} - \beta_0}{\text{SE}_{\hat{\beta}_\text{F}}}. \tag{4.13}$$

Under the null hypothesis, z has an approximate standard normal distribution. We reject the null hypothesis at $\alpha = 0.05$ when the absolute value of z is equal or larger than 1.96. Alternatively, we can also construct a Wald CI as discussed in Section 2.4.5.

4.2.2 Testing the homogeneity of effect sizes

Under the fixed-effects model, the differences in the observed effect sizes are only due to the sampling error. We test whether the data are consistent with the fixed-effects model by computing a Q statistic (Cochran, 1954):

$$Q = \sum_{i=1}^{k} w_i (y_i - \hat{\beta}_\text{F})^2. \tag{4.14}$$

Under the null hypothesis of the homogeneity of effect sizes $H_0 : \beta_\text{F} = f_1 = f_2 = \cdots = f_k$, the Q statistic has a chi-square distribution with $(k - 1)$ degrees of freedom (dfs). When the weight w_i is a *true* constant, the Q statistic has an exact chi-square distribution. As w_i is estimated in a meta-analysis and most effect sizes are approximately distributed as normal distributions, Equation 4.14 is only approximately distributed as a chi-square distribution.

On the basis of computer simulation studies, the Q statistic has been generally found to have Type I error rates that are close to the nominal α value, for example, 0.05 for $\alpha = 0.05$, whereas the power of the test in detecting the heterogeneity of effect sizes is low, especially when the number of studies is small, the sample size per study is small, and the degree of the heterogeneity of the effect sizes is low (e.g., Harwell, 1997; Huedo-Medina et al., 2006; Oswald and Johnson, 1998; Sánchez-Meca and Marín-Martínez, 1997; Viechtbauer, 2007).

4.2.3 Treating the sampling variance as known versus as estimated

Before moving to the random-effects model, we discuss the implications of treating v_i as known rather than as estimated. Let us illustrate the differences between the two approaches by using the sample mean as an effect size. Suppose that we are conducting a meta-analysis on the intelligence quotient (IQ); the sample mean and the standard deviation for one study ($n = 100$) are $\bar{x}_{\text{IQ}} = 110$, and $s_{\text{IQ}} = 14$. On the basis of the central limit theorem, the sampling variance of $\bar{x}_{\text{IQ}}$ is s^2_{IQ}/n. Thus, the effect size (the raw mean) and its sampling variance for this study is $y_{\text{IQ}} = 110$ and $v_{\text{IQ}} = 14^2/100 = 1.96$. $v_{\text{IQ}} = 1.96$, which is treated as known in a meta-analysis, is used to quantify the precision of $y_{\text{IQ}} = 110$.

But how accurate is v_{IQ}? When the data are normally distributed, the sampling variance of the variance s^2_{IQ} is $\mathsf{Var}(s^2_{\text{IQ}}) = 2\sigma^4_{\text{IQ}}/(n-1)$ (e.g., Tamhane and Dunlop, 2000). As we rarely know the population variance, we replace it with its sample statistic, that is, $\mathsf{Var}(s^2_{\text{IQ}}) = 2s^4_{\text{IQ}}/(n-1)$. As $v_{\text{IQ}} = s^2_{\text{IQ}}/n$, it follows that $\mathsf{Var}(v_{\text{IQ}}) = \mathsf{Var}(s^2_{\text{IQ}})/n^2$. In our example, the sampling variance of v_{IQ} is $2(14^4/99)/100^2 = 0.0776$. Therefore, v_{IQ} in the meta-analysis is actually a random variable with an estimated variance of 0.0776. When the sample sizes are large enough, it is reasonable to treat v_i as known. When the sample sizes are small, however, this assumption may not be appropriate. Treating v_i as fixed may then affect the accuracy of the meta-analysis.

On the other hand, it should be noted that the formula to estimate the sampling variance (and SE) on v_i is not very stable unless the sample sizes are huge. Thus, it has limited applications. Besides the sample size, the other factors affecting the accuracy of treating v_i as known are (i) the types of effect sizes and (ii) the values of the effect sizes if v_i depends on y_i. Because v_i is usually estimated, the accuracy of Equation 4.12 depends on how accurate the estimated value of v_i is (e.g., Hedges, 2007).

v_i is usually a function of the *true* effect size and the sample size. As the *true* effect size is unknown, we replace it with the observed effect size y_i in calculating v_i. Sometimes, it may be easier to justify treating v_i as known in a fixed-effects model than in a random-effects model. Under a fixed-effects model, all population effect sizes are assumed the same. Some authors prefer to obtain a better estimate on the *true* effect size by using the average effect size, for example, $\breve{y} = \frac{\sum_{i=1}^{k} y_i}{k}$ (or $\acute{y} = \frac{\sum_{i=1}^{k} n_i y_i}{\sum_{i=1}^{k} n_i}$) to calculate v_i. We may estimate $\breve{v}_i$ (or $\acute{v}_i$) by replacing y_i with $\breve{y}$ (or $\acute{y}$) (e.g., Hafdahl, 2007; Schmidt and Hunter, 2015). Please note that all studies use the same $\breve{y}$ (or $\acute{y}$) to calculate the individual sampling variances. As $\breve{y}$ (or $\acute{y}$) is based on a larger sample size than y_i does, the calculated value will be more accurate than v_i which is based on y_i. In principle, this approach is not appropriate for random-effects model because f_i varies across studies. However, Schmidt and Hunter (2015) argued that the above approach is still valid in estimating the sampling variances under a random-effects model.

Ideally, we may develop models that quantify the accuracy in estimating v_i. However, this is challenging because such models are effect size specific. Researchers would, therefore, need to develop different models to handle different effect sizes. For example, Malzahn et al. (2000) developed a nonparametric approach to analyze standardized mean differences by treating v_i as estimates rather than known. For binary data, it is also possible to directly apply a generalized mixed-effects model on frequency counts (e.g., Brown and Prescott, 2006). The first stage of the two-stage structural equation modeling (TSSEM) based on a fixed-effects model in synthesizing correlation or covariance matrices takes the sampling covariance matrix of the effect sizes into account (see Section 7.3.1).

4.3 Random-effects model

The fixed-effects model assumes that all studies share the same common effect. Studies may vary in terms of samples, measures, and quality. It may not be reasonable to expect all studies to have the same common effect. A random-effects model allows studies to have their own population effect sizes. We may write the equations for the observed effect size y_i and its *true* population effect size f_i in the ith study as a two-level model:

$$\begin{aligned} \text{Level 1: } y_i &= f_i + e_i, \\ \text{Level 2: } f_i &= \beta_R + u_i, \end{aligned} \tag{4.15}$$

where β_R is the average population effect size under a random-effects model and $u_i \sim \mathcal{N}(0, \tau^2)$ is the heterogeneity variance that has to be estimated. As τ^2 is the variance of the *true* effect sizes, it is not affected by the sampling error in theory. Alternatively, we may combine the two levels into one single-level equation:

$$y_i = \beta_R + u_i + e_i. \tag{4.16}$$

When u_i and e_i are independent, which is the usual assumption in a meta-analysis, it is clear from Equation 4.16 that

$$\mathsf{Var}(y_i) = \tau^2 + v_i. \tag{4.17}$$

It should be noted that v_i depends on the sample size, whereas τ^2 does not. Given the *true* effect size f_i, the observed effect size y_i varies around f_i with the sampling variance v_i or $\mathsf{Var}(y_i|f_i) = v_i$. If we randomly select one effect size and do not know the *true* effect size, the best guess is β_R. We expect that the observed effect size y_i is centered around β_R with a sampling variance of the sum of τ^2 and v_i. In the literature, $(\tau^2 + v_i)$ and v_i are known as the unconditional and the conditional variances, respectively. There are two unknown parameters, β_R and τ^2, in a random-effects model, whereas there is only one parameter β_F in a fixed-effects model.

It is of relevance to discuss the roles of the normality assumption in the random-effects model. The observed effect size y_i is conditionally distributed around f_i with a variance of v_i. When the sample sizes in the studies are large enough, y_i tends to be normally distributed as shown by the central limit theorem regardless of what the distribution of the original data was. On the other hand, the assumption of normality on the random effects u_i is a true assumption. Increasing the number of studies and/or sample sizes in the studies does not help to improve the normality assumption of the random effects.

In estimating the heterogeneity variance, some estimation methods, for example, unweighted method of moments (UMM) and weighted method of moments

(WMM), do not required the normality assumption of the random effects, whereas the ML and the REML estimation methods do. Although estimating the variance component with the method of moments does not assume normality of the effect sizes, testing and constructing CIs on the variance component require such an assumption. Moreover, the testing and constructing of CIs on the averaged effect size also rely on the assumption of normality on the part of the effect sizes. Therefore, most methods, including the method of moments, ML, and REML, implicitly or explicitly assume normality of the random effects.

4.3.1 Estimation and hypothesis testing

Except for ML estimation, two steps are usually required in fitting a random-effects model. In the first step, the variance component τ^2 is first estimated. In the second step, the fixed effect β_R is estimated by treating the estimated τ^2 as known. There are several common methods for estimating τ^2. We briefly review some of them here.

4.3.1.1 Unweighted method of moments

Hedges (1983) proposed an UMM approach to estimate τ^2. The idea of the method of moments is to equate the expected value of a statistic to its sample value. We may define the variance of y_i as

$$s_y^2 = \frac{\sum_{i=1}^{k}(y_i - \bar{y})^2}{k - 1}, \tag{4.18}$$

where $\bar{y} = \sum_{i=1}^{k} y_i/k$. From Equation 4.17, Hedges noted that the expected value of s_y^2 is

$$\mathsf{E}(s_y^2) = \tau^2 + \frac{\sum_{i=1}^{k} v_i}{k}. \tag{4.19}$$

By equating the moments (Equation 4.18 with Equation 4.19), an unbiased estimator of τ^2 based on the UMM method is

$$\hat{\tau}^2_{\text{UMM}} = s_y^2 - \frac{\sum_{i=1}^{k} v_i}{k}. \tag{4.20}$$

If $\hat{\tau}^2_{\text{UMM}}$ is negative, it is usually truncated to 0. As this estimator is easy to calculate, it looks attractive. However, when s_y^2 is less than $\frac{\sum_{i=1}^{k} v_i}{k}$, $\hat{\tau}^2_{\text{UMM}}$ can be negative even though the Q statistic is large or even statistically significant (Friedman, 2000). Therefore, $\hat{\tau}^2_{\text{UMM}}$ may be inconsistent with the inferences based on the Q statistic.

4.3.1.2 Weighted method of moments

DerSimonian and Laird (1986) proposed a WMM approach to estimate τ^2. These authors noted that the expected value of Q is

$$\mathsf{E}(Q) = c\tau^2 + k - 1, \tag{4.21}$$

where $c = \sum_{i=1}^{k} w_i - \frac{\sum_{i=1}^{k} w_i^2}{\sum_{i=1}^{k} w_i}$. When $\tau^2 = 0$, the expected value of Q is $k - 1$. Thus, the Q statistic has an approximate chi-square distribution with $k - 1$ df under the assumption of $\tau^2 = 0$ in Equation 4.14. We may derive an unbiased estimator of τ^2 based on the WMM as

$$\hat{\tau}^2_{\text{WMM}} = \frac{Q - (k - 1)}{c}. \tag{4.22}$$

If $\hat{\tau}^2_{\text{WMM}}$ is negative, it is usually truncated to 0. An attractive feature of the WMM approach is that it is consistent with the significance test on the Q statistic. When the Q statistic is larger than its df, $\hat{\tau}^2_{\text{WMM}}$ is positive.

4.3.1.3 Maximum likelihood estimation

On the basis of the model in Equation 4.16, the -2LL_i for the random-effects meta-analysis is

$$-2\text{LL}_i(\beta_{\text{R}}, \tau^2) = \log(2\pi) + \log(\tau^2 + v_i) + \frac{(y_i - \beta_{\text{R}})^2}{\tau^2 + v_i}. \tag{4.23}$$

The ML estimates of $\beta_{\text{R(ML)}}$ and τ^2_{ML} are obtained by solving the partial derivative of $-2\text{LL}(\theta) = -\sum_{i=1}^{n} 2\text{LL}_i(\theta)$ set to zero (Hardy and Thompson, 1996). It should be noted that both $\beta_{\text{R(ML)}}$ and τ^2_{ML} are estimated simultaneously in the ML estimation method. One attractive property of $\hat{\beta}_{\text{R(ML)}}$ is that it is unbiased even in small samples (Demidenko, 2013). However, there are still issues surrounding the use of ML estimation in a meta-analysis. The first one is that v_i is treated as known versus estimated (see Section 4.2.3). The second issue is that $\hat{\tau}^2_{\text{ML}}$ is slightly negatively biased (see Section 8.1). The SEM-based meta-analysis mainly uses the ML estimation as the estimation method.

4.3.1.4 Restricted maximum likelihood estimation

The ML estimator of τ^2 tends to underestimate the population heterogeneity in finite samples because it does not take into account the fact that $\hat{\beta}_{\text{R}}$ is estimated rather than known when estimating τ^2. The REML estimation attempts to correct this bias by taking into account the loss of df in estimating the fixed effect. More details on the REML estimation is provided in Section 8.1.

4.3.1.5 Comparing among the estimation methods

Several authors have compared the empirical performance of these estimators. Friedman (2000) compared the efficiency of the UMM and WMM estimators. She found that the WMM estimator is more efficient when the population heterogeneity is small, while the UMM estimator is more efficient when the population heterogeneity is large. Similarly, Demidenko (2013) found no estimator to be uniformly better, in terms of mean squares error, over the whole range of τ^2 even in the case of normal distribution. Viechtbauer (2005) compared the performance of the above four estimators and another estimator proposed by Hunter and Schmidt (1990) by deriving the mean squares error of the estimators and by a computer simulation. He found that there is no estimator that is universally better than others in terms of bias, efficiency, and mean squares error. Generally, Viechtbauer (2005) recommended the REML estimator, as it shows a good balance between unbiasedness and efficiency.

4.3.1.6 Estimation of the average population effect

After estimating the heterogeneity variance, we can compute the unconditional sampling variance using $\tilde{v}_i = v_i + \hat{\tau}^2$ as the weight. We treat $\tilde{v}_i$ as known and minimize the criterion F_{WLS} with the WLS estimation method:

$$F_{\text{WLS}} = \sum_{i=1}^{k} \tilde{w}_i (y_i - \beta_{\text{R}})^2, \tag{4.24}$$

where $\tilde{w}_i = 1/\tilde{v}_i$.

The solution, which is similar to that for the fixed-effects model, is the weighted mean of y_i using $\tilde{w}_i$ (the reciprocal of the precision) as the weight:

$$\hat{\beta}_{\text{R}} = \frac{\sum_{i=1}^{k} \tilde{w}_i y_i}{\sum_{i=1}^{k} \tilde{w}_i}. \tag{4.25}$$

The sampling variance of $\hat{\beta}_{\text{R}}$ may also be derived in a similar way as that in Equation 4.12:

$$\text{Var}(\hat{\beta}_{\text{R}}) = \frac{1}{\sum_{i=1}^{k} \tilde{w}_i}. \tag{4.26}$$

As $\hat{\tau}^2$ is nonnegative, v_i in the fixed-effects model cannot be larger than $\tilde{v}_i$ in the random-effects model. This suggests that $\text{Var}(\hat{\beta}_{\text{R}}) \geq \text{Var}(\hat{\beta}_{\text{F}})$. In other words, the calculated CIs under a fixed-effects model are usually shorter than those calculated under a random-effects model. If a fixed-effects model is *incorrectly* applied to studies where τ^2 is not zero, both the SEs and CIs are likely to be underestimated. Schmidt et al. (2009) found that most meta-analyses published in *Psychological Bulletin* used the fixed-effects model rather than the random-effects model. These authors argued that the random-effects model should be routinely used. If the estimated τ^2 is zero, the random-effects model automatically becomes a fixed-effects model. In estimating $\text{Var}(\hat{\beta}_{\text{R}})$, it should also be noted that $\hat{\tau}^2$ is treated as known rather than as estimated. Therefore, $\text{Var}(\hat{\beta}_{\text{R}})$ is smaller than the true variability of $\hat{\beta}_{\text{R}}$ (see Sánchez-Meca and Marín-Martínez (2008) for some alternative approaches).

4.3.2 Testing the variance component

The fixed-effects model in Equation 4.5 is nested within the random-effects model by setting $\tau^2 = 0$. Statistically, we may compare a fixed-effects model and a random-effects meta-analysis by testing $H_0 : \tau^2 = 0$. As the difference on the numbers of parameters is 1, it is reasonable to expect that the difference between the likelihood ratio (LR) statistics of these two models follows a chi-square distribution with 1 df when $H_0 : \tau^2 = 0$ is true. However, this is not true because the null hypothesis $H_0 : \tau^2 = 0$ is tested on the boundary.

Let us illustrate the concept of testing on a boundary condition by considering the random-effects model in Equation 4.16. Suppose that we want to test $H_0 : \beta_R = 0$; we fit two models—one with β_R being free and the another with the constraint $\beta_R = 0$. When $H_0 : \beta_R = 0$ is true, there is a 50:50 chance that the sample estimates on β_R are positive (or negative) because of a sampling error. The difference in the LR statistics between the two models asymptotically follows a chi-square distribution with df = 1 when $H_0 : \beta_R = 0$ is true.

This is not the case when we are testing $H_0 : \tau^2 = 0$. As τ^2 cannot be negative, there is a 50% chance that the sample estimates on τ^2 will be positive and another 50% chance that $\hat{\tau}^2$ is 0. When $\hat{\tau}^2 = 0$, the difference in the LR statistics between these two models is 0. Therefore, the difference in the LR statistics between these two models is distributed as a 50:50 mixture of a degenerate random variable with all of its probability mass concentrated at 0 ($\chi^2_{\text{df}=0}$) and $\chi^2_{\text{df}=1}$ (Self and Liang, 1987).

Figure 4.2 shows the distributions of $\chi^2_{\text{df}=1}$ and a 50:50 mixture of $\chi^2_{\text{df}=0}$ and $\chi^2_{\text{df}=1}$. As shown in the figure, the critical value of $\chi^2_{\text{df}=1}$ and a 50:50 mixture of $\chi^2_{\text{df}=0}$ and $\chi^2_{\text{df}=1}$ are 3.84 and 2.71, respectively. If we *incorrectly* refer the test statistic to $\chi^2_{\text{df}=1}$, the empirical (or true) Type I error will be smaller than 0.05 if $\alpha = 0.05$ is used. If it is really necessary to test whether $\tau^2 = 0$, we should use 2.71 as the critical value for $\alpha = 0.05$. An alternative strategy is to use 2α instead of α as the

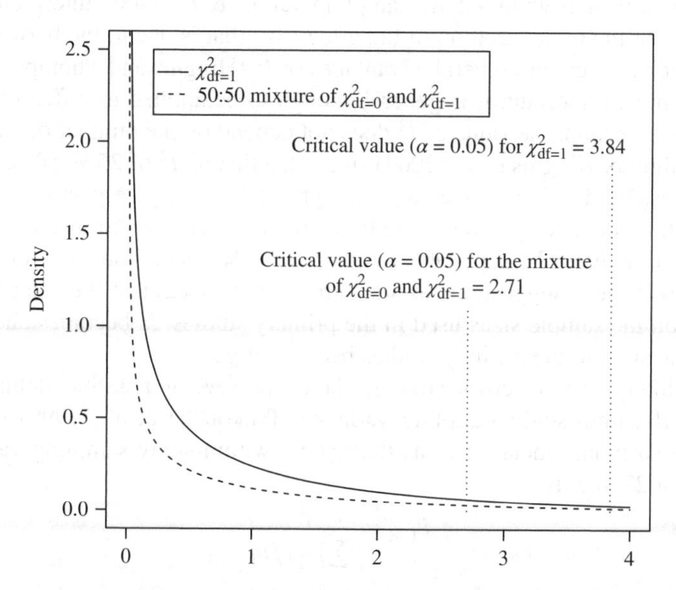

Figure 4.2 Distributions of $\chi^2_{\text{df}=1}$ and a 50:50 mixture of $\chi^2_{\text{df}=0}$ and $\chi^2_{\text{df}=1}$.

alpha level (Pinheiro and Bates, 2000). That is, we may reject the null hypothesis when the observed p value is larger than 0.10 for $\alpha = 0.05$. Readers may refer to Viechtbauer (2007) and Stoel et al. (2006) for a discussion on this issue in the context of a meta-analysis and an SEM.

4.3.3 Quantifying the degree of the heterogeneity of effect sizes

Although the Q statistic may be used to test the homogeneity of effect sizes, it does not indicate the degree of heterogeneity. The Q statistic may be significant simply because of the large number of studies involved. Conversely, a large Q statistic may be nonsignificant because of the small number of studies involved. The heterogeneity variance τ^2 can be used to indicate the heterogeneity of studies. One major limitation of τ^2 is that it depends on the types of effect sizes. For example, $\hat{\tau}^2 = 0.1$ may mean different degrees of heterogeneity in a raw mean difference or a correlation coefficient.

Higgins and Thompson (2002) proposed three indices to quantify the heterogeneity of effect sizes that do not depend on the types of effect sizes. They are the H, R, and I^2 indices. H is the square root of the Q statistic divided by its dfs; R is the ratio of the SE of the $\hat{\beta}_\text{R}$ to the SE of $\hat{\beta}_\text{F}$, and I^2 is the between-study heterogeneity to the total variation of studies.

Among these three indices, I^2 is the most popular. We focus on it here. Its general formula is

$$I^2 = \frac{\hat{\tau}^2}{\hat{\tau}^2 + \tilde{v}}, \tag{4.27}$$

where $\tilde{v}$ is a *typical* within-study sampling variance. I^2 can be interpreted as the proportion of the total variation of the effect size that is due to the between-study heterogeneity. There are several advantages of I^2 (Higgins and Thompson, 2002). First and most important, it is general enough to be applied to different types of effect sizes. Second, the value of I^2 does not depend on the number of studies. As a rule of thumb, Higgins et al. (2003) suggested that an I^2 of 25%, 50%, and 75% can be considered as low, moderate, and high heterogeneity, respectively. It should be noted that these suggestions were based on meta-analyses in medical research. They may or may not be applicable to other fields. Readers should exercise caution and not treat these suggestions as standards. As the extent of the sampling error depends on the sample sizes used in the primary studies, I^2 becomes larger when the sample sizes in the primary studies become larger.

As v_i likely varies across studies, there are several possible definitions of the *typical* within-study sampling variance. Takkouche et al. (1999) suggested using the harmonic mean of v_i as the *typical* within-study sampling variance in Equation 4.27, that is,

$$\tilde{v}_\text{HM} = \frac{k}{\sum_{i=1}^{k} 1/v_i}. \tag{4.28}$$

The I^2 calculated using $\tilde{v}_\text{HM}$ is called R_I in Takkouche et al. (1999, 2013).

Higgins and Thompson (2002) preferred to define the *typical* within-study sampling variance in Equation 4.27 using the Q statistic:

$$\tilde{v}_{\mathrm{Q}} = \frac{(k-1)\sum_{i=1}^{k} 1/v_i}{(\sum_{i=1}^{k} 1/v_i)^2 - \sum_{i=1}^{k} 1/v_i^2}. \tag{4.29}$$

One advantage of using $\tilde{v}_{\mathrm{Q}}$ as the *typical* within-study sampling variance is that I^2 can be simplified to $I_{\mathrm{Q}}^2 = (Q - (k-1))/Q$.

Besides these two estimators, Xiong et al. (2010) also discussed an estimator of I^2 that is based on the arithmetic mean of v_i:

$$\tilde{v}_{\mathrm{AM}} = \sum_{i=1}^{k} \frac{v_i}{k}. \tag{4.30}$$

It should be noted that, in general, I^2 is not used to estimate any population quantities. This is because it contains the $\tilde{v}$ that is calculated based on specific studies. Were a different set of studies to be selected, $\tilde{v}$ would likely be different. As the sample sizes per study get larger and larger, I^2 approaches 1 as $\tilde{v}$ approaches 0. Therefore, I^2 is used as a descriptive statistic rather than an inferential statistic. Even though we are not testing any population parameters, CIs on I^2 may still be useful for quantifying the precision of I^2. Higgins and Thompson (2002) and Takkouche et al. (2013) discussed several methods of constructing CIs on I^2. As I^2 is a function of $\hat{\tau}^2$ by treating $\tilde{v}$ as a constant in Equation 4.27, the `metaSEM` package has implemented the likelihood-based confidence interval (LBCI) on I^2.

4.4 Comparisons between the fixed- and the random-effects models

After introducing both the fixed-effects and the random-effects models, this section discusses their similarities and differences. The definitions of fixed and random effects are not consistent in the literature. According to Gelman (2005, p. 20), "different—in fact, incompatible—definitions are used in different contexts." Gelman further listed five different definitions of fixed and random effects. We use the conventional definitions in a meta-analysis here.

4.4.1 Conceptual differences

It is generally recommended that the decision on whether to use the fixed- or random-effects model should depend on the research questions. The fixed-effects model is used when we are only interested in studies with the same characteristics as those selected in the meta-analysis. For example, a medical researcher may only be interested in the effectiveness of a new drug in a few well-controlled studies rather than in all possible studies. A fixed-effects model may be used to synthesize a few well-controlled studies.

By using a random-effects model, researchers may estimate the average effect on all possible studies and the variability of the effect sizes. A common assumption in the random-effects model is that the k studies that are included in the meta-analysis are randomly sampled from a large superpopulation. For many reasons, for instance, publication bias, the selected studies are unlikely to have been randomly sampled from the so-called superpopulation. Therefore, this random sampling assumption will almost never be satisfied in practice.

On the basis of the above criticism, Bonett (2008, 2009) proposed a fixed-effects model without this random sampling assumption. He argued that this model is preferable to the random-effects model. The usual assumption under the fixed-effects model is the homogeneity of effect sizes, $\beta_F = f_1 = f_2 = \cdots = f_k$. Because of this assumption, this (fixed-effects) model is sometimes called the *common effect model*. Bonett's fixed-effects model does not assume that there is a common effect. He proposed to estimate the *average* population effect by $\beta_A = \frac{f_1+f_2+\cdots+f_k}{k}$. There are two key features to quantify this. First, it is a fixed-effects model because researchers are only interested in these k studies. Second, no random sampling assumption is required for these k studies. Readers may refer to Bonett (2008, 2009) for details.

On the other hand, Higgins et al. (2009) and Raudenbush (2009) argued from a Bayesian perspective that the random sampling assumption is not necessary for the random-effects model. These authors argued that only the concept of the *exchangeability* of the studies is required. Exchangeability represents a judgment of the researchers that the effect sizes may be nonidentical, but their magnitudes cannot be differentiated before the data analysis is conducted. In other words, the researchers have no reason to believe that the effect of one study is larger than that of other studies.

4.4.2 Statistical differences

Figure 4.3 illustrates the concepts of fixed- and random-effects meta-analysis. Let us first consider the fixed-effects model in Figure 4.3a. All studies have the same common effect, which is β_F in the figure. As studies are likely to have different v_i, the sampling variances of different effect sizes may be different. Because of the sampling error, it happens that $y_1 \sim \mathcal{N}(\beta_F, v_1)$ and $y_2 \sim \mathcal{N}(\beta_F, v_2)$ are observed by the researchers even though their population values are the same.

Regarding the random-effects model in Figure 4.3b, the average population effect is β_R. Two studies are randomly selected from $\mathcal{N}(\beta_R, \tau^2)$ (the shaded area). It happens that f_1 and f_2 (not observable to the researchers) were selected. Given that f_1 and f_2 are the true population values, it happens that $y_1 \sim \mathcal{N}(f_1, v_1)$ and $y_2 \sim \mathcal{N}(f_2, v_2)$ are observed by the researchers.

There are several points to note when comparing the random-effects model in Equation 4.16 and the fixed-effects model in Equation 4.5. First, the conditional sampling variance v_i is the same, regardless of whether a fixed-effects or a random-effects model is used. The conditional sampling variance v_i is usually a function of the *true* effect size and the sample size. As the *true* effect size is

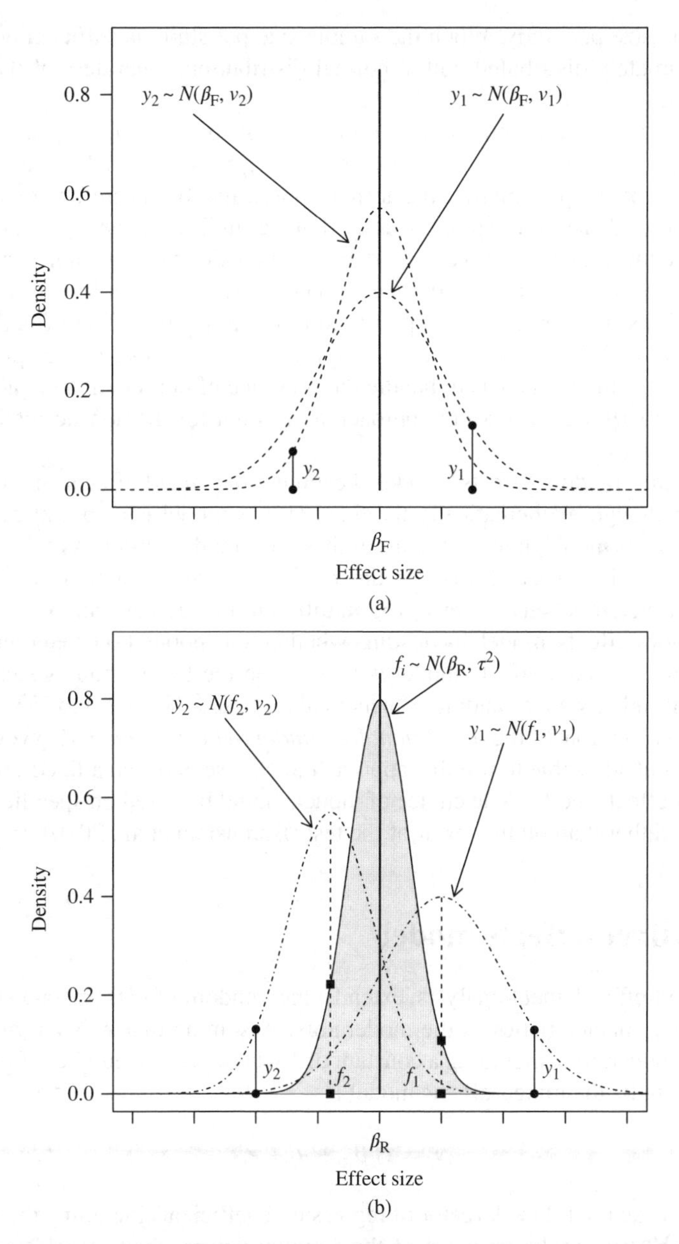

Figure 4.3 (a) Fixed-effects versus (b) random-effects meta-analyses.

usually unknown, the sample effect size is used to replace the *true* effect sizes in calculating the conditional sampling variances. Therefore, v_i is usually estimated based on the sample effect size. Second, the validity of the normality assumption on y_i depends on the type of effect size, the population value of the effect size, and the sample size per study. When the sample size per study is sufficiently large, y_i is approximately distributed with a normal distribution, regardless of the type of effect size.

However, the assumption that f_i is normally distributed under a random-effects model is a true assumption. If the distribution of f_i is non-normal, collecting more studies does not help to improve the normal approximation. Therefore, this assumption is independent of the sample sizes used in the studies and the number of studies used in the meta-analysis. Several models based on alternative parametric distributions have been suggested for the *true* effect sizes. For example, a t-distribution or skewed distribution have been proposed to minimize the effect of outliers (see Higgins et al., 2009) under a Bayesian approach. Demidenko (2013) proposed a robust/median meta-analysis to handle the presence of outliers in a meta-analysis. Böhning (2000) used a mixture approach to account for the non-normality of the *true* effect sizes.

Third, the random-effects model becomes the fixed-effects model when $\tau^2 = 0$. Although mathematically the fixed-effects model is a special case of the random-effects model, interpretations of these two models differ. As a fixed-effects meta-analysis is a special case of the random-effects meta-analysis when the population heterogeneity is zero, it is intuitive to decide between a fixed-effects or a random-effects model by testing whether the population heterogeneity is zero. That is, a fixed-effects model is used when the test of homogeneity is not significant; otherwise, a random-effects model is used. Hedges and Vevea (1998) termed this approach the *conditionally random-effects model*. However, it is generally not advisable to use this approach to choose between a fixed-effects and a random-effects model. The choice of model should be based on specific research questions rather than on the result of the test (Borenstein et al., 2010).

4.5 Mixed-effects model

The mixed-effects meta-analysis extends the random-effects meta-analysis by using study characteristics as the moderators. Assuming that $\boldsymbol{x}_i$ is an $(m+1) \times 1$ vector of moderators, including a constant of 1 where m is the number of predictors in the ith study, the mixed-effects model is

$$y_i = \boldsymbol{x}_i^{\mathrm{T}} \boldsymbol{\beta}_{\mathrm{R}} + u_i + e_i, \tag{4.31}$$

where $\boldsymbol{\beta}_{\mathrm{R}}$ is an $(m+1) \times 1$ vector of regression coefficients including the intercept and $\tau^2 = \mathsf{Var}(u_i)$ is the variance of the residual heterogeneity (Goldstein, 2011; Raudenbush and Bryk, 2002; Stram, 1996). Although $\boldsymbol{x}_i$ (excluding the constant) *predict* the outcome variable y_i, $\boldsymbol{x}_i$ are usually called *moderators* in the literature of meta-analysis. It is because $\boldsymbol{x}_i$ *moderate* the strength of the effect on y_i at the study level.

If we assume that the between-study heterogeneity is purely due to differences in the study characteristics (e.g., Hedges and Olkin, 1983), Equation 4.31 can be simplified to

$$y_i = \boldsymbol{x}_i^{\mathrm{T}} \boldsymbol{\beta}_{\mathrm{F}} + e_i. \tag{4.32}$$

This model is a fixed-effects model, as it includes only the regression coefficients without the random effects. Conceptually, it allows studies with different population effects. After controlling for the predictors $\boldsymbol{x}_i$, the only uncertainty left is the sampling error e_i. As this model is a special case of the more general mixed-effects model listed in Equation 4.31, we will only focus on the mixed-effects model.

4.5.1 Estimation and hypotheses testing

In order to estimate the parameters of the mixed-effects meta-analysis, it is easier to stack all data into vectors and matrices for ease of calculations. The model in Equation 4.31 becomes

$$\boldsymbol{y} = \boldsymbol{X}\boldsymbol{\beta}_{\mathrm{R}} + \boldsymbol{u} + \boldsymbol{e}, \tag{4.33}$$

where $\boldsymbol{y}$ is a $k \times 1$ vector of effect sizes, $\boldsymbol{X}$ is a $k \times (m+1)$ design matrix including 1 in the first column, $\boldsymbol{u}$ is a $k \times 1$ vector of the random effects of residual heterogeneity, and $\boldsymbol{e}$ is a $k \times 1$ vector of sampling error. Consider a meta-analysis with one predictor x; the model is

$$\overbrace{\begin{bmatrix} y_1 \\ y_2 \\ \vdots \\ y_k \end{bmatrix}}^{\boldsymbol{y}} = \overbrace{\begin{bmatrix} 1 & x_1 \\ 1 & x_2 \\ \vdots & \vdots \\ 1 & x_k \end{bmatrix}}^{\boldsymbol{X}} \overbrace{\begin{bmatrix} \beta_0 \\ \beta_1 \end{bmatrix}}^{\boldsymbol{\beta}_{\mathrm{R}}} + \overbrace{\begin{bmatrix} u_1 \\ u_2 \\ \vdots \\ u_k \end{bmatrix}}^{\boldsymbol{u}} + \overbrace{\begin{bmatrix} e_1 \\ e_2 \\ \vdots \\ e_k \end{bmatrix}}^{\boldsymbol{e}}, \tag{4.34}$$

where β_0 is the intercept and β_1 is the expected change in y when x increases by 1 unit.

As the studies are independent, $\boldsymbol{T}^2 = \mathsf{Var}(\boldsymbol{u}) = \mathrm{Diag}(\tau^2, \tau^2, \ldots, \tau^2) = \begin{bmatrix} \tau^2 & 0 & \cdots & 0 \\ 0 & \tau^2 & \ddots & \vdots \\ \vdots & \ddots & \ddots & 0 \\ 0 & \cdots & 0 & \tau^2 \end{bmatrix}$ and $\boldsymbol{V} = \mathrm{Diag}(v_1, v_2, \ldots, v_k)$ are both diagonal matrices where the off-diagonal elements are all 0. The estimation methods discussed in Section 4.3.1 can be modified to the mixed-effects model. Generally, the heterogeneity variance τ^2 in Equation 4.31 is first estimated (for the calculations, see Raudenbush (2009) and Viechtbauer (2008)).

By treating the unconditional sampling variance $\boldsymbol{V}_{\mathrm{R}} = \boldsymbol{T}^2 + \boldsymbol{V}$ as known, we estimate $\boldsymbol{\beta}_{\mathrm{R}}$ by minimizing the criterion F_{WLS} with the WLS estimation method (Hedges and Olkin, 1985):

$$F_{\mathrm{WLS}} = (\boldsymbol{y} - \boldsymbol{X}\boldsymbol{\beta}_{\mathrm{R}})^{\mathrm{T}} \boldsymbol{V}_{\mathrm{R}}^{-1} (\boldsymbol{y} - \boldsymbol{X}\boldsymbol{\beta}_{\mathrm{R}}). \tag{4.35}$$

The solutions for $\hat{\boldsymbol{\beta}}_\mathrm{R}$ and its asymptotic sampling variance are

$$\begin{aligned}\hat{\boldsymbol{\beta}}_\mathrm{R} &= (\boldsymbol{X}^\mathrm{T}\boldsymbol{V}_\mathrm{R}^{-1}\boldsymbol{X})^{-1}\boldsymbol{X}^\mathrm{T}\boldsymbol{V}_\mathrm{R}^{-1}\boldsymbol{y} \quad \text{and} \\ \hat{\boldsymbol{\Omega}}_\mathrm{R} &= \mathsf{Var}(\hat{\boldsymbol{\beta}}_\mathrm{R}) = (\boldsymbol{X}^\mathrm{T}\boldsymbol{V}_\mathrm{R}^{-1}\boldsymbol{X})^{-1}.\end{aligned} \tag{4.36}$$

The diagonals of $\hat{\boldsymbol{\Omega}}_\mathrm{R}$ indicates the sampling variances of the parameter estimates. We may test the significance of the ith parameter $\hat{\boldsymbol{\beta}}_{\mathrm{R}[i]}$ by

$$z_{\hat{\boldsymbol{\beta}}_{\mathrm{R}[i]}} = \frac{\hat{\boldsymbol{\beta}}_{\mathrm{R}[i]} - \beta_0}{\sqrt{\hat{\boldsymbol{\Omega}}_{\mathrm{R}[i,i]}}}, \tag{4.37}$$

which has an approximate standard normal distribution under $H_0 : \boldsymbol{\beta}_{\mathrm{R}[i]} = \beta_0$ (see Huizenga et al. (2011), for other approaches to test the moderator effects). If we want to test whether some of the parameters in $\hat{\boldsymbol{\beta}}_\mathrm{R}$ are zero, we may construct a Wald test on a linear contrast based on the $\hat{\boldsymbol{\beta}}_\mathrm{R}$ and $\hat{\boldsymbol{\Omega}}_\mathrm{R}$ (see Fox (2008), for details). An alternative approach is to use an LR statistic to compare two models with and without the constraints. This approach is discussed Section 4.6.3 under the SEM approach. Approximate CIs may also be constructed based on the estimated SEs.

4.5.2 Explained variance

Besides testing whether the moderators are significant, researchers may want to quantify the degree of prediction with R^2 like indices. As the sampling variance v_i is known in a meta-analysis, the variation owing to v_i should not be included in calculating the R^2 (e.g., Raudenbush, 2009). To simplify the presentation, we take the intercept out of the regression coefficients. With a slight abuse of notation, we consider two models—Model 0 without any moderator and Model 1 with the moderators:

$$\begin{aligned}\text{Model 0:} \, \boldsymbol{y}_0 &= \beta_0 + \boldsymbol{u}_0 + \boldsymbol{e}, \\ \text{Model 1:} \, \boldsymbol{y}_1 &= \beta_0 + \boldsymbol{X}\boldsymbol{\beta} + \boldsymbol{u}_1 + \boldsymbol{e},\end{aligned} \tag{4.38}$$

where β_0 is the intercept, $\boldsymbol{\beta}$ is the vector of regression coefficients, $\boldsymbol{X}$ is a design matrix for the moderators, $\tau_0^2 = \mathsf{Var}(\boldsymbol{u}_0)$ is the heterogeneity variance of the random effects without the moderator, $\tau_1^2 = \mathsf{Var}(\boldsymbol{u}_1)$ is the residual variance of the random effects with the moderators, and $\boldsymbol{e}$ is the sampling error.

The population percentage of the explained variance P^2 by including the predictors can be calculated by comparing the τ_0^2 and τ_1^2 (Aloe et al., 2010; Borenstein et al., 2009; Raudenbush, 2009),

$$P^2 = \frac{\tau_0^2 - \tau_1^2}{\tau_0^2}. \tag{4.39}$$

As the population values are rarely known, we estimate them by their sample statistics,

$$R^2 = \frac{\hat{\tau}_0^2 - \hat{\tau}_1^2}{\hat{\tau}_0^2}. \tag{4.40}$$

In a regression analysis, the computed R^2 is always non-negative. However, the calculated R^2 can be negative. If this happens, it is truncated to zero. When there are missing values in the moderators, these studies are removed in calculating $\hat{\tau}_1^2$. It is of importance also to remove the studies with missing data in calculating $\hat{\tau}_0^2$; otherwise, $\hat{\tau}_0^2$ and $\hat{\tau}_1^2$ are calculated based on different numbers of studies.

López-López et al. (2014) conducted a simulation study to evaluate the empirical performance of Equation 4.40. As different estimators may lead to different estimates of τ_0^2 and τ_1^2, these authors evaluated seven estimates of R^2. On the basis of their results, none of the studied estimators performed accurately when the number of studies was small ($k < 20$). When the number of studies was moderate (20–40 studies), they found that the REML and the empirical Bayes methods performed best when the bias and efficiency criteria were jointly considered.

An alternative way to define the percentage of variance explained is based on Model 1 only:

$$\tilde{P}^2 = \frac{\boldsymbol{\beta}^{\mathrm{T}}\boldsymbol{\Sigma}_X\boldsymbol{\beta}}{\boldsymbol{\beta}^{\mathrm{T}}\boldsymbol{\Sigma}_X\boldsymbol{\beta} + \tau_1^2}, \tag{4.41}$$

where $\boldsymbol{\Sigma}_X$ is the population covariance matrix of the moderators. Its sample estimate is

$$\tilde{R}^2 = \frac{\hat{\boldsymbol{\beta}}^{\mathrm{T}}\boldsymbol{S}_X\hat{\boldsymbol{\beta}}}{\hat{\boldsymbol{\beta}}^{\mathrm{T}}\boldsymbol{S}_X\hat{\boldsymbol{\beta}} + \hat{\tau}_1^2}, \tag{4.42}$$

where $\boldsymbol{S}_X$ is the sample covariance matrix of the moderators. $\tilde{R}^2$ is always non-negative as long as $\hat{\tau}_1^2$ is nonnegative. As $\tilde{R}^2$ is rarely used in meta-analysis, it is unclear about its empirical properties. Further research may compare the empirical performance of these two estimators.

4.5.3 A cautionary note

Although a mixed-effects meta-analysis looks similar to a typical regression analysis, special care has to be taken to avoid an ecological fallacy—incorrectly interpreting findings from the aggregated statistics as findings from the individuals. Take the proportion of females as an example. If it positively predicts the odds ratio of having a disease, it is tempting to take the interpretation that females are more likely to get the disease than males do. This interpretation would be incorrect. As the predictor (or the moderator) is the proportion of females (not gender at the individual level), there are still males in studies with a high proportion of females. What the result suggests is that studies with more females are more likely to have larger odds ratios. This effect may or may not be attributed to the female participants in these studies. Findings at the study level cannot directly be translated into findings at the individual level. Readers may refer to Petkova et al. (2013) for some discussion on this issue.

4.6 Structural equation modeling approach

After reviewing the meta-analytic models, this section introduces how the SEM-based meta-analysis can be used to model the univariate meta-analytic models (Cheung, 2008). The SEM-based meta-analysis is mathematically equivalent to that of the conventional meta-analysis using ML (or possibly REML) estimation. The main difference is that the meta-analytic models are formulated as structural equation models. Therefore, the power of SEM can be extended to meta-analysis. As ML estimation is used in the SEM-based meta-analysis, both the fixed and the random effects are simultaneously estimated. We first begin with the fixed-effects model and then extend the discussion to the random-effects and the mixed-effects models. The techniques introduced in the previous sections can be applied in the SEM-based meta-analysis.

4.6.1 Fixed-effects model

As indicated in the previous section, there is only one parameter β_F under a fixed-effects model, while the sampling variance v_i is assumed to be known. The fixed-effects meta-analytic model can be formulated as a structural equation model displayed in Figure 4.4. The observed effect size in the ith study is represented by the variable y_i, while the variance of the measurement error is fixed as the known sampling variance v_i. As v_i is fixed in the model, the only parameter in this model is β_F.

When comparing the fixed-effects meta-analytic model and the model with only one variable in Figure 4.1, it is clear that the fixed-effects meta-analytic model is a very simple model with only one parameter. Although the fixed-effects meta-analytic model is a simple model, it is not straightforward to handle it under conventional SEM packages. This is because v_i varies across subjects (or across studies in a meta-analysis), whereas most SEM packages were not developed to handle subject-specific models.

There are two approaches to incorporate the known v_i under the SEM framework (Cheung, 2013). The first approach is to transform the effect sizes so that they have a common known sampling distribution with a variance of 1 (see Cheung, 2008). As all effect sizes have the same known sampling distribution, conventional SEM packages may be used to model the meta-analytic models by fixing the error variance at 1. The second approach is to assign the known v_i as the known sampling

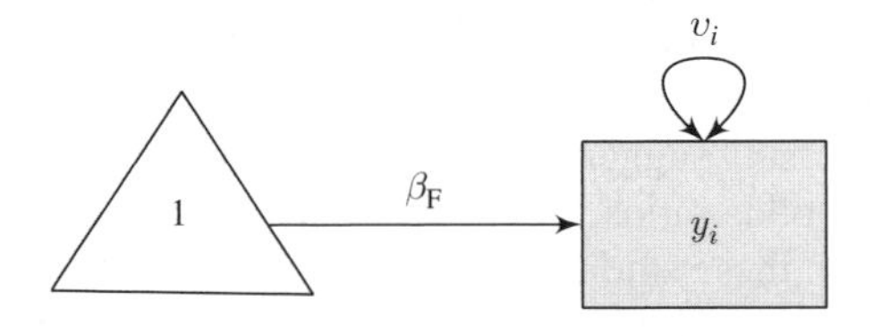

Figure 4.4 A univariate fixed-effects meta-analytic model.

error with the use of definition variables (see Section 2.5.2). Each subject (or each study in a meta-analysis) may has its own known v_i in the model in Figure 4.4. The definition variable approach is introduced here, while Chapter 9 introduces the transformed variables approach that can be implemented in M*plus*.

To conduct the fixed-effects meta-analysis in SEM, we fit the following model-implied mean and variance matrix:

$$\begin{aligned} \mu_i(\theta) &= \beta_\mathrm{F} \quad \text{and} \\ \Sigma_i(\theta) &= v_i. \end{aligned} \tag{4.43}$$

After fitting this model, $\hat{\beta}_\mathrm{F}$ is reported. We test the significance of $H_0 : \beta_\mathrm{F} = \beta_0$ by

$$z_{\hat{\beta}_\mathrm{F}} = \frac{\hat{\beta}_\mathrm{F} - \beta_0}{SE(\hat{\beta}_\mathrm{F})}, \tag{4.44}$$

which has an approximate standard normal distribution under H_0. We reject the null hypothesis $H_0 : \beta_\mathrm{F} = 0$ at $\alpha = 0.05$ if the absolute value of $z_{\hat{\beta}_\mathrm{F}}$ is larger than or equal to 1.96. We may construct an approximate 95% Wald CI on β_F using $\hat{\beta}_\mathrm{F} \pm 1.96 * SE_{\hat{\beta}_\mathrm{F}}$. Alternatively, we may construct an LBCI on $\hat{\beta}_\mathrm{F}$ (see Section 2.4.5).

4.6.2 Random-effects model

The random-effects model allows studies to have their own study-specific random effects. Study-specific random effects are unobservable and latent in nature. In the context of SEM, we may treat the study-specific random effects (f_i in Equation 4.15) as a latent variable. Figure 4.5 shows two equivalent graphical representations of a random-effects model. Figure 4.5a shows the graphical model of the random-effects model. The latent variable $f_i \sim \mathcal{N}(\beta_\mathrm{R}, \tau^2)$ represents the *true* effect size in the ith study, while the error variance v_i is known and varies across studies. Conceptually, the random-effects model can be viewed as a one-factor CFA model with only one indicator. The measurement error in conventional CFA models is used to represent the sampling error in a meta-analysis. This model is similar to the CFA model used to model multilevel data (e.g., Mehta and Neale, 2005).

Unlike conventional CFA models, the variances of the measurement (sampling) error are fixed according to the studies in a meta-analysis. One attractive feature of this presentation is that it mirrors Equation 4.15 by showing all of the elements—$\beta_\mathrm{R}, f_i, e_i, \tau^2$, and v_i in the figure. Level 1 and level 2 can be considered as the measurement and structural models in conventional SEM. As the *true* effect size in the ith study is represented by the latent factor f_i, the *true* effect size may be used as the independent variable or the dependent variable in more complex analyses, such as the mediation and moderation analyses in Section 5.6.

On the basis of Equations 4.16 and 4.17, we may write $y_i \sim \mathcal{N}(\beta_\mathrm{R}, \tau^2 + v_i)$. All studies have the same expected mean (β_R), whereas the expected variance ($\tau^2 + v_i$) varies across studies. Figure 4.5b shows a model that is equivalent to the random-effects model. The latent variable f_i is not shown in the figure. The key

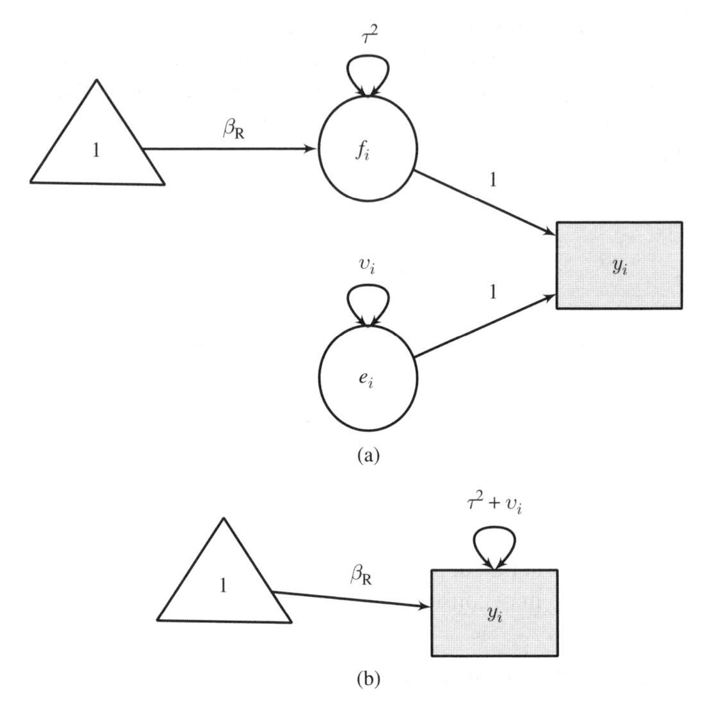

Figure 4.5 A univariate random-effects meta-analytic model.

feature of this presentation is that each study has its own distribution ($\tau^2 + v_i$). The advantage of this presentation is its simplicity, as the latent variable f_i and the residual e_i are not shown in the figure.

To conduct the univariate random-effects meta-analysis in SEM, we fit the following model-implied moments:

$$\begin{aligned} \mu_i(\boldsymbol{\theta}) &= \beta_R \quad \text{and} \\ \Sigma_i(\boldsymbol{\theta}) &= \tau^2 + v_i. \end{aligned} \tag{4.45}$$

We have to estimate both β_R and τ^2 simultaneously. After the estimation, the SEs on $\hat{\beta}_R$ and $\hat{\tau}^2$ may be used to test the significance of these estimates. As $\hat{\tau}^2$ is unlikely to be normally distributed, a Wald CI is not recommended. LBCI is preferred for quantifying the precision of $\hat{\tau}^2$ (Hardy and Thompson, 1996).

4.6.3 Mixed-effects model

We may easily extend the random-effects model to a mixed-effects model by including study characteristics as moderators. Similar to a regression analysis (see Section 2.5.2), there are two representations for a mixed-effects meta-analysis. We may treat the moderators as either variables or design matrices (Cheung, 2008, 2013).

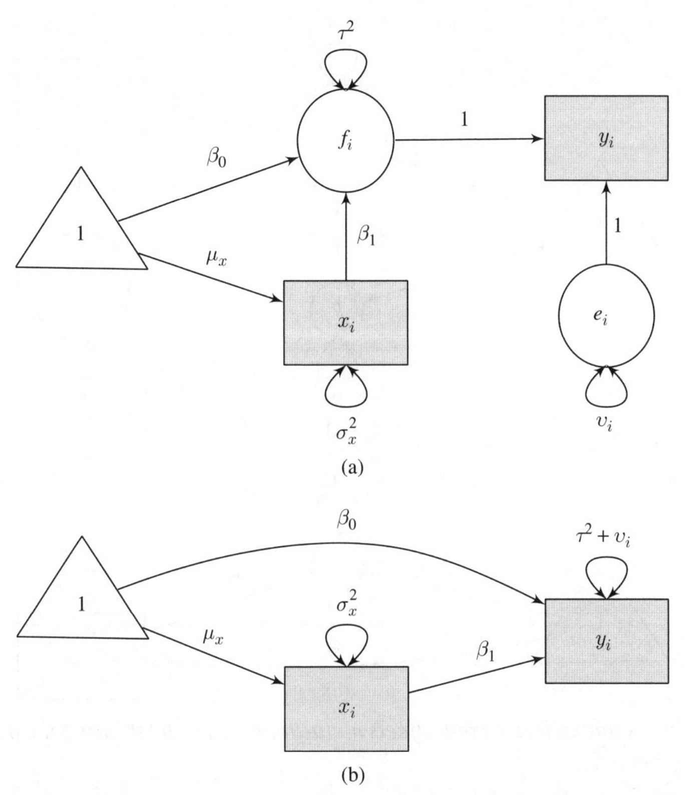

Figure 4.6 A univariate mixed-effects meta-analytic model treating the moderator as a variable.

We first discuss the approach by treating the moderators as variables. Figure 4.6 shows two equivalent models of a mixed-effects model with one moderator. Figure 4.6a explicitly displays the *true* effect size, which is represented by $f_i = \beta_0 + \beta_1 x_i + u_i$. The observed effect size is then defined by $y_i = f_i + e_i$. Figure 4.6b displays the same model without showing f_i. As the moderator x_i is a variable, its mean μ_x and its variance σ_x^2 have to be estimated. The expected mean vectors and the expected covariance matrix are

$$\mathsf{E}\left(\begin{bmatrix} y_i \\ x_i \end{bmatrix}\right) = \begin{bmatrix} \beta_0 + \beta_1 \mu_x \\ \mu_x \end{bmatrix} \quad \text{and}$$

$$\mathsf{Cov}\left(\begin{bmatrix} y_i \\ x_i \end{bmatrix}\right) = \begin{bmatrix} \beta_1^2 \sigma_x^2 + \tau^2 + v_i & \\ \beta_1 \sigma_x^2 & \sigma_x^2 \end{bmatrix}. \tag{4.46}$$

The second approach is to treat the moderators as a design matrix or fixed values (Cheung, 2010, 2013). Figure 4.7 shows two equivalent models of the mixed-effects meta-analysis. Figure 4.7a includes the *true* effect size. A phantom variable P with a variance of 0 is created to implement the predictors. The mean of P is fixed at x_i by

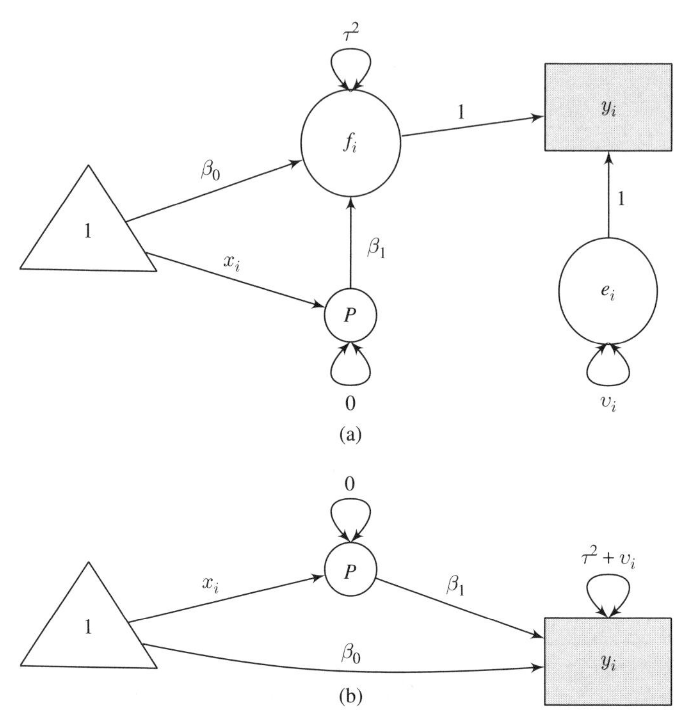

Figure 4.7 A univariate mixed-effects meta-analytic model treating the moderator as a design matrix.

using a definition variable, meaning that each study has a different value of x_i. The effect of regressing f_i on P is β_1. Therefore, $f_i = \beta_0 + \beta_1 x_i + u_i$. As x_i is treated as a fixed value, there is no estimate of the mean and variance of x_i. Figure 4.7b shows the same model without the *true* effect size or the latent variable. The following are the model-implied conditional mean and variance for m moderators:

$$\begin{aligned} \mu_i(\boldsymbol{\theta}|\boldsymbol{x}_i) &= \boldsymbol{x}_i^{\mathrm{T}}\boldsymbol{\beta} \quad \text{and} \\ \Sigma_i(\boldsymbol{\theta}|\boldsymbol{x}_i) &= \tau^2 + v_i. \end{aligned} \tag{4.47}$$

When there is more than one moderator, we may want to test the composite hypothesis that all of the regression coefficients (excluding the intercept) are zero, $H_0 : \beta_1 = \beta_2 = \cdots = \beta_m = 0$. When the scales of the predictors are comparable, for example, dummy or standardized variables, we may want to test whether some or all of the regression coefficients are the same, for example, $H_0 : \beta_1 = \beta_2 = \cdots = \beta_m$. To test the above hypotheses, we may fit two models—one with the constraint that all regression coefficients are fixed at zero or that all or some regression coefficients are equal and the other without any constraint. As these two models are nested, we may use an LR test to test the above null hypothesis.

4.7 Illustrations using R

This section illustrates how to conduct the univariate meta-analysis using the metaSEM package. Readers may refer to Appendix A for an introduction on the metaSEM package and on how to read external data in R. Two examples are used to demonstrate the fixed-, random-, and mixed-effects meta-analyses. The fixed-effects meta-analysis is a special case of the random-effects model by fixing the heterogeneity variance at 0 in the metaSEM package. Therefore, we first show how to conduct a random-effects meta-analysis. Then we illustrate how to conduct a fixed-effects meta-analysis by imposing the appropriate constraints. Moreover, examples on applying constraints and constructing LBCI are also demonstrated. The statistics reported in the illustrations were captured by using the Sweave function in R. The numbers of decimal places may be slightly different for those reported in the selected output and in the text.

4.7.1 Odds ratio of atrial fibrillation between bisphosphonate and non-bisphosphonate users

Eight studies were selected from Mak et al. (2009). The objective in Mak et al. (2009) was to compare the incidents of atrial fibrillation among bisphosphonate and non-bisphosphonate users. The odds ratio between the bisphosphonate and non-bisphosphonate users was calculated as the effect size. There are substantial differences in data and statistical analyses between Mak et al. (2009) and the present illustration. As an illustration, we combine studies from both randomized controlled trials and observational studies (RCT and Obs in the variable type in the data set) (see also Cheung et al., 2012). It is generally not advisable to do this. A preferred approach is to use a Bayesian meta-analysis, which was employed by Mak et al. (2009). Under a Bayesian approach, results from the observational studies are used as the prior for analyzing the randomized controlled trials (e.g., Sutton and Abrams, 2001). The findings of the present study may not be directly comparable to those reported by Mak et al. (2009).

The data was stored in the object Mak09. We first show the full data set by typing the name of the data set. The effect size and its sampling variance are yi (log of the odds ratio) and vi, respectively.

```
R> #### Load the metaSEM library
R> library("metaSEM")
R> ## Display the full dataset
R> Mak09
```

```
              Study type AF.BP Tot.BP AF.non.BP Tot.non.BP       yi
1        Black (2007)  RCT    94   3862        73       3852  0.25575
2     Cummings (2007)  RCT    81   3236        71       3223  0.13082
3        Karam (2007)  RCT   189  10018        94       5048  0.01331
4        Lyles (2007)  RCT    29   1054        27       1057  0.07633
5 Papapoulous (2008)  RCT    57   6830        18       1924 -0.11526
```

```
6  Abrahamsen (2009)  Obs   797  14302       1280       28731  0.23558
7    Heckbert (2008)  Obs    47     87        672        1598  0.48188
8    Sorensen (2008)  Obs   724   3862      12862       77643  0.15019
        vi age.mean study.duration
1 0.024867     73.0            3.0
2 0.027064     69.3            3.6
3 0.016233     73.5             NA
4 0.073466     74.5            5.0
5 0.073772     66.9            2.5
6 0.002146     74.3           10.0
7 0.048845     72.7            3.0
8 0.001793     76.1            6.0
```

4.7.1.1 Random-effects model

We conduct a random-effects meta-analysis by using the `meta()` command with `y` and `v` as the arguments for the effect size and its sampling variance. After running the analysis, we may store the results in an R object, say `mak1`. We extract the results by calling up the `summary()` function.

```
R> mak1 <- meta(y=yi, v=vi, data=Mak09)
R> summary(mak1)
R> #### Alternative specification without storing the results
R> summary( meta(y=yi, v=vi, data=Mak09) )
```

```
-----------------------  Selected output  -----------------------

95% confidence intervals: z statistic approximation
Coefficients:
            Estimate Std.Error    lbound    ubound z value Pr(>|z|)
Intercept1  1.81e-01  2.89e-02  1.24e-01  2.38e-01    6.25  4.2e-10
Tau2_1_1    9.98e-11  2.22e-03 -4.34e-03  4.34e-03    0.00        1

Intercept1 ***
Tau2_1_1
---
Signif. codes:  0 '***' 0.001 '**' 0.01 '*' 0.05 '.' 0.1 ' ' 1

Q statistic on the homogeneity of effect sizes: 7.16
Degrees of freedom of the Q statistic: 7
P value of the Q statistic: 0.4124

Heterogeneity indices (based on the estimated Tau2):
                             Estimate
Intercept1: I2 (Q statistic)        0

Number of studies (or clusters): 8
Number of observed statistics: 8
Number of estimated parameters: 2
```

```
Degrees of freedom: 6
-2 log likelihood: -10.27
OpenMx status1: 0 ("0" or "1": The optimization is considered fine.
Other values indicate problems.)

-----------------------  Selected output  -----------------------
```

Before interpreting the results, we should check whether there are any estimation problems in the analysis. We check the `OpenMx status1` at the end of the output. If the status is either 0 or 1, the optimization is fine; otherwise, the results are not trustworthy. If there are problems in the optimization, we may try to rerun the analysis with the `rerun()` function (see Section 7.6.3 for an example).

The test statistic of the homogeneity of effect sizes is $Q(\text{df} = 7) = 7.1598$, $p = 0.4124$, which is statistically nonsignificant. Both $\hat{\tau}^2$ (`Tau2_1_1` in the output) and I^2 are approximately 0. The results suggest that the between-study heterogeneity is trivial. Results based on the fixed- and the random-effects models are nearly identical.

4.7.1.2 Fixed-effects model

As a demonstration, we fitted a fixed-effects model. A fixed-effects meta-analysis is a special case of the random-effects meta-analysis by fixing the heterogeneity variance at zero. We may fix the heterogeneity to any value by using the `RE.constraints` argument in `meta()`. It expects a matrix as input. In this example, the dimension of the matrix is 1×1. If we specify a scalar, the value will be automatically converted into a matrix.

```
R> ( mak2 <- summary(meta(y=yi, v=vi, data=Mak09,
                     RE.constraints=0)) )
R> #### Alternative specification
R> ( mak2 <- summary(meta(y=yi, v=vi, data=Mak09,
                     RE.constraints=matrix(0, ncol=1, nrow=1))) )
```

```
-----------------------  Selected output  -----------------------

95% confidence intervals: z statistic approximation
Coefficients:
           Estimate Std.Error lbound ubound z value Pr(>|z|)
Intercept1   0.1808    0.0287 0.1245 0.2372    6.29  3.2e-10 ***
---
Signif. codes:  0 '***' 0.001 '**' 0.01 '*' 0.05 '.' 0.1 ' ' 1

Heterogeneity indices (based on the estimated Tau2):
                              Estimate
Intercept1: I2 (Q statistic)         0

-----------------------  Selected output  -----------------------
```

There is no estimate on τ^2 because it is fixed at zero. The estimated average effects (`Intercept1` at the output) based on either a fixed- or a random-effects model are the same. The estimated common effect with their 95% Wald CIs are equal: 0.1808 (0.1245, 0.2372). The results are identical (up to several decimal places) because $\hat{\tau}^2 \approx 0$. The analyses were based on a log of the odds ratio. After the analyses, we can convert the estimated `Intercept1` and its 95% `lbound` and `ubound` back to an odds ratio.

```
R> ## Get the estimate and its 95% CI
R> ( Est <- mak2$coefficients["Intercept1",
                              c("Estimate","lbound","ubound")] )
```

```
           Estimate lbound ubound
Intercept1   0.1808 0.1245 0.2372
```

```
R> ## Convert them into odds ratio
R> exp(Est)
```

```
           Estimate lbound ubound
Intercept1    1.198  1.133  1.268
```

From the above calculations, the transformed odds ratio with its 95% Wald CI is 1.1982 (1.1326, 1.2677).

4.7.2 Correlation between organizational commitment and salesperson job performance

Jaramillo et al. (2005, Table 1) conducted a meta-analysis of 61 studies on the relationship between organizational commitment and salesperson job performance. The effect size was a correlation coefficient. Jaramillo et al. (2005) corrected for unreliability before conducting the analysis. As an illustration, we use the reported (uncorrected) correlation coefficients here. We first show the first few cases by using the `head()` command.

```
R> head(Jaramillo05)
```

```
                         Author Sample_size    Sales Country IDV
1         Aryee et al. (2002)         179    mixed   India  48
2 Balfour and Wechsler (1991)         232 nonsales     USA  91
3      Bashaw and Grant (1994)        560    sales     USA  91
4            Benkhoff (1997)         181    sales Germany  67
5         Brett et al. (1995)         156    sales     USA  91
6         Brett et al. (1995)         180    sales     USA  91
         OC_scale OC_alpha JP_alpha     r       r_v
```

```
1 Porter or Mowday      0.87      0.89 0.02 0.005582
2             other     0.82        NA 0.12 0.004187
3 Porter or Mowday      0.83      0.76 0.09 0.001757
4 Porter or Mowday        NA      1.00 0.20 0.005092
5 Porter or Mowday      0.83        NA 0.08 0.006328
6 Porter or Mowday      0.83        NA 0.04 0.005538
```

The effect size and its sampling variance are `r` and `r_v`, respectively. We use `r` as the effect size in this illustration. If a Fisher's z score is required, we can calculate it by

```
R> z <- with( Jaramillo05, 0.5*log((1+r)/(1-r)) )
R> z.v <- with( Jaramillo05, 1/(Sample.size-3) )
```

4.7.2.1 Random-effects model

We employ a random-effects model with the following syntax. By default, the I^2 is calculated based on the Q statistic (with the `I2="I2q"` argument in calling the `meta()` function). Readers can also use either the harmonic mean (`I2="I2hm"`) or the arithmetic mean (`I2="I2am"`) to calculate the I^2 (see Section 4.3.3).

```
R> summary( meta(y=r, v=r_v, data=Jaramillo05) )
```

```
----------------------- Selected output -----------------------

95% confidence intervals: z statistic approximation
Coefficients:
           Estimate Std.Error  lbound  ubound z value Pr(>|z|)
Intercept1  0.18662   0.01933 0.14874 0.22451    9.65  < 2e-16 ***
Tau2_1_1    0.01703   0.00414 0.00893 0.02514    4.12  3.8e-05 ***
---
Signif. codes:  0 '***' 0.001 '**' 0.01 '*' 0.05 '.' 0.1 ' ' 1

Q statistic on the homogeneity of effect sizes: 339.4
Degrees of freedom of the Q statistic: 60
P value of the Q statistic: 0

Heterogeneity indices (based on the estimated Tau2):
                               Estimate
Intercept1: I2 (Q statistic)       0.81

Number of studies (or clusters): 61
Number of observed statistics: 61
Number of estimated parameters: 2
Degrees of freedom: 59
-2 log likelihood: -55.44
```

```
OpenMx status1: 0 ("0" or "1": The optimization is considered fine.
Other values indicate problems.)

----------------------- Selected output -----------------------
```

The homogeneity test of effect sizes is statistically significant with $Q(\text{df} = 60) = 339.3886, p < 0.001$. The $\hat{\tau}^2 = 0.0170$ and $I^2 = 0.8144$. These indicate that there is a high degree of heterogeneity. The between-study effect explains 81.44% of the total variation. The estimated average population correlation coefficient (with its 95% Wald CI) based on a random-effects model is 0.1866 (0.1487, 0.2245).

4.7.2.2 Likelihood-based CI

The above CIs are based on the Wald approximation (labeled as a `z statistic approximation` in the output). When the number of studies is small, LBCI (labeled as a `Likelihood-based statistic` in the output; see Section 2.4.5) is preferred. We may request the LBCI by specifying the `intervals.type="LB"` argument. As I^2 is a function of $\hat{\tau}^2$ (see Section 4.3.3), LBCI on I^2 is also reported.

```
R> summary( meta(y=r, v=r_v, data=Jaramillo05,
                 intervals.type="LB") )
```

```
----------------------- Selected output -----------------------

95% confidence intervals: Likelihood-based statistic
Coefficients:
           Estimate Std.Error lbound ubound z value Pr(>|z|)
Intercept1   0.1866        NA 0.1480 0.2251      NA       NA
Tau2_1_1     0.0170        NA 0.0106 0.0276      NA       NA

Heterogeneity indices (I2) and their 95% likelihood-based CIs:
                            lbound Estimate ubound
Intercept1: I2 (Q statistic)  0.732    0.814   0.88

----------------------- Selected output -----------------------
```

The 95% LBCIs on $\hat{\tau}^2$ and I^2 are (0.0106, 0.0276), and (0.7319, 0.8767), respectively.

4.7.2.3 Mixed-effects model

The moderators can be included by specifying the `x` argument in the `meta()` function. When there is more than one moderators, they can be combined using the `cbind()` command. The explained variance R^2 on the effect size with Equation 4.40 is also reported.

4.7.2.4 One moderator

Let us consider the individualism index of the studies (IDV in the data set) as a moderator. Individualism refers to the degree to which individuals are integrated into groups (Hofstede, 2001). People in individualistic societies tend to focus on personal achievements and individual rights, while people in collectivist societies tend to think and act as members of a group. This variable is often used to explain cross-cultural behavioral and psychological differences. We use the individualism index to predict the variation of the effect sizes. We usually center the predictors to improve numerical stability and ease of interpretations of the results.

```
R> ## Center IND: scale(IND, scale=FALSE)
R> ##  scale=TRUE: standardize the variable
R> ##  scale=FALSE: not standardize the variable
R> summary( meta(y=r, v=r_v, x=scale(IDV, scale=FALSE),
                 data=Jaramillo05) )
```

```
----------------------- Selected output -----------------------

95% confidence intervals: z statistic approximation
Coefficients:
            Estimate Std.Error    lbound    ubound z value Pr(>|z|)
Intercept1  0.185963  0.019030  0.148664  0.223262    9.77  < 2e-16
Slope1_1   -0.001321  0.000977 -0.003236  0.000593   -1.35     0.18
Tau2_1_1    0.016341  0.004020  0.008461  0.024220    4.06  4.8e-05
---
Signif. codes:  0 '***' 0.001 '**' 0.01 '*' 0.05 '.' 0.1 ' ' 1

Explained variances (R2):
                          y1
Tau2 (no predictor)     0.02
Tau2 (with predictors) 0.02
R2                      0.04

----------------------- Selected output -----------------------
```

The estimated regression coefficient for IDV (Slope1_1 in the output) is not statistically significant with $\hat{\beta}_{\text{IDV}} = -0.0013$, SE $= 0.0010$ and $p = 0.1762$. The $\hat{\tau}^2$s for the models with and without IDV are 0.0163, and 0.0170, respectively. Thus, $R^2 = 1 - 0.0163/0.0170 = 0.0407$. The IDV can only explain about 4% of the variation on the effect size of the correlation between organizational commitment and job performance.

4.7.2.5 Two moderators

The data set includes the coefficient alpha of the scales on measuring organizational commitment and job performance (OC_alpha and JP_alpha in the data set). Besides correcting the correlations for unreliability with the approach advocated by Schmidt and Hunter (2015), Hox (2010) proposed an alternative approach to adjust for measurement unreliability. Hox argued that the relationship between the reliability coefficient (e.g., coefficient alpha) and the observed correlation can be approximated by an additive model as long as the reliability is not too low (e.g., coefficient alpha > 0.70). The model for the ith study is

$$r_i = \beta_0 + \beta_1 r_{(xx')i} + \beta_2 r_{(yy')i} + u_i + e_i, \tag{4.48}$$

where r_i is the observed correlation, $r_{(xx')i}$ is the reliability coefficient of the predictor, $r_{(yy')i}$ is the reliability coefficient of the dependent variable, u_i is the study-specific random effects, and e_i is the sampling variance. One advantage of this approach is that we can statistically test whether the reliabilities are significant in predicting the effect size.

If the above model is correct, we may estimate the average population correlation ρ_{cor} corrected for measurement unreliability by using $r_{(xx')i} = r_{(yy')i} = 1.0$, that is,

$$\hat{\rho}_{\text{cor}} = \hat{\beta}_0 + \hat{\beta}_1(1.0) + \hat{\beta}_2(1.0). \tag{4.49}$$

As an illustration, we include both `OC_alpha` and `JP_alpha` in the model. We save the results to an object, say `model1`, that will be used later for model comparisons.

```
R> model1 <- meta(y=r, v=r_v, x=cbind(OC_alpha, JP_alpha),
                  data=Jaramillo05,
                  model.name="Unequal coefficients")
R> summary(model1)
```

```
----------------------- Selected output -----------------------

95% confidence intervals: z statistic approximation
Coefficients:
            Estimate Std.Error    lbound    ubound z value Pr(>|z|)
Intercept1 -0.57554   0.50152 -1.55849  0.40742   -1.15  0.25114
Slope1_1    0.13110   0.45872 -0.76797  1.03018    0.29  0.77503
Slope1_2    0.80442   0.43038 -0.03912  1.64796    1.87  0.06161 .
Tau2_1_1    0.01873   0.00565  0.00765  0.02981    3.31  0.00092 ***
---
Signif. codes:  0 '***' 0.001 '**' 0.01 '*' 0.05 '.' 0.1 ' ' 1

Explained variances (R2):
                          y1
Tau2 (no predictor)     0.02
Tau2 (with predictors)  0.02
R2                      0.00

----------------------- Selected output -----------------------
```

The estimated regression coefficients for `OC_alpha` and `JP_alpha` (`Slope1_1` and `Slope1_2` in the output) are $\hat{\beta}_{\text{OC}_{\text{alpha}}} = 0.1311$, $\text{SE}_{\text{OC}_{\text{alpha}}} =$

0.4587, $p_{OC_{alpha}} = 0.7750$, and $\hat{\beta}_{JP_{alpha}} = 0.8044$, $SE_{JP_{alpha}} = 0.4304$, $p_{JP_{alpha}} = 0.0616$, respectively. Neither of them are statistically significant at $\alpha = 0.05$ and $R^2 = 0$. Therefore, there is no evidence indicating that the reliabilities of the measures are correlated with the effect size. Although the above approach is simple and intuitive, there are limited studies comparing its performance against the conventional approaches in correcting for unreliability. The main objective of this example was to demonstrate how to impose equality constraints on the regression coefficients in the `metaSEM` package. Further studies may investigate how useful this approach is.

4.7.2.6 Testing the equality of all coefficients

Although both coefficients are nonsignificant in the above analysis, we test $H_0 : \beta_{equal} = \beta_{OC_{alpha}} = \beta_{JP_{alpha}}$ as an illustration. First, we need to fit a model with an equality constraint on the regression coefficients by specifying the `coef.constraints` argument. The argument expects a $p \times m$ matrix, where p is the number of effect sizes and m is the number of predictors. In this example, it is 1×2 matrix, where the first and second elements refer to the regression coefficients of `OC_alpha` and `JP_alpha`, respectively.

We may impose the equality constraint by using the same label in the constraint. In this example, `0*` represents the starting value for the regression coefficients while `Slope_equal` is the name of both coefficients. We further call this model `model.name="Equal slopes"` for ease of comparison and save the results to an R object called `model2`.

```
R> ( constraint <- matrix(c("0*Slope_equal", "0*Slope_equal"),
                           nrow=1, ncol=2) )
```

```
         [,1]            [,2]
[1,] "0*Slope_equal" "0*Slope_equal"
```

```
R> model2 <- meta(y=r, v=r_v, x=cbind(OC_alpha, JP_alpha),
                 data=Jaramillo05, coef.constraints=constraint,
                 model.name="Equal coefficients")
R> summary(model2)
```

```
----------------------- Selected output -----------------------

95% confidence intervals: z statistic approximation
Coefficients:
             Estimate Std.Error    lbound    ubound z value Pr(>|z|)
Intercept1   -0.60367   0.50745 -1.59826   0.39092   -1.19  0.23420
Slope_equal   0.48630   0.29531 -0.09249   1.06509    1.65  0.09961 .
Tau2_1_1      0.01937   0.00582  0.00796   0.03079    3.33  0.00088 ***
---
```

```
Signif. codes:  0 '***' 0.001 '**' 0.01 '*' 0.05 '.' 0.1 ' ' 1

Explained variances (R2):
                           y1
Tau2 (no predictor)      0.02
Tau2 (with predictors) 0.02
R2                         0.00

---------------------- Selected output ----------------------
```

The estimated constrained regression coefficient is $\hat{\beta}_{\text{equal}} = 0.4863, \text{SE}_{\text{equal}} = 0.2953, p_{\text{equal}} = 0.0996$. To test $H_0 : \beta_{\text{OC}_{\text{alpha}}} = \beta_{\text{JP}_{\text{alpha}}}$, we compare `model1` against `model2` with the `anova()` function.

```
R> anova(model1, model2)
```

```
                    base              comparison ep minus2LL df    AIC
1 Unequal coefficients                     <NA>  4   -31.01 31 -93.01
2 Unequal coefficients Equal coefficients  3   -30.02 32 -94.02
  diffLL diffdf      p
1     NA     NA     NA
2 0.9943      1 0.3187
```

The LR statistic is $\Delta\chi^2(\text{df} = 1) = 0.9943, p = 0.3187$. Therefore, there is not enough evidence to reject the null hypothesis of equal regression coefficients.

4.7.2.7 Testing categorical predictors

There are three types of samples: `sales`, `nonsales`, and `mixed` in the variable `Sales`. A typical approach is to use one group, say `nonsales`, as the reference group. We may then create two dummy variables ($\boldsymbol{D}_{\text{sales}}$ and $\boldsymbol{D}_{\text{mixed}}$) with only 0 and 1 for `sales` and `mixed` to represent the differences between these groups to the reference group, the `nonsales`. The model

$$\boldsymbol{y} = \beta_0 + \beta_1\boldsymbol{D}_{\text{sales}} + \beta_2\boldsymbol{D}_{\text{mixed}} + \boldsymbol{u} + \boldsymbol{e}, \tag{4.50}$$

where β_0 is the population effect size for `nonsales`, β_1 is the difference between `sales` and `nonsales`, and β_2 is the difference between `mixed` and `nonsales`.

Although the model can be used to test the differences among the groups, it does not provide the estimates for all groups. An alternative approach is to create three indicator variables. We may fit a model without an intercept:

$$\boldsymbol{y} = \beta_1\boldsymbol{D}_{\text{sales}} + \beta_2\boldsymbol{D}_{\text{mixed}} + \beta_3\boldsymbol{D}_{\text{nonsales}} + \boldsymbol{u} + \boldsymbol{e}, \tag{4.51}$$

where β_1, β_2, and β_3 now represent the average population effect sizes for `sales`, `mixed`, and `nonsales`, respectively. In order to estimate the means for all three groups, the intercept must be fixed at 0; otherwise, the model is not identified. To test whether all group means are the same, we compare the above model against the intercept model. Under the null hypothesis $H_0 : \beta_1 = \beta_2 = \beta_3$, the test statistic has a chi-square distribution with df = 2.

First, we show the frequency table of the variable `Sales`. Then, we create three indicator variables by using the `ifelse()` command.

```
R> table(Jaramillo05$Sales)
```

```
   mixed nonsales    sales
       6       27       28
```

```
R> sales <- ifelse(Jaramillo05$Sales=="sales", yes=1, no=0)
R> nonsales <- ifelse(Jaramillo05$Sales=="nonsales", yes=1, no=0)
R> mixed <- ifelse(Jaramillo05$Sales=="mixed", yes=1, no=0)
```

To fit the model without an intercept, we fix the intercept at 0 by specifying the `intercept.constraints=matrix(0, ncol=1, nrow=1)` argument. As the original starting values assume that there is an intercept, there were estimation problems in the model without the intercept. We provide starting values for the regression coefficients by using the `coef.constraints` argument:

```
R> ( startvalues <- matrix(c("0*Slope1_1", "0*Slope1_2",
                          "0*Slope1_3"), nrow=1, ncol=3) )
```

```
     [,1]         [,2]         [,3]
[1,] "0*Slope1_1" "0*Slope1_2" "0*Slope1_3"
```

```
R> model3 <- meta(y=r, v=r_v, x=cbind(sales, mixed, nonsales),
                  data=Jaramillo05, coef.constraints=startvalues,
                  intercept.constraints=matrix(0, ncol=1, nrow=1),
                  model.name="Indicator variables")
R> summary(model3)
```

```
----------------------- Selected output -----------------------

95% confidence intervals: z statistic approximation
Coefficients:
          Estimate Std.Error  lbound  ubound z value Pr(>|z|)
Slope1_1  0.22830   0.02759 0.17421 0.28238    8.27  2.2e-16 ***
Slope1_2  0.14659   0.06328 0.02257 0.27061    2.32    0.021 *
Slope1_3  0.15196   0.02794 0.09720 0.20672    5.44  5.4e-08 ***
Tau2_1_1  0.01573   0.00385 0.00818 0.02328    4.08  4.4e-05 ***
---
```

```
Signif. codes:  0 '***' 0.001 '**' 0.01 '*' 0.05 '.' 0.1 ' ' 1

Explained variances (R2):
                            y1
Tau2 (no predictor)       0.02
Tau2 (with predictors) 0.02
R2                        0.08

----------------------- Selected output -----------------------
```

The estimated average effects and their 95% Wald CIs for the `sales`, `mixed`, and `nonsales` are 0.2283 (0.1742, 0.2824), 0.1466 (0.0226, 0.2706), and 0.1520 (0.0972, 0.2067), respectively. All are statistically significant at $\alpha = 0.05$.

When the null hypothesis $H_0 : \beta_1 = \beta_2 = \beta_3$ is true, this model is equivalent to the model with only an intercept. As the model with only an intercept `model4` is nested within the model with predictors `model3`, we compare them with the following code:.

```
R> model4 <- meta(y=r, v=r_v, data=Jaramillo05)
R> anova(model3, model4)
```

```
                     base            comparison ep minus2LL df    AIC
1 Indicator variables                      <NA>  4   -59.56 57 -173.6
2 Indicator variables Meta analysis with ML  2   -55.44 59 -173.4
  diffLL diffdf      p
1     NA     NA     NA
2  4.114      2 0.1278
```

The LR statistic is $\Delta\chi^2(\text{df} = 2) = 4.1140, p = 0.1278$. Therefore, there is not enough evidence to reject the null hypothesis of equal population correlations.

4.8 Concluding remarks and further readings

This chapter introduced and compared the basic ideas of fixed- and random-effects meta-analyses. A fixed-effects model is used when researchers are mainly interested in the collected studies, whereas a random-effects model is used when researchers want to generalize the findings to other potential studies. The conventional assumption of common effects (homogeneity of effect sizes) is used for the fixed-effects model. Readers may refer to, for example, Bonett (2008, 2009) and Shuster (2010) and the comments followed for alternative models without the assumption of a common effect. The meta-analytic models can be formulated as structural equation models by fixing the known sampling variances as variance of measurement error via definition variable. Future studies may explore how the techniques in SEM can be applied in the SEM-based meta-analysis.

References

Aloe AM, Becker BJ and Pigott TD 2010. An alternative to R2 for assessing linear models of effect size. *Research Synthesis Methods* **1**(3-4), 272–283.

Böhning D 2000. *Computer-assisted analysis of mixtures and applications: meta-analysis, disease mapping and others*. Chapman & Hall/CRC, Boca Raton, FL.

Bonett DG 2008. Meta-analytic interval estimation for bivariate correlations. *Psychological Methods* **13**(3), 173–181.

Bonett DG 2009. Meta-analytic interval estimation for standardized and unstandardized mean differences. *Psychological Methods* **14**(3), 225–238.

Borenstein M, Hedges LV, Higgins JP and Rothstein HR 2009. *Introduction to meta-analysis*. John Wiley & Sons, Ltd, Chichester, West Sussex, Hoboken, NJ.

Borenstein M, Hedges LV, Higgins JP and Rothstein HR 2010. A basic introduction to fixed-effect and random-effects models for meta-analysis. *Research Synthesis Methods* **1**(2), 97–111.

Brown H and Prescott R 2006. *Applied mixed models in medicine*. John Wiley & Sons, Inc., New York.

Cheung MWL 2008. A model for integrating fixed-, random-, and mixed-effects meta-analyses into structural equation modeling. *Psychological Methods* **13**(3), 182–202.

Cheung MWL 2010. Fixed-effects meta-analyses as multiple-group structural equation models. *Structural Equation Modeling: A Multidisciplinary Journal* **17**(3), 481–509.

Cheung MWL 2013. Multivariate meta-analysis as structural equation models. *Structural Equation Modeling: A Multidisciplinary Journal* **20**(3), 429–454.

Cheung MWL, Ho RCM, Lim Y and Mak A 2012. Conducting a meta-analysis: Basics and good practices. *International Journal of Rheumatic Diseases* **15**(2), 129–135.

Cochran W 1954. The combination of estimates from different experiments. *Biometrics* **10**(1), 101–129.

Demidenko E 2013. *Mixed models: theory and applications with R*, 2nd edn. Wiley-Interscience, Hoboken, NJ.

DerSimonian R and Laird N 1986. Meta-analysis in clinical trials. *Controlled Clinical Trials* **7**(3), 177–188.

Fox J 2008. *Applied regression analysis and generalized linear models*, 2nd edn. Sage Publications, Inc., Thousand Oaks, CA.

Friedman L 2000. Estimators of random effects variance components in meta-analysis. *Journal of Educational and Behavioral Statistics* **25**(1), 1–12.

Gelman A 2005. Analysis of variance-why it is more important than ever. *Annals of Statistics* **33**(1), 1–53.

Goldstein H 2011. *Multilevel statistical models*, 4th edn. John Wiley & Sons, Inc., Hoboken, NJ.

Hafdahl AR 2007. Combining correlation matrices: simulation analysis of improved fixed-effects methods. *Journal of Educational and Behavioral Statistics* **32**(2), 180–205.

Hardy RJ and Thompson SG 1996. A likelihood approach to meta-analysis with random effects. *Statistics in Medicine* **15**(6), 619–629.

Harwell M 1997. An empirical study of hedge's homogeneity test. *Psychological Methods* **2**(2), 219–231.

Hedges LV 1983. A random effects model for effect sizes. *Psychological Bulletin* **93**(2), 388–395.

Hedges LV 2007. Meta-analysis In *Psychometrics* (ed. Rao CR and Sinharay S), vol. 26 of *Handbook of statistics*. Russell Sage Foundation, New York, pp. 919–953.

Hedges LV and Olkin I 1983. Regression-models in research synthesis. *American Statistician* **37**(2), 137–140.

Hedges LV and Olkin I 1985. *Statistical methods for meta-analysis*. Academic Press, Orlando, FL.

Hedges LV and Vevea JL 1998. Fixed- and random-effects models in meta-analysis. *Psychological Methods* **3**(4), 486–504.

Higgins JPT and Thompson SG 2002. Quantifying heterogeneity in a meta-analysis. *Statistics in Medicine* **21**(11), 1539–1558.

Higgins JPT, Thompson SG, Deeks JJ and Altman DG 2003. Measuring inconsistency in meta-analyses. *British Medical Journal* **327**(7414), 557–560.

Higgins JPT, Thompson SG and Spiegelhalter DJ 2009. A re-evaluation of random-effects meta-analysis. *Journal of the Royal Statistical Society: Series A (Statistics in Society)* **172**(1), 137–159.

Hofstede GH 2001. *Culture's consequences: comparing values, behaviors, institutions, and organizations across nations*, 2nd edn. Sage Publications, Inc., Thousand Oaks, CA.

Hox JJ 2010. *Multilevel analysis: techniques and applications*, 2nd edn. Routledge, New York.

Huedo-Medina TB, Sanchez-Meca J, Marin-Martinez F and Botella J 2006. Assessing heterogeneity in meta-analysis: Q statistic or I2 index? *Psychological Methods* **11**(2), 193–206.

Huizenga HM, Visser I and Dolan CV 2011. Testing overall and moderator effects in random effects meta-regression. *British Journal of Mathematical and Statistical Psychology* **64**(1), 1–19.

Hunter JE and Schmidt FL 1990. *Methods of meta-analysis: correcting error and bias in research findings*. Sage Publications, Inc., Newbury Park, CA.

Jaramillo F, Mulki JP and Marshall GW 2005. A meta-analysis of the relationship between organizational commitment and salesperson job performance: 25 years of research. *Journal of Business Research* **58**(6), 705–714.

López-López JA, Marín-Martínez F, Sánchez-Meca J, Van den Noortgate W and Viechtbauer W 2014. Estimation of the predictive power of the model in mixed-effects meta-regression: a simulation study. *British Journal of Mathematical and Statistical Psychology* **67**(1), 30–48.

Mak A, Cheung MWL, Ho RCM, Cheak AAC and Lau CS 2009. Bisphosphonates and atrial fibrillation: Bayesian meta-analyses of randomized controlled trials and observational studies. *BMC Musculoskeletal Disorders* **10**(1), 1–12.

Malzahn U, Böhning D and Holling H 2000. Nonparametric estimation of heterogeneity variance for the standardised difference used in meta-analysis. *Biometrika* **87**(3), 619–632.

Mehta PD and Neale MC 2005. People are variables too: multilevel structural equations modeling. *Psychological Methods* **10**(3), 259–284.

Oswald FL and Johnson JW 1998. On the robustness, bias, and stability of statistics from meta-analysis of correlation coefficients: some initial Monte Carlo findings. *Journal of Applied Psychology* **83**(2), 164–178.

Petkova E, Tarpey T, Huang L and Deng L 2013. Interpreting meta-regression: application to recent controversies in antidepressants' efficacy. *Statistics in Medicine* **32**(17), 2875–2892.

Pinheiro JC and Bates D 2000. *Mixed-effects models in S and S-Plus*. Springer-Verlag, New York.

Raudenbush SW 2009. Analyzing effect sizes: random effects models In *The handbook of research synthesis and meta-analysis* (ed. Cooper HM, Hedges LV and Valentine JC), 2nd edn. Russell Sage Foundation, New York, pp. 295–315.

Raudenbush SW and Bryk AS 2002. *Hierarchical linear models: applications and data analysis methods*. Sage Publications, Inc., Thousand Oaks, CA.

Sánchez-Meca J and Marín-Martínez F 1997. Homogeneity tests in meta-analysis: a Monte Carlo comparison of statistical power and Type I error. *Quality and Quantity* **31**(4), 385–399.

Sánchez-Meca J and Marín-Martínez F 2008. Confidence intervals for the overall effect size in random-effects meta-analysis. *Psychological Methods* **13**(1), 31–48.

Schmidt FL and Hunter JE 2015. *Methods of meta-analysis: correcting error and bias in research findings*, 3rd edn. Sage Publications, Inc., Thousand Oaks, CA.

Schmidt FL, Oh IS and Hayes TL 2009. Fixed- versus random-effects models in meta-analysis: model properties and an empirical comparison of differences in results. *British Journal of Mathematical and Statistical Psychology* **62**(1), 97–128.

Self SG and Liang KY 1987. Asymptotic properties of maximum likelihood estimators and likelihood ratio tests under nonstandard conditions. *Journal of the American Statistical Association* **82**(398), 605–610.

Shuster JJ 2010. Empirical vs natural weighting in random effects meta-analysis. *Statistics in Medicine* **29**(12), 1259–1265.

Stoel RD, Garre FG, Dolan C and van den Wittenboer G 2006. On the likelihood ratio test in structural equation modeling when parameters are subject to boundary constraints. *Psychological Methods* **11**(4), 439–455.

Stram DO 1996. Meta-analysis of published data using a linear mixed-effects model. *Biometrics* **52**(2), 536–544.

Sutton AJ and Abrams KR 2001. Bayesian methods in meta-analysis and evidence synthesis. *Statistical Methods in Medical Research* **10**(4), 277 –303.

Takkouche B, Cadarso-Suárez C and Spiegelman D 1999. Evaluation of old and new tests of heterogeneity in epidemiologic meta-analysis. *American Journal of Epidemiology* **150**(2), 206–215.

Takkouche B, Khudyakov P, Costa-Bouzas J and Spiegelman D 2013. Confidence intervals for heterogeneity measures in meta-analysis. *American Journal of Epidemiology* **178**(6), 993–1004.

Tamhane AC and Dunlop DD 2000. *Statistics and data analysis: from elementary to intermediate*. Prentice Hall, Upper Saddle River, NJ.

Viechtbauer W 2005. Bias and efficiency of meta-analytic variance estimators in the random-effects model. *Journal of Educational and Behavioral Statistics* **30**(3), 261–293.

Viechtbauer W 2007. Hypothesis tests for population heterogeneity in meta-analysis. *British Journal of Mathematical and Statistical Psychology* **60**(1), 29–60.

Viechtbauer W 2008. Analysis of moderator effects in meta-analysis In *Best practices in quantitative methods* (ed. Osborne J). Sage Publications, Inc., Thousand Oaks, CA, pp. 471–487.

Xiong C, Miller JP and Morris JC 2010. Measuring study-specific heterogeneity in meta-analysis: application to an antecedent biomarker study of Alzheimer's disease. *Statistics in Biopharmaceutical Research* **2**(3), 300–309.

5

Multivariate meta-analysis

This chapter extends univariate meta-analysis to a multivariate meta-analysis that allows researchers to analyze more than one effect size per study. We begin the chapter by discussing different types of dependence in the effect sizes and the need for a multivariate meta-analysis to handle multiple effect sizes. Several conventional approaches to conducting multivariate meta-analysis are briefly mentioned. The structural equation modeling (SEM) approach to conducting fixed-, random-, and mixed-effects multivariate meta-analyses is introduced. We then extend the multivariate meta-analysis to mediation and moderation models among the *true* effect sizes. Several examples are used to illustrate the procedures in the R statistical environment.

5.1 Introduction

Most meta-analytic procedures assume independence among the effect sizes. Because of the research design of the primary studies, many effect sizes reported in publications are not independent. The assumption of independence among the effect sizes may not be tenable in many research settings. Moreover, many research questions are multivariate in nature. A single effect size may not be sufficient to summarize the outcome effect. Multivariate meta-analysis is required to address the complexity of the research questions.

5.1.1 Types of dependence

There are several types of dependence in a meta-analysis (e.g., Hedges et al., 2010). The first type is the dependence owing to sampling error. This type of dependence is introduced using the same participants to calculate the effect sizes, for example, the standardized mean differences (SMDs) on the verbal subtest (SAT-verbal) and

Meta-Analysis: A Structural Equation Modeling Approach, First Edition. Mike W. -L. Cheung.

Companion Website: www.wiley.com/go/cheung/meta_analysis

the math subtest (SAT-math) of the Scholastic Aptitude Test (SAT) calculated from the same participants (Kalaian and Raudenbush, 1996) and the correlation matrices calculated from the same participants (Cheung and Chan, 2004). The key feature of this type of dependence is that we may estimate the degree of dependence for *each study* when sufficient information, such as the correlation among the original variables, is given (see Section 3.3 for details). Besides the sampling variances, we also need to estimate the sampling covariances. The estimated dependence (the conditional sampling covariance matrix $\boldsymbol{V}_i$ in Equation 5.4 introduced later) is treated as known values in a multivariate meta-analysis. This type of dependence is always present even though a fixed-effects model is used.

The second type of dependence is the dependence on the *true* effect sizes. Using the SMDs on SAT-verbal and SAT-math calculated from the same participants as an example, the *true* effect sizes on SAT-verbal and SAT-math may be positively correlated at the population level. That is, studies with a larger *true* effect on the SAT-verbal also tend to have a larger *true* effect on the SAT-math. This type of dependence (correlation) is the nature of the phenomenon being studied, which is not caused by using the same participants to calculate the effect sizes. It is represented by the between-study variance component $\boldsymbol{T}^2$ in Equation 5.13. When we are only given the information on *one study*, we are not able to estimate this type of dependence even if the raw data are given. It has to be estimated with a random-effects model based on *all studies*. Even the effect sizes are conditionally independent (the off-diagonal elements of $\boldsymbol{V}_i$ are zero), the *true* effect sizes can still be correlated (the off-diagonal elements of $\boldsymbol{T}^2$ are nonzero) (see Section 5.7.1 for an example).

The third type of dependence is due to the observed effect sizes nested within some hierarchies. For example, multiple effect sizes are reported by each study. Unlike the types of dependence introduced earlier, the degree of dependence is usually unknown. When we are synthesizing the multiple effect sizes, we need to take the dependence into account by estimating the degree of dependence.

Both this chapter and Chapter 6 address nonindependent effect sizes. This chapter applies a multivariate meta-analysis to address the first and second types of dependence when the conditional sampling covariances $\boldsymbol{V}_i$ among the effect sizes are known (e.g., Arends et al., 2003; Becker, 1992, 1995, 2007; Cheung, 2013b; Jackson et al., 2011; Kalaian and Raudenbush, 1996; Raudenbush et al., 1988). When the sampling covariances in $\boldsymbol{V}_i$ are not available, it may be difficult to apply the multivariate meta-analysis. Chapter 6 uses a three-level model to address the third type of dependence when the conditional sampling covariances among the effect sizes are unknown (Cheung, 2014c).

5.1.2 Univariate meta-analysis versus multivariate meta-analysis

Before introducing the multivariate meta-analysis, one obvious but crucial question is whether the multivariate approach is always better than the univariate approach. We compare these two approaches from a missing data perspective. As the primary studies are conducted by different researchers, it is reasonable to expect that the numbers of effect sizes reported in each study may be different. Let us consider the meta-analysis reported by Kalaian and Raudenbush (1996) as an example. These authors reported 47 studies on the SMDs of SAT-verbal and SAT-math. Among these studies, 20 studies reported both effect sizes, while 18 and 9 studies reported only the effect sizes on SAT-verbal and SAT-math, respectively (see Table 1 in Kalaian and Raudenbush (1996) for the data).

One option to handle the data is to conduct two separate univariate meta-analyses on the SAT-verbal and the SAT-math. This approach is similar to the pairwise deletion in a primary data analysis, where the means and variances of the variables are independently calculated for each variable (effect size in a meta-analysis). The correlation between SAT-verbal and SAT-math is ignored in separate univariate analyses. The univariate approach assumes that the mechanism of the missing effect sizes in each study is missing completely at random (MCAR) (see Section 8.2 for more details on missing data). This assumption implies that the missingness is unrelated to the values of the missing data and other effect sizes in the data set.

MCAR is rather restrictive with regard to applied research. A less restrictive assumption is missing at random (MAR), which assumes that the missing data can be related to other effect sizes included in the analysis. In other words, the missingness on SAT-verbal can be related to the observed effect size on SAT-math (and vice versa). When there are moderators in a mixed-effects meta-analysis, the missing values can be related to the moderators, such as the year of publication and the study characteristics. The MAR is assumed on the missing values under a multivariate meta-analysis. The missing data can be handled by the use of the maximum likelihood (ML; also known as full information maximum likelihood or FIML in SEM literature) estimation method or multiple imputation (MI). The literature on analyzing missing data analysis consistently shows that ML and MI are better than pairwise deletion in handling missing data when the missing data mechanism is either MCAR or MAR (Enders, 2010). Section 8.2 compares and contrasts the ML and MI in handling missing data. We mainly focus on ML in this chapter.

When the missing value is related to the value of the missing data, which is similar to publication bias in a meta-analysis, it is known as *not missing at random* (NMAR) or *nonignorable missingness*. Neither the pairwise deletion nor ML is unbiased (e.g., Schafer, 1997). However, the bias of the ML is still less than that on the pairwise deletion (Jamshidian and Bentler, 1999); (Muthén et al., 1987). Therefore, ML is generally recommended to be used to handle missing data (e.g., Enders, 2010; Schafer and Graham, 2002). When we put these findings in the background of a meta-analysis, it is clear that the multivariate meta-analysis should be used to handle multiple effect sizes.

On the other hand, Ishak et al. (2008) argued that we may conduct several univariate meta-analyses without any significant risk of bias or loss of precision in the estimates if our interest is only on estimating the fixed effects. However, Riley (2009) showed that the estimated fixed effects in a multivariate meta-analysis generally have smaller standard errors (SEs) and mean square errors than those based on separate univariate meta-analyses. Demidenko (2013) also demonstrated similar results. The SEs for the multivariate meta-analysis and several univariate analyses will only be equal when either (i) the known sampling variances are the same for all studies, that is, $\boldsymbol{V}_1 = \boldsymbol{V}_2 = \cdots = \boldsymbol{V}_k$, or (ii) the effect sizes are conditionally independent, that is, $\boldsymbol{V}_i$ is a diagonal matrix. Hafdahl (2007) showed another condition for the equivalence of both the approaches under a fixed-effects model. Besides these scenarios, the estimates based on the multivariate meta-analysis are more precise than those based on the separate univariate meta-analysis. Moreover, the multivariate meta-analysis allows us to estimate the degree of dependence (correlation) between the *true* effect sizes that is also informative in a meta-analysis. Therefore, a multivariate meta-analysis is generally recommended over several univariate meta-analyses (Jackson et al., 2011).

5.2 Fixed-effects model

When all of the studies share the same population or *true* effect sizes, a fixed-effects model may be used. Let p be the number of effect sizes per study involved in a multivariate meta-analysis and p_i be the number of the observed effect sizes in the ith study. As the primary studies are conducted by different researchers, it is likely that different numbers of effect sizes are reported in each study. When there is no missing effect size, p_i is the same as p; otherwise, p_i is smaller than p. It is more convenient to use the vector notation to handle multiple effect sizes. The multiple effect sizes in the ith study may be stacked into a $p_i \times 1$ vector of $\boldsymbol{y}_i$. The model for the ith study is

$$\underset{p_i \times 1}{\boldsymbol{y}_i} = \underset{p_i \times p}{\boldsymbol{X}_i} \underset{p \times 1}{\boldsymbol{f}_i} + \underset{p_i \times 1}{\boldsymbol{e}_i}, \tag{5.1}$$

where $\boldsymbol{y}_i$ is the $p_i \times 1$ vector of the observed effect sizes, $\boldsymbol{X}_i$ is a $p_i \times p$ design matrix with 0 and 1 to select the observed effect sizes, $\boldsymbol{f}_i$ is a $p \times 1$ vector of *true* or population effect sizes, and $\boldsymbol{e}_i$ is a $p_i \times 1$ vector of the sampling error. When the sample sizes are reasonably large, $\boldsymbol{e}_i$ is assumed to be multivariate normally distributed with a mean vector of zero and a known covariance matrix $\boldsymbol{V}_i$, that is, $\boldsymbol{e}_i \sim \mathcal{N}(0, \boldsymbol{V}_i)$ (see Section 3.3 for the formulas for the common effect sizes for multivariate meta-analysis).

Suppose that there are k studies involved in a meta-analysis; all the *true* effect sizes are assumed the same, that is, $\beta_\text{F} = \boldsymbol{f}_1 = \boldsymbol{f}_2 = \cdots = \boldsymbol{f}_k$. This model is also known as the *common effects model*. The model for the ith study is

$$\boldsymbol{y}_i = \boldsymbol{X}_i \beta_\text{F} + \boldsymbol{e}_i. \tag{5.2}$$

Suppose that we are conducting a multivariate meta-analysis with two effect sizes. Study 1 is complete without missing data, while Studies 2 and 3 report only the first and second effect sizes, respectively. The model is

$$\text{Study 1:} \qquad \begin{bmatrix} y_1 \\ y_2 \end{bmatrix}_{(1)} = \begin{bmatrix} 1 & 0 \\ 0 & 1 \end{bmatrix}_{(1)} \begin{bmatrix} \beta_1 \\ \beta_2 \end{bmatrix}_\text{F} + \begin{bmatrix} e_1 \\ e_2 \end{bmatrix}_{(1)},$$

$$\text{Study 2:} \qquad \begin{bmatrix} y_1 \end{bmatrix}_{(2)} = \begin{bmatrix} 1 & 0 \end{bmatrix}_{(2)} \begin{bmatrix} \beta_1 \\ \beta_2 \end{bmatrix}_{\text{F}} + \begin{bmatrix} e_1 \end{bmatrix}_{(2)},$$

$$\text{Study 3:} \qquad \begin{bmatrix} y_2 \end{bmatrix}_{(3)} = \begin{bmatrix} 0 & 1 \end{bmatrix}_{(3)} \begin{bmatrix} \beta_1 \\ \beta_2 \end{bmatrix}_{\text{F}} + \begin{bmatrix} e_2 \end{bmatrix}_{(3)},$$

where the subscripts outside the parenthesis indicate studies. We may stack all effect sizes and write the model as

$$\boldsymbol{y} = \boldsymbol{X}\boldsymbol{\beta}_{\text{F}} + \boldsymbol{e}, \tag{5.3}$$

where $\boldsymbol{y} = \begin{bmatrix} \boldsymbol{y}_1 \\ \boldsymbol{y}_2 \\ \vdots \\ \boldsymbol{y}_k \end{bmatrix}$, $\boldsymbol{X} = \begin{bmatrix} \boldsymbol{X}_1 \\ \boldsymbol{X}_2 \\ \vdots \\ \boldsymbol{X}_k \end{bmatrix}$, and $\boldsymbol{e} = \begin{bmatrix} \boldsymbol{e}_1 \\ \boldsymbol{e}_2 \\ \vdots \\ \boldsymbol{e}_k \end{bmatrix}$. As the effects are likely to be correlated within a study, the off-diagonals of $\boldsymbol{V}_i$ are usually nonzero, while $\boldsymbol{e}_i$ and $\boldsymbol{e}_j$ from two studies are independent, that is, $\text{Cov}(\boldsymbol{e}_i, \boldsymbol{e}_j) = 0$. Therefore, the conditional sampling covariance matrix $\boldsymbol{V}$ is a block diagonal (symmetric) matrix,

$$\boldsymbol{V} = \begin{bmatrix} \boldsymbol{V}_1 & & & \\ 0 & \boldsymbol{V}_2 & & \\ \vdots & \ddots & \ddots & \\ 0 & \cdots & 0 & \boldsymbol{V}_k \end{bmatrix}. \tag{5.4}$$

In our example, the stacked matrices with a study index in parenthesis are

$$\overbrace{\begin{bmatrix} y_{1(1)} \\ y_{2(1)} \\ y_{1(2)} \\ y_{2(3)} \end{bmatrix}}^{y} = \overbrace{\begin{bmatrix} 1 & 0 \\ 0 & 1 \\ 1 & 0 \\ 0 & 1 \end{bmatrix}}^{X} \overbrace{\begin{bmatrix} \beta_1 \\ \beta_2 \end{bmatrix}}^{\beta_{\text{F}}} + \overbrace{\begin{bmatrix} e_{1(1)} \\ e_{2(1)} \\ e_{1(2)} \\ e_{2(3)} \end{bmatrix}}^{e} \quad \text{and}$$

$$V = \begin{bmatrix} V_{11(1)} & V_{21(1)} & & \\ V_{21(1)} & V_{22(1)} & & \\ 0 & 0 & V_{11(2)} & \\ 0 & 0 & 0 & V_{22(3)} \end{bmatrix}, \tag{5.5}$$

where the numbers in parentheses in the subscripts indicate the studies.

5.2.1 Testing the homogeneity of effect sizes

We may generalize the Q statistic proposed by Cochran (1954) in the univariate meta-analysis to multivariate meta-analysis (see Becker, 1992; Demidenko, 2013; Hedges and Olkin, 1985). For a multivariate meta-analysis, the Q statistic is defined as

$$Q = (\boldsymbol{y} - \boldsymbol{X}\hat{\boldsymbol{\beta}}_{\text{F}})^{\text{T}} \boldsymbol{V}^{-1} (\boldsymbol{y} - \boldsymbol{X}\hat{\boldsymbol{\beta}}_{\text{F}}), \tag{5.6}$$

where $\hat{\beta}_\text{F}$ is estimated common effect sizes under a fixed-effects model by Equation 5.9. As there are likely missing effect sizes in a multivariate meta-analysis, we may filter out the missing effect sizes before testing the null hypothesis $H_0 : \beta_\text{F} = X_1 f_1 = X_2 f_2 = \cdots = X_k f_k$. Under the null hypothesis, the Q statistic is approximately distributed as a chi-square distribution with $(\sum_{i=1}^{k} p_i - p)$ degrees of freedom (dfs) in large samples.

5.2.2 Estimation and hypotheses testing

By treating V in Equation 5.4 as known, we may estimate β_F by minimizing the criterion F_GLS with the generalized least squares (GLS), which is a direct generalization of the univariate meta-analysis in Equation 4.6 (Becker, 1992, 1995; Hedges and Olkin, 1985):

$$\begin{aligned} F_\text{GLS} &= (y - X\beta_\text{F})^\text{T} V^{-1} (y - X\beta_\text{F}) \\ &= y^\text{T} V^{-1} y - y^\text{T} V^{-1} X\beta_\text{F} - \beta_\text{F}^\text{T} X^\text{T} V^{-1} y + \beta_\text{F}^\text{T} X^\text{T} V^{-1} X\beta_\text{F} \\ &= y^\text{T} V^{-1} y - 2\beta_\text{F}^\text{T} X^\text{T} V^{-1} y + \beta_\text{F}^\text{T} X^\text{T} V^{-1} X\beta_\text{F}. \end{aligned} \tag{5.7}$$

To solve the above equation, we take the partial derivative of F_GLS with respect to β_F,

$$\begin{aligned} \frac{\partial F_\text{GLS}}{\partial \beta_\text{F}} &= \frac{\partial (y^\text{T} V^{-1} y - 2\beta_\text{F}^\text{T} X^\text{T} V^{-1} y + \beta_\text{F}^\text{T} X^\text{T} V^{-1} X\beta_\text{F})}{\partial \beta_\text{F}} \\ &= -2X^\text{T} V^{-1} y + 2X^\text{T} V^{-1} X\beta_\text{F}. \end{aligned} \tag{5.8}$$

By setting $\frac{\partial F_\text{GLS}}{\partial \beta_\text{F}} = 0$, the solution for $\hat{\beta}_\text{F}$ is

$$\hat{\beta}_\text{F} = (X^\text{T} V^{-1} X)^{-1} X^\text{T} V^{-1} y. \tag{5.9}$$

As $\frac{\partial^2 F_\text{GLS}}{\partial \beta_\text{F}^2} = 2X^\text{T} V^{-1} X$, which is positive definite, the solution is a local minimum.

Similar to the asymptotic sampling variance in the univariate meta-analysis (Equation 4.12), we may derive the asymptotic sampling covariance matrix $\hat{\Omega}_\text{F} = \text{Cov}(\hat{\beta}_\text{F})$ by treating $(X^\text{T} V^{-1} X)^{-1} X^\text{T} V^{-1}$ as a constant matrix:

$$\begin{aligned} \hat{\Omega}_\text{F} &= \text{Cov}((X^\text{T} V^{-1} X)^{-1} X^\text{T} V^{-1} y) \\ &= (X^\text{T} V^{-1} X)^{-1} X^\text{T} V^{-1} \text{Cov}(y) V^{-1} X (X^\text{T} V^{-1} X)^{-1} \\ &= (X^\text{T} V^{-1} X)^{-1} X^\text{T} V^{-1} V V^{-1} X (X^\text{T} V^{-1} X)^{-1} \quad \text{as} \quad \text{Cov}(y) = V \\ &= (X^\text{T} V^{-1} X)^{-1} (X^\text{T} V^{-1} X)(X^\text{T} V^{-1} X)^{-1} \\ &= (X^\text{T} V^{-1} X)^{-1}. \end{aligned} \tag{5.10}$$

The diagonal elements of $\hat{\Omega}_\text{F}$ represent the sampling variances of the parameter estimates in $\hat{\beta}_\text{F}$, while the off-diagonal elements represent the sampling covariances among the parameter estimates. We may use the sampling variances to test the significance of the parameter estimates. Suppose that we want to test the ith effect size in $\hat{\beta}_{\mathrm{F}}$, that is, $H_0 : \beta_{\mathrm{F}[i]} = \beta_0$; we may use a Wald statistic to test it:

$$z_{\hat{\beta}_{\mathrm{F}[i]}} = \frac{\hat{\beta}_{\mathrm{F}[i]} - \beta_0}{\sqrt{\hat{\Omega}_{\mathrm{F}[i,i]}}}, \tag{5.11}$$

where $\hat{\Omega}_{\mathrm{F}[i,i]}$ is the ith diagonal element in $\hat{\Omega}_{\mathrm{F}}$. Under the null hypothesis, $z_{\hat{\beta}_{\mathrm{F}[i]}}$ has approximate standard normal distribution. Approximate Wald confidence intervals (CIs) on the parameter estimates may also be constructed.

5.3 Random-effects model

In applied research, it is reasonable to expect that each study may have its own study-specific effect sizes because of the differences in samples, design, and measures. Besides the sampling error, the random-effects model includes the random effects. We may use two levels to represent the model:

$$\begin{aligned} &\text{Level 1:} \quad && \boldsymbol{y}_i = \boldsymbol{f}_i + \boldsymbol{e}_i, \\ &\text{Level 2:} \quad && \boldsymbol{f}_i = \boldsymbol{\beta}_{\mathrm{R}} + \boldsymbol{u}_i, \end{aligned} \tag{5.12}$$

where $\boldsymbol{\beta}_{\mathrm{R}}$ is the average population effect vector under the random-effects model and $\boldsymbol{u}_i \sim \mathcal{N}(0, \boldsymbol{T}^2)$ is the heterogeneity variance–covariance matrix that has to be estimated.

The above two-level model can be combined into a single equation with all of the data stacked together. Two selection matrices ($\boldsymbol{X}$ for the fixed effects and $\boldsymbol{Z}$ for the random effects) are required to filter the missing effect sizes. The random-effects model is

$$\boldsymbol{y} = \boldsymbol{X}\boldsymbol{\beta}_{\mathrm{R}} + \boldsymbol{Z}\boldsymbol{u} + \boldsymbol{e}, \tag{5.13}$$

where $\mathbf{Z} = \mathrm{Diag}(\mathbf{Z}_1, \mathbf{Z}_2, \ldots, \mathbf{Z}_k)$ is selection matrix of 1 and 0 to select the random effects, $\boldsymbol{u} = [\boldsymbol{u}_1^{\mathrm{T}} | \boldsymbol{u}_2^{\mathrm{T}} | \cdots | \boldsymbol{u}_i^{\mathrm{T}}]^{\mathrm{T}}$ is the stacked random effects for all studies, $\boldsymbol{\beta}_{\mathrm{R}}$ is the average population effect sizes under the random-effects model, and $\boldsymbol{y}$, $\boldsymbol{X}$, and $\boldsymbol{e}$ are defined in Equation 5.3. $\boldsymbol{u}_i \sim \mathcal{N}(0, \boldsymbol{T}^2)$ is the study-specific random effects in the ith study, where $\boldsymbol{T}^2$ is a $p \times p$ nonnegative definite matrix. Loosely speaking, a nonnegative definite matrix means that the diagonals of $\boldsymbol{T}^2$ (variances) cannot be negative and the off-diagonals in terms of correlations must stay within the meaningful range, that is, -1 to $+1$. Moreover, the values of the correlations are constrained by the triangular inequality condition. The triangular inequality condition states the possible range of correlation between variables x and y when the correlations with a third variable, say z, are fixed (Wothke, 1993). We will address the issue of nonpositive definite matrix later. Using our previous example as an illustration, the random-effects model with a study index in parenthesis is

$$\overbrace{\begin{bmatrix} y_{1(1)} \\ y_{2(1)} \\ \hdashline y_{1(2)} \\ \hdashline y_{2(3)} \end{bmatrix}}^{y} = \overbrace{\begin{bmatrix} 1 & 0 \\ 0 & 1 \\ \hdashline 1 & 0 \\ \hdashline 0 & 1 \end{bmatrix}}^{X} \overbrace{\begin{bmatrix} \beta_1 \\ \beta_2 \end{bmatrix}}^{\beta_R} + \overbrace{\left[\begin{array}{cc:cc:cc} 1 & 0 & 0 & 0 & 0 & 0 \\ 0 & 1 & 0 & 0 & 0 & 0 \\ \hdashline 0 & 0 & 1 & 0 & 0 & 0 \\ \hdashline 0 & 0 & 0 & 0 & 0 & 1 \end{array}\right]}^{Z} \overbrace{\begin{bmatrix} u_{1(1)} \\ u_{2(1)} \\ \hdashline u_{1(2)} \\ u_{2(2)} \\ \hdashline u_{1(3)} \\ u_{2(3)} \end{bmatrix}}^{u} + \overbrace{\begin{bmatrix} e_{1(1)} \\ e_{2(1)} \\ \hdashline e_{1(2)} \\ \hdashline e_{2(3)} \end{bmatrix}}^{e}, \tag{5.14}$$

In fixed-effects models, there is only one source of variation—the conditional sampling covariance $\boldsymbol{V}_i$. Besides the conditional sampling covariance, random-effects models include an extra between-study variance component $\boldsymbol{T}^2$. Although both $\boldsymbol{V}_i$ and $\boldsymbol{T}^2$ are related to the dependence of the effect sizes, their meanings are different. The sampling covariances (dependence) in $\boldsymbol{V}_i$ are inherited from the measures and sampling error. The correlation among the effect sizes in $\boldsymbol{V}_i$ is due to the use of the same samples to calculate the multiple effect sizes in the same studies or due to the use of the same control group to calculate the treatment effect sizes. Even under a fixed-effects model, where the population effect sizes are fixed (i.e., constant) at the population, the observed effect sizes are still correlated. When the sample sizes get larger and larger, $\boldsymbol{V}_i$ approaches zero. On the other hand, $\boldsymbol{T}^2$ refers to the dependence at the population level, regardless of the sample sizes.

5.3.1 Structure of the variance component of random effects

The off-diagonals of $\boldsymbol{T}^2$ indicate the covariance among the *true* effect sizes. When the covariances are positive, it indicates that studies with a positive effect on one effect size tend to have a positive effect on the other effect size at the population level. There are several choices for the structure of $\boldsymbol{T}^2$. The most obvious choice is that $\boldsymbol{T}^2$ is simply a nonnegative definite *unstructured* matrix,

$$\boldsymbol{T}^2_{\text{Un}} = \begin{bmatrix} \tau^2_{11} & & & \\ \tau^2_{21} & \tau^2_{22} & & \\ \vdots & \ddots & \ddots & \\ \tau^2_{p1} & \cdots & \tau^2_{p(p-1)} & \tau^2_{pp} \end{bmatrix}. \tag{5.15}$$

This is the default choice for most multivariate meta-analyses, especially when the number of effect sizes per study is small. The main advantage of this structure is that all covariances among the *true* effect sizes are empirically estimated. This gives us useful information on how the *true* effect sizes are correlated. The heterogeneity variances can be different for different effect sizes.

The use of an unstructured matrix in $\boldsymbol{T}^2$ may not always be feasible. When the number of studies is small, there may not be sufficient data to estimate all $p(p+1)/2$ elements in $\boldsymbol{T}^2$. $\hat{\boldsymbol{T}}^2$ may be negative definite, which is not acceptable in theory. This may happen quite often in meta-analytic structural equation modeling (MASEM; see Chapter 7), where the number of effect sizes per study is the number of correlation coefficients. For example, Cheung (2014a) illustrated a random-effects MASEM on the higher-order factor structure of the five-factor model from a data set from Digman (1997). The effect sizes were 10 correlation coefficients (a 5 × 5 correlation matrix). Thus, there were a total of 55 elements in $\boldsymbol{T}^2$. As there were only 14 studies, the estimated $\boldsymbol{T}^2$ was negative definite when an unstructured matrix was used in $\boldsymbol{T}^2$.

One way of working around this is to use a *diagonal* structure on $\boldsymbol{T}^2$, meaning that the *true* multiple effect sizes are uncorrelated at the population,

$$\boldsymbol{T}^2_{\text{Diag}} = \begin{bmatrix} \tau^2_{11} & & & \\ 0 & \tau^2_{22} & & \\ \vdots & \ddots & \ddots & \\ 0 & \cdots & 0 & \tau^2_{pp} \end{bmatrix}. \tag{5.16}$$

This reduces the number of parameters in $\boldsymbol{T}^2$ from $p(p+1)/2$ elements to only p elements. Although this model is similar to running several univariate meta-analyses on individual effect sizes, there are still advantages in using it. First, the multivariate meta-analysis still takes the conditional sampling covariance $\boldsymbol{V}_i$ into account, whereas the univariate meta-analyses further assume that the diagonals of $\boldsymbol{V}_i$ are zero. Second, the multivariate meta-analysis, including all multivariate effect sizes in the model, allows comparisons of models involving the fixed or the random effects. For example, researchers may compare whether the average effect sizes are the same or whether the heterogeneity variances are the same in different effect sizes by using likelihood ratio (LR) statistic. It should be noted, however, that the covariance structure is misspecified because the off-diagonal elements are fixed at zero. The test statistics may be inaccurate. Further studies may address the empirical performance of this approach.

Another possible choice is the *compound symmetry* structure. This structure may be useful if we want to test whether the heterogeneity variances are the same for the *true* multiple effect sizes and whether the correlations among the *true* multiple effects are the same,

$$\boldsymbol{T}^2_{\text{CS}} = \begin{bmatrix} \tau^2_{11} & & & \\ \tau^2_{21} & \tau^2_{11} & & \\ \vdots & \ddots & \ddots & \\ \tau^2_{21} & \cdots & \tau^2_{21} & \tau^2_{11} \end{bmatrix}. \tag{5.17}$$

The compound symmetry structure may also be used to analyze a three-level meta-analysis within the multivariate meta-analysis framework (see Section 6.4). A final option is a *zero* matrix, $\boldsymbol{T}^2_{\text{Zero}} = 0$. The random-effects model is then equivalent to a fixed-effects model.

5.3.2 Nonnegative definite of the variance component of random effects

When we calculate a covariance matrix $\boldsymbol{S}$ from p observed variables with $n > p$, where n is the sample size and there is no missing data, $\boldsymbol{S}$ is usually nonnegative definite, that is, $x^{\mathrm{T}}Sx \geq 0$ for any $p \times 1$ vector of x with $x \neq 0$. A nonnegative covariance matrix means that

(i) the variances are always nonnegative;

(ii) the correlation coefficients are within the ± 1 boundary; and

(iii) the combinations of the correlations meet some conditions so that they are *real* correlations (e.g., Wothke, 1993).

As $\hat{\boldsymbol{T}}^2$ is estimated from the *true* effect sizes (latent scores in the SEM framework) and there are usually incomplete effect sizes, some or all of the above conditions may not meet. If $\hat{\boldsymbol{T}}^2$ is negative definite, it may not be appropriate to be interpreted as a covariance matrix. Several approaches have been suggested to handle this situation (e.g., Demidenko, 2013; Pinheiro and Bates, 1996). There are two general approaches—the parameterized approach and the constrained approach.

The parameterized approach uses a set of new parameters in the optimization. After the optimization, $\boldsymbol{T}^2$ is calculated. One popular choice is based on the Cholesky decomposition using

$$\boldsymbol{T}^2 = \boldsymbol{L}\boldsymbol{L}^{\mathrm{T}}, \tag{5.18}$$

where $\boldsymbol{L}$ is a lower triangular matrix. Let us illustrate the idea of the Cholesky decomposition on a single variable. If we want to ensure that the estimated variance is nonnegative, we may use the standard deviation σ as the parameter in the model. This is an unconstrained optimization because we do not put any restriction on σ. After the optimization, $\hat{\sigma}$ can be positive, zero, or even negative. As we take the squared of $\hat{\sigma}$ as the estimate of σ^2, $\hat{\sigma}^2$ is always nonnegative. Conceptually, the Cholesky decomposition is similar to taking the square root of the covariance matrix. In the analysis, we replace $\boldsymbol{T}^2$ by $\boldsymbol{L}\boldsymbol{L}^{\mathrm{T}}$. After fitting the models, $\hat{\boldsymbol{L}}$ is estimated instead of $\hat{\boldsymbol{T}}^2$. We compute $\hat{\boldsymbol{T}}^2$ from Equation 5.18. By using the Cholesky decomposition, we can guarantee that $\hat{\boldsymbol{T}}^2$ is always nonnegative.

Although the Cholesky decomposition is attractive in ensuring $\hat{\boldsymbol{T}}^2$ to be nonnegative definite, this approach has several limitations. First, there is no unique solution on the Cholesky decomposition. Similar to the case that $\hat{\sigma}$ can be negative, there may be multiple solutions for the Cholesky decomposition. Although the presence of multiple solutions may not affect the calculated $\hat{\boldsymbol{T}}^2$, the computer algorithm may switch between different solutions. This may make it slow to converge to the final solution. Another issue is that $\hat{\boldsymbol{L}}$ bears no meaning. Therefore, the constructed SEs and CIs on $\hat{\boldsymbol{L}}$ are basically useless. Extra steps are required to translate these SEs (or CIs) into the correspondent elements in $\hat{\boldsymbol{T}}^2$. Finally, the analysis becomes complicated when there are equality constraints on some of the elements in $\boldsymbol{T}^2$. For example, extra care has to be applied to the Cholesky decomposition on $\boldsymbol{T}^2$ for the compound symmetry structure.

A second approach is to impose nonlinear constraints on $\boldsymbol{T}^2$ such that the estimated matrix is always nonnegative. One option is to parameterize $\boldsymbol{T}^2$ as

$$\boldsymbol{T}^2 = \boldsymbol{DPD}, \tag{5.19}$$

where $\boldsymbol{P}$ and $\boldsymbol{D}$ are the correlation matrix and the matrix of standard deviations of the random effects, respectively. Then, we set the upper and lower bounds on $\boldsymbol{P}$ within the ± 1 boundary. This approach can ensure conditions (i) and (ii) listed earlier. Although the correlation coefficients are within the ± 1 boundary, the matrix can still be negative definite (e.g., Wothke, 1993). Similar to the Cholesky decomposition, this method cannot be easily implemented when there are constraints in $\boldsymbol{T}^2$.

Another option is to set the lower bounds on the diagonals of $\boldsymbol{T}^2$ to be nonnegative, that is, $\text{Diag}(\boldsymbol{T}^2) \geq 0$. One advantage of this option is that it works fine even when there are equality constraints on $\boldsymbol{T}^2$. Condition (i) can be easily fulfilled. However, there is no guarantee that conditions (ii) and (iii) can be met. The current version of the `metaSEM` package (0.9-0) has only implemented this option. Future versions may explore other methods to ensure the positive definiteness of $\hat{\boldsymbol{T}}^2$.

5.3.3 Estimation and hypotheses testing

Similar to the univariate random-effects meta-analysis, there are several methods to estimate the heterogeneity variance component matrix of $\boldsymbol{T}^2$. Becker (1992) proposed an unweighted and weighted methods of moment approach to estimate the variance components for data without missing values. Demidenko (2013) discussed the methods of moment approach to estimate the variance component in the presence of missing values. The popular DerSimonian–Laird (DL) estimator (DerSimonian and Laird, 1986) in a univariate meta-analysis has been extended to a multivariate meta-analysis, (Chen et al., 2012; Jackson et al., 2010). Both ML and restricted (or residual) maximum likelihood estimation (REML) estimation methods implemented in a conventional multilevel modeling framework (e.g., Arends et al., 2003; Kalaian and Raudenbush, 1996; van Houwelingen et al., 2002) and an SEM framework (Cheung, 2013a, b) can also be used to conduct the multivariate meta-analysis. Hafdahl (2008) investigated the misspecified multivariate case of fixed-effects methods for heterogeneous correlation matrices. Besides the frequentist approach, the Bayesian approach may also be used (e.g., Bujkiewicz et al., 2013; Nam et al., 2003; Wei and Higgins, 2013).

Assuming that we have estimated $\boldsymbol{T}^2$, we may calculate the unconditional sampling covariance matrix for the ith study $\tilde{\boldsymbol{V}}_i = \boldsymbol{Z}_i \hat{\boldsymbol{T}}^2 \boldsymbol{Z}_i^{\text{T}} + \boldsymbol{V}_i$, where $\boldsymbol{Z}_i$ is used to select the random effects. We stack the conditional sampling covariance matrix $\tilde{\boldsymbol{V}}_i$ together as we did in Equation 5.4:

$$\tilde{\boldsymbol{V}} = \begin{bmatrix} \tilde{\boldsymbol{V}}_1 & & & \\ 0 & \tilde{\boldsymbol{V}}_2 & & \\ \vdots & \ddots & \ddots & \\ 0 & \cdots & 0 & \tilde{\boldsymbol{V}}_k \end{bmatrix}. \tag{5.20}$$

We treat $\tilde{V}$ as known and estimate β_R by minimizing the criterion F_{GLS} with GLS:

$$F_{GLS} = (y - X\beta_R)^T \tilde{V}^{-1} (y - X\beta_R). \tag{5.21}$$

Similar to the solutions in Equation 5.9, the fixed-effects and their asymptotic sampling covariance matrix for the random-effects model can be estimated by

$$\begin{aligned} \hat{\beta}_R &= (X^T \tilde{V}^{-1} X)^{-1} X^T \tilde{V}^{-1} y \quad \text{and} \\ \hat{\Omega}_R &= (X^T \tilde{V}^{-1} X)^{-1}. \end{aligned} \tag{5.22}$$

$\hat{\Omega}_R$ can be used to test the significance of the individual parameters in $\hat{\beta}_R$.

$$z_{\hat{\beta}_{R[i]}} = \frac{\hat{\beta}_{R[i]} - \beta_0}{\sqrt{\hat{\Omega}_{R[i,i]}}} \tag{5.23}$$

has an approximate standard normal distribution in testing $H_0 : \hat{\beta}_{R[i]} = \beta_0$. Approximate Wald CIs on the parameter estimates can also be constructed based on the estimated SEs.

5.3.3.1 Testing the variance component

Under the random-effects model, we sometimes want to test $H_0 : T^2 = 0$. We may create two models for comparison—one model with no restriction on T^2 and the other model with $T^2 = 0$. As these two models are nested, an LR statistic may be used to compare these two models. Because of the boundary condition as discussed in Section 4.3.2, the test statistic is not distributed as a chi-square distribution. Generally, it is difficult to derive the exact distribution because it is a mixture of chi-square variates (see Stoel et al., 2006).

When the multivariate effect sizes are similar in scale, for example, SMDs on SAT-verbal and SAT-math in Kalaian and Raudenbush (1996), it is of interest to test whether the heterogeneity variances are the same in the multivariate effect sizes. This can be tested by comparing two nested models—one model without any constraint on the diagonals and the other model with equality constraints on the diagonals. Under the null hypothesis $H_0 : \tau_{11}^2 = \tau_{22}^2 = \cdots = \tau_{pp}^2$, the difference on the LR statistics has a chi-square distribution with $(p - 1)$ dfs. We illustrate some examples in Section 5.7.

5.3.4 Quantifying the degree of heterogeneity of effect sizes

Equation 4.27 defines an I^2 (Higgins and Thompson, 2002) based on the estimated heterogeneity with the *typical* within-study sampling variance that can be used to quantify the degree of between-study heterogeneity to the total variance of the effect sizes in univariate meta-analysis. In the case of the multivariate meta-analysis, the estimated variance component of the heterogeneity is T^2. Jackson et al. (2012) proposed several multivariate extensions to the I^2 that may be used to quantify the degree of heterogeneity in a multivariate meta-analysis. The $I^2_{\text{Q(Uni)}}$ in a univariate meta-analysis based on the Q statistic is $I^2_{\text{Q(Uni)}} = 1 - \text{df}_{\text{(Uni)}}/Q_{\text{(Uni)}}$, where $Q_{\text{(Uni)}}$ and $\text{df}_{\text{(Uni)}}$ are the Q statistic and its degrees of freedom in testing the homogeneity of effect sizes in Equation 4.14, respectively. One of them is the multivariate generalization of the $I^2_{\text{Q(Uni)}}$:

$$I^2_{\text{Q(Mul)}} = 1 - \frac{\text{df}_{\text{(Mul)}}}{Q_{\text{(Mul)}}}, \tag{5.24}$$

where $Q_{\text{(Mul)}}$ and $\text{df}_{\text{(Mul)}}$ are the Q statistic and its dfs in testing the homogeneity of effect sizes in Equation 5.6, respectively. When the estimated $I^2_{\text{Q(Mul)}}$ is negative, it is truncated to zero.

When the heterogeneity varies across the effect sizes, it is not clear how useful a single index on the multiple effect sizes is. Alternatively, we may apply the I^2 on each effect size. This gives us insights on the heterogeneity of the individual effect sizes. There may be a high degree of heterogeneity on some effect sizes, while the heterogeneity on the other effect sizes may be very low. The `metaSEM` package (Cheung, 2014b) uses this approach to quantify the degree of heterogeneity in a multivariate meta-analysis.

5.3.5 When the sampling covariances are not known

Models for multivariate meta-analyses usually assume that the sampling variances and covariances of the effect sizes are known. The sampling variances of the effect sizes are usually available, whereas the sampling covariances or correlations of the effect sizes may be missing in some studies. Several strategies have been suggested for handling the missing correlations among the effect sizes (see Jackson et al., (2011) for a discussion).

The first approach is to conduct univariate meta-analyses on separate effect sizes by ignoring the dependence of the effect sizes (see Section 5.1.2). The second approach is to estimate the sampling correlation from other sources. For example, Raudenbush et al. (1988) illustrated a multivariate meta-analysis on the SMDs of the SAT-verbal and SAT-math. As the correlation between SAT-verbal and SAT-math was not reported in the original studies, the authors assumed that $\rho = 0.66$, the population correlation between SAT-verbal and SAT-math reported by the Educational Testing Service. Another approach is to use the correlation between the observed effect sizes to approximate the within-study correlation (Hox, 2010). However, Hox (2010) reminded readers that the estimated correlation is based on an ecological analysis that may confound the within-study correlation and the between-study correlation.

The fourth approach is to provide a range of correlations for a sensitivity analysis. If the results are similar for different values of correlation used, the results are robust for the correlation used in the analysis. The fifth approach is to apply an alternative model proposed by Riley et al. (2008). These authors proposed a model that combines both the within- and between-study correlations into a single parameter. Therefore, there is no need to provide the within-study correlation in the analysis. There are two limitations to this approach, however. First, we cannot estimate the between-study heterogeneity variances, although some estimates can be used to approximate them. Second, this approach only works for models with two effect sizes per study. It has yet to see how it can be extended to meta-analysis with more than two effect sizes per study. The other approaches are to handle the dependence with a robust variance estimation (Hedges et al., 2010) or to approximate the multivariate meta-analysis with a three-level model (see Section 6.4.2).

5.4 Mixed-effects model

When there is substantial heterogeneity, it is of interest to test how the effect sizes vary by including the study characteristics as moderators. The random-effects model in Equation 5.13 can be extended to the mixed-effects model by defining suitable matrices for $\boldsymbol{X}$ and β_{R}. There are two formats for representing multivariate data: the long format and the wide format. The long format is usually used in the multilevel modeling approach or the GLS approach, while the wide format is used in the SEM-based meta-analysis.

Tables 5.1 and 5.2 display some sample data for the long and the wide formats, respectively. The moderators can be either study or effect size specific. For example, if year of publication is used as a moderator, the moderator is the same for both effect sizes in a study. An example is the x_1 in Tables 5.1 and 5.2. The moderators may also be effect size specific. For example, the moderator may refer to different types of intervention for y_1 and y_2. An example is the x_2 in Tables 5.1 and 5.2.

The following model shows the design matrices for our previous example in Tables 5.1 and 5.2 with x_2 as the moderator by stacking all studies together.

$$\overbrace{\begin{bmatrix} y_{1(1)} \\ y_{2(1)} \\ \hdashline y_{1(2)} \\ \hdashline y_{2(3)} \end{bmatrix}}^{y} = \overbrace{\begin{bmatrix} 1 & 0 & x_{2,1(1)} & 0 \\ 0 & 1 & 0 & x_{2,2(1)} \\ \hdashline 1 & 0 & x_{2,1(2)} & 0 \\ \hdashline 0 & 1 & 0 & x_{2,2(3)} \end{bmatrix}}^{X} \overbrace{\begin{bmatrix} \beta_{1,0} \\ \beta_{2,0} \\ \beta_{1,1} \\ \beta_{2,1} \end{bmatrix}}^{\beta_{\mathrm{R}}}$$

$$+ \overbrace{\begin{bmatrix} 1 & 0 & 0 & 0 & 0 & 0 \\ 0 & 1 & 0 & 0 & 0 & 0 \\ \hdashline 0 & 0 & 1 & 0 & 0 & 0 \\ \hdashline 0 & 0 & 0 & 0 & 0 & 1 \end{bmatrix}}^{Z} \overbrace{\begin{bmatrix} u_{1(1)} \\ u_{2(1)} \\ \hdashline u_{1(2)} \\ u_{2(2)} \\ \hdashline u_{1(3)} \\ u_{2(3)} \end{bmatrix}}^{u} + \overbrace{\begin{bmatrix} e_{1(1)} \\ e_{2(1)} \\ \hdashline e_{1(2)} \\ \hdashline e_{2(3)} \end{bmatrix}}^{e}, \tag{5.25}$$

where β_{R} is the vector of the intercepts and regression coefficients. In this example, $\beta_{1,0}$ and $\beta_{2,0}$ are the intercepts for the effect sizes 1 and 2, respectively, whereas $\beta_{1,1}$ and $\beta_{2,1}$ are the regression coefficients from $\boldsymbol{x}_2$ on the effect sizes 1 and 2,

Table 5.1 Long format data for a multivariate meta-analysis.

Study	Effect size	x_1 (Study level)	x_2 (Effect size level)
1	$y_{1(1)}$	$x_{1(1)}$	$x_{2,1(1)}$
1	$y_{2(1)}$	$x_{1(1)}$	$x_{2,2(1)}$
2	$y_{1(2)}$	$x_{1(2)}$	$x_{2,1(2)}$
3	$y_{2(3)}$	$x_{1(3)}$	$x_{2,2(3)}$

Table 5.2 Wide format data for a multivariate meta-analysis.

Study	y_1	y_2	x_1 (Study level)	x_2 for y_1	x_2 for y_2
1	$y_{1(1)}$	$y_{2(1)}$	$x_{1(1)}$	$x_{2,1(1)}$	$x_{2,2(1)}$
2	$y_{1(2)}$	NA	$x_{1(2)}$	$x_{2,1(2)}$	NA
3	NA	$y_{2(3)}$	$x_{1(3)}$	NA	$x_{2,2(3)}$

Abbreviation: NA, Not available.

respectively. $\boldsymbol{T}^2 = \mathsf{Cov}(\boldsymbol{u}_i)$ is the variance component of the residual heterogeneity after controlling for $\boldsymbol{x}_2$.

Several points are worth mentioning here. $\boldsymbol{x}_i$ represents the moderators for the effect sizes. x_1 and x_2 can be different if they are representing different values for different effect sizes. If $\boldsymbol{x}_i$ is a study-level covariate, such as year of publication, the values will be the same for the same study. Second, it may be of interest to test $H_0 : \beta_{1,1} = \beta_{1,2}$. This null hypothesis states that the effects of the moderators are the same on the effect sizes. The null hypothesis can be tested by formulating two models—one model without any constraint and the other model with the constraint $\beta_{1,1} = \beta_{1,2}$. As these two models are nested, the difference on the LR statistics follows a chi-square distribution with 1 df under the null hypothesis.

The unweighted and weighted methods of moment approach for the random-effects model has been extended to the multivariate mixed-effects model (e.g., Demidenko, 2013). Once the variance component matrix of the residual heterogeneity $\hat{\boldsymbol{T}}^2$ has been estimated, we can estimate the fixed effects with Equations 5.20 and 5.22 and a design matrix $\boldsymbol{X}$ that includes the moderators. Moreover, ML or REML implemented in multilevel modeling (e.g., Kalaian and Raudenbush, 1996) and SEM (Cheung, 2013a, b) may be used to obtain the estimates on the fixed-effects and the random-effects.

5.4.1 Explained variance

When there are moderators, it is of interest to see how much variation the moderators explain. Jackson et al. (2012) extended the concept of explained variance to multivariate meta-analyses. Alternatively, the concept of R^2 in the univariate meta-analysis in Section 4.5.2 can be applied to individual effect sizes. We may calculate R^2 for each effect size that indicates the percentage of explained variance on that effect size. This may provide information on how useful the predictors are in explaining the heterogeneity of the effect sizes.

5.5 Structural equation modeling approach

In this section, we extend the univariate SEM-based meta-analysis to the multivariate meta-analysis. The key idea is to treat the *true* effect sizes as the latent variables under the SEM framework. The known sampling covariance matrices are imposed as the covariance matrices of the measurement errors via definition variables. We first begin with the fixed-effects model and then extend the discussion to the random- and mixed-effects models.

5.5.1 Fixed-effects model

Similar to the SEM approach conducting a univariate meta-analysis, we treat the observed effect sizes as raw data. Figure 5.1 shows the model for the multivariate fixed-effects model with two effect sizes per study. The two observed effect sizes in the *i*th study are represented by the variables $y_{1,i}$ and $y_{2,i}$. $\beta_{1,0}$ and $\beta_{2,0}$ represent the common population effect sizes for the first and second effect sizes under a fixed-effects model. The elements of the known sampling variance–covariance matrix of the effect sizes in the *i*th study are imposed as the known covariance matrix, $v_{1,1,i}$, $v_{2,1,i}$, and $v_{2,2,i}$ via the definition variables.

When there are missing effect sizes, the missing values are handled using the FIML estimation method. The treatment of the missing values in $\boldsymbol{V}_i$ deserves some explanation. Suppose that there are missing values in some of the effect sizes, say $y_{1,4}$ in the fourth study, and that of its associated variance and covariances are also

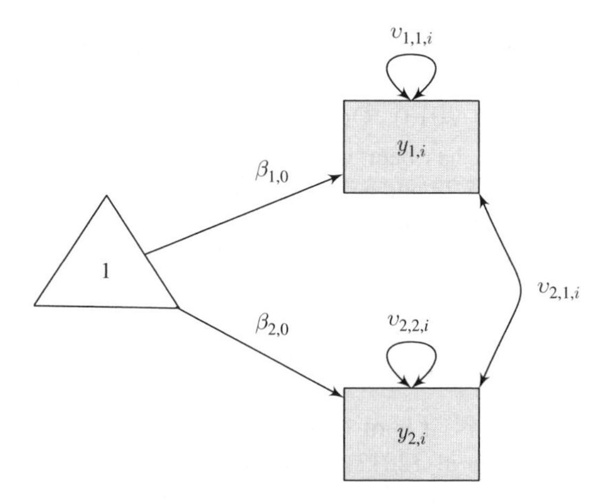

Figure 5.1 A multivariate fixed-effects model with two effect sizes.

missing, that is, $v_{1,1,4} = \text{NA}$ and $v_{2,1,4} = \text{NA}$. As the elements of $\boldsymbol{V}_4$ are imposed as known values via definition variables in Figure 5.1, missing values are not allowed in the definition variables. To address this issue, we have to replace the missing values, for example, $v_{1,1,4}$ and $v_{2,1,4}$ in this example, with some arbitrary values. As the missing value $y_{1,4}$ in this example will be filtered out, the values on $v_{1,1,4}$ and $v_{2,1,4}$ will not be entered into the analysis.

To fit the multivariate fixed-effects meta-analysis in SEM, we use the following model-implied conditional mean vector and the covariance matrix:

$$\begin{aligned} \mu_i(\theta) &= \beta_\text{F} \quad \text{and} \\ \Sigma_i(\theta) &= \boldsymbol{V}_i. \end{aligned} \tag{5.26}$$

After fitting this model, the parameter estimates on β_F and its asymptotic covariance matrix $\mathsf{Cov}(\hat{\beta}_\text{F})$ are estimated. Hypothesis testing on β_F may be carried out as discussed in Section 5.2.2.

5.5.2 Random-effects model

The fixed-effects model can be extended to the random-effects model by including study-specific random effects. Figure 5.2 shows two graphical models that are equivalent to the random-effects model with two effect sizes per study. Figure 5.2a displays the model using Equation 5.12. The *true* effect sizes are represented by the latent variables $f_{1,i}$ and $f_{2,i}$. The heterogeneity of the effect sizes $\boldsymbol{T}^2$ is represented by the variance–covariance matrix of $\boldsymbol{f}$, while the conditional known error variance–covariance matrix $\boldsymbol{V}_i$ is treated as the known variance–covariance matrix of the measurement errors. Conceptually, the random-effects meta-analysis can be viewed as a confirmatory factor analytic (CFA) model—the *true* effect sizes and the known sampling variances in a meta-analysis are treated as the latent variables and the measurement errors with known variances in the CFA. As the observed effect sizes are correlated, the measurement errors are also correlated.

Figure 5.2b, which is equivalent to Equation 5.13, shows the same model by skipping the latent variables. $\boldsymbol{T}^2$ and $\boldsymbol{V}_i$ are combined to form the unconditional variance–covariance matrix. Therefore, the model-implied conditional mean vector and the variance–covariance matrix are

$$\begin{aligned} \mu_i(\theta) &= \beta_\text{R} \quad \text{and} \\ \Sigma_i(\theta) &= \boldsymbol{T}^2 + \boldsymbol{V}_i. \end{aligned} \tag{5.27}$$

After fitting this model, the parameter estimates on β_R and $\boldsymbol{T}^2$, and their asymptotic covariance matrix, are available. Hypothesis testing on $\hat{\beta}_\text{R}$ may be carried out. For instance, we may test the significance of the individual parameter estimates by using either an SE or an LR test. We can also test the composite hypothesis of several parameter estimates by comparing the models with and without the constraints with an LR test. When an SEM approach is used, it is straightforward to impose a special covariance structure on $\boldsymbol{T}^2$. For example, we may impose a diagonal matrix on $\boldsymbol{T}^2$ if there are too many effect sizes per study and/or there are not enough studies to estimate an unstructured $\boldsymbol{T}^2$.

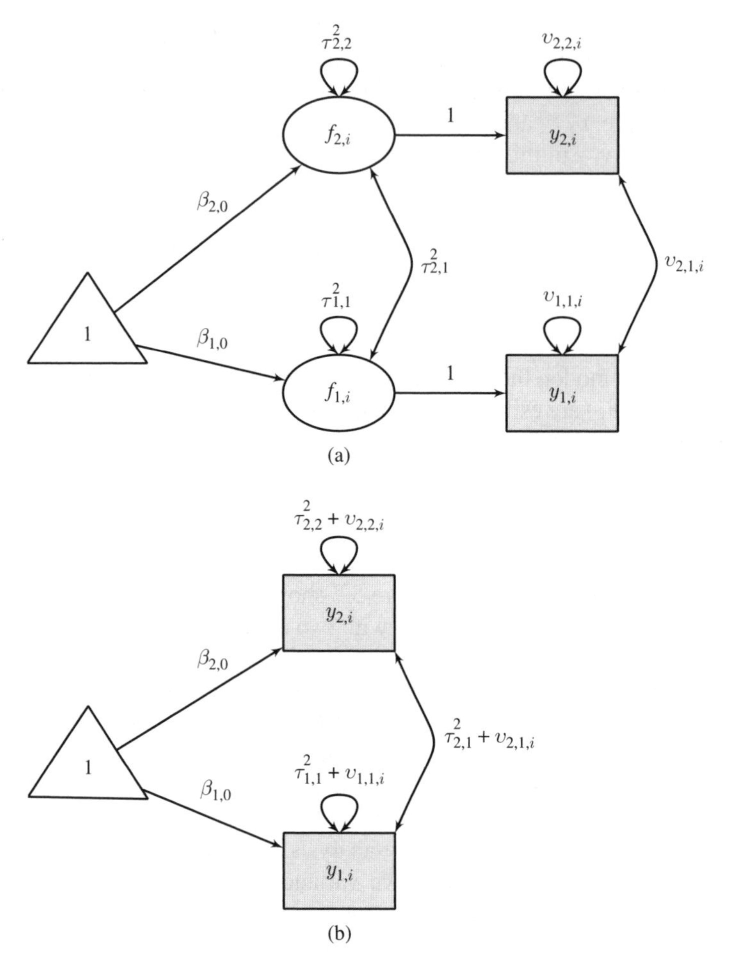

Figure 5.2 A multivariate random-effects model with two effect sizes.

5.5.3 Mixed-effects model

We may extend the random-effects model to a mixed-effects model by using study characteristics as moderators. The moderators are treated as either variables or a design matrix (see Section 4.6.3). Following the practice in meta-analysis, we treat the predictors as a design matrix here. Figure 5.3 shows two equivalent models of a mixed-effects model with two effect sizes per study and one moderator. Without the loss of generality, the moderator x_i is the same for both effect sizes in the ith study. Figure 5.3a explicitly shows the *true* effect sizes or the latent variables $f_{1,i}$ and $f_{2,i}$, while Figure 5.3b skips the *true* effect sizes.

A phantom variable P is created to store the values of the moderators. $\beta_{1,1}$ and $\beta_{2,1}$ are the regression coefficients indicating the effects from x_i to $y_{1,i}$ and $y_{2,i}$ in

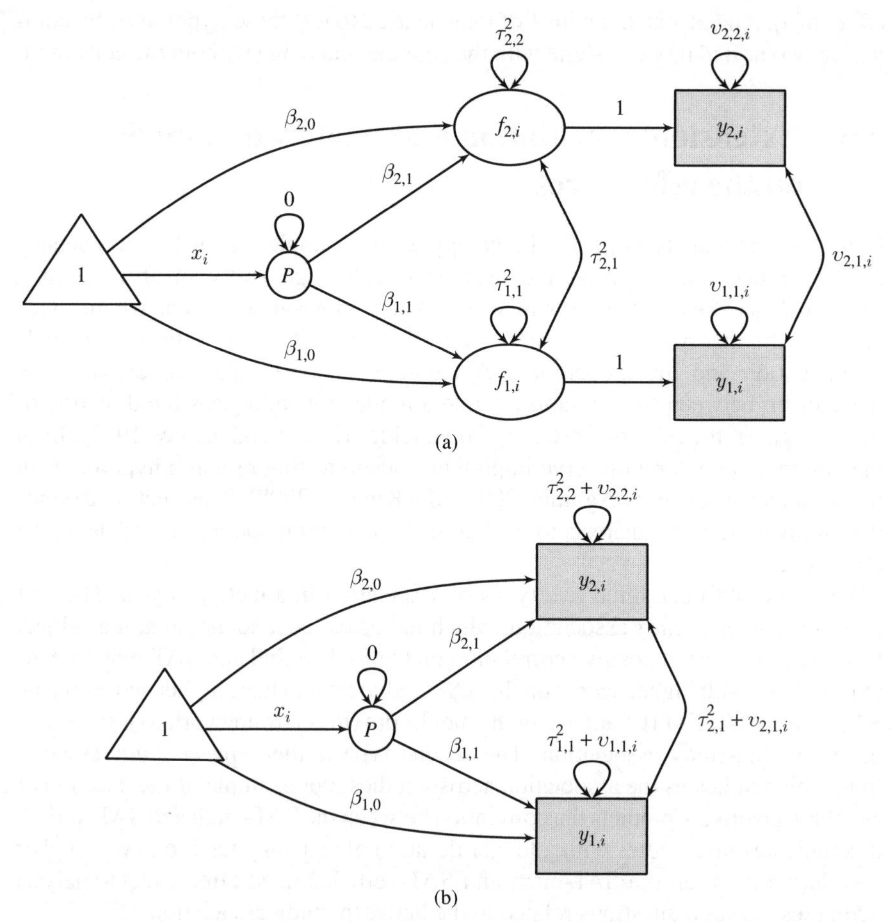

Figure 5.3 A multivariate mixed-effects model with two effect sizes and one moderator.

the ith study, respectively. $\beta_{1,0}$ and $\beta_{2,0}$ are the intercepts of the first and the second effect sizes, respectively, when $x_i = 0$.

By using the tracing rules (e.g., Mulaik, 2009), it can be shown that the conditional implied mean vector and covariance matrix for the model in Figure 5.3 are

$$\begin{aligned}\mu_i(\theta|x_i) &= \begin{bmatrix}\beta_{1,0} + \beta_{1,1}x_i\\ \beta_{2,0} + \beta_{2,1}x_i\end{bmatrix} \quad \text{and}\\ \Sigma_i(\theta|x_i) &= \boldsymbol{T}^2 + \boldsymbol{V}_i.\end{aligned} \tag{5.28}$$

Several research hypotheses on both the fixed effects and the random effects discussed in previous sections can be easily tested under the SEM approach. For example, we may test whether the regression coefficients are the same $H_0 : \beta_{1,1} = \beta_{2,1}$, or whether the intercepts are the same $H_0 : \beta_{1,0} = \beta_{2,0}$ after controlling for the effect of x_i. An LR statistic with 1 df may be used to test these hypotheses by comparing two nested models—one with the constraint and one without the constraint.

5.6 Extensions: mediation and moderation models on the effect sizes

Using the meta-analysis in psychotherapy as an example, Shadish and Sweeney (1991) argued that knowing the average effect is generally insufficient in a meta-analysis. Researchers would like to know more about when, where, why, and how the theory works. These last questions are related to testing the models of mediation and the moderation. A mediator is a variable that explains the mechanism between two variables, while a moderator influences the direction or the strength of the relation between two variables (Baron and Kenny, 1986). Both mediation and moderation serve important roles in testing research hypotheses in the social sciences (e.g., Aguinis, 2004; MacKinnon, 2008). This section extends the multivariate meta-analysis to testing mediation and moderation models on the effect sizes.

Shadish (1996) classified two types of association in a meta-analysis. The first type is the *within-study* association, which indicates the association at the subject level. For example, a positive correlation on the SAT-verbal and SAT-math means that subjects with higher scores on the SAT-verbal tend to have higher scores on the SAT-math. MASEM (Chapter 7 in this book) provides a framework to test models on the within-study association. The second type is the *between-study* association, which indicates the association across studies. For example, if the duration of coaching positively predicts the correlation between the SAT-math and SAT-verbal, this indicates that studies with a longer duration of coaching tend to have a higher correlation between the SAT-math and SAT-verbal. Mixed-effects meta-analysis addresses research questions related to the between-study association.

Shadish (1992, 1996) and Shadish and Sweeney (1991) further extended the above mediation models on correlation matrices and moderation models in mixed-effects meta-analysis to *any* generic effect size, such as the SMD. He also provided theoretical justifications and examples for these models. In his work, Shadish used the coded study characteristics as the variables in testing mediation, moderation, and even structural equation models. We show in this section that the *true* effect sizes may also be used as mediators and moderators. This section provides the necessary statistical techniques to model the mediation and moderation on the effect sizes.

As both mediation and moderation involve a regression analysis of the effect sizes, it is of importance to discuss why it is not preferable to conduct a mixed-effects meta-analysis on one effect size, say y_1, by treating the other effect size, say y_2, as the predictor (or moderator). In a regression analysis, the predictors are assumed to be measured without measurement error. When y_2 is treated as a predictor, this practice violates this assumption because the precision of y_2 is $v_{2,2}$. There is a clear consensus in the literature on regression analysis and SEM that the parameter estimates are biased when the predictors are measured with measurement error (e.g., Bollen, 1989; Buonaccorsi, 2010; Fuller, 1987). Findings of these studies are directly applicable to meta-analysis. The issue is even more complicated in a meta-analysis where the effect sizes are measured with different degrees of precision. For example, the baseline risk y_2 is sometimes used as a predictor to control for possible differences across studies in synthesizing the treatment effect y_1 in a meta-analysis (e.g., Arends et al., 2000; Ghidey et al., 2013). The parameter estimates are likely to be biased when the observed baseline risk is directly used to predict the treatment effect.

It is well known that SEM may be used to correct for the unreliability of the measurement error. When there is only one indicator, say y_1, we may fix the error variance of the item to the estimated error variance (e.g., Hayduk, 1987). Suppose that there are two variables y_1 and y_2 with their estimated reliability coefficients (coefficient alphas) $\hat{\alpha}_{y1}$ and $\hat{\alpha}_{y2}$. We may fix the error variances as shown in Figure 5.4. The latent factors f_1 and f_2 represent the *true* scores without measurement error. Thus, $\phi_{1,2}$ is the covariance (correction) corrected for unreliability.

The main drawback of this approach is that the error variances $\text{Var}(y_1)(1-\hat{\alpha}_{y1})$ and $\text{Var}(y_2)(1-\hat{\alpha}_{y2})$ are considered to be fixed rather than estimated. Thus, the SEs of the parameter estimates are likely to be underestimated when the sample sizes are small. Oberski and Satorra (2013) proposed a method to adjust for this bias.

The above approach may also be applied in the SEM-based meta-analysis to conduct mediation and moderation analyses. The key point is to treat the *true* effect sizes as latent variables in SEM, whereas the known sampling covariance matrix is fixed as known values on the measurement errors via definition variables. As the variances of the sampling errors are treated as known in a meta-analysis, there is no need to adjust for the uncertainty by treating the variances of the sampling errors as known.

5.6.1 Regression model

We first illustrate a regression model between two effect sizes. After showing how to model the *true* effect size as a predictor, we extend the analysis to the mediation and the moderation models. Let us consider $y_{1,i}$ and $y_{2,i}$ as the dependent and the

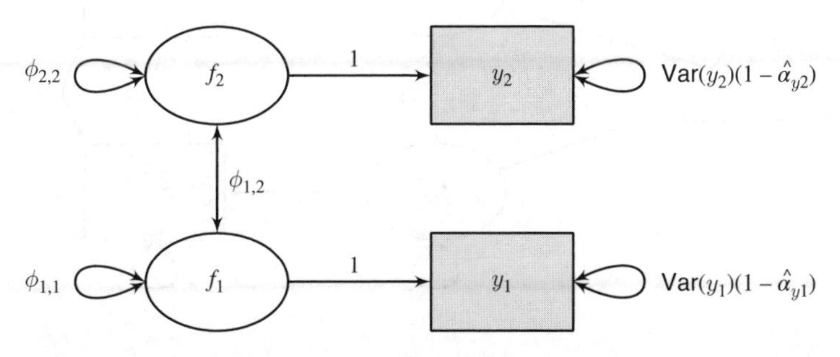

Figure 5.4 A structural equation model to correct for unreliabilities.

independent variables, respectively. The measurement (or level 1) model and the structural (or level 2) model for the ith study are

$$\begin{aligned} \text{Measurement model: } & \begin{bmatrix} y_{1,i} \\ y_{2,i} \end{bmatrix} = \begin{bmatrix} f_{1,i} \\ f_{2,i} \end{bmatrix} + \begin{bmatrix} e_{1,i} \\ e_{2,i} \end{bmatrix} \quad \text{and} \\ \text{Structural model: } & \begin{bmatrix} f_{1,i} \\ f_{2,i} \end{bmatrix} = \begin{bmatrix} \beta_{1,0} \\ \beta_{2,0} \end{bmatrix} + \begin{bmatrix} 0 & \beta_{1,2} \\ 0 & 0 \end{bmatrix} \begin{bmatrix} f_{1,i} \\ f_{2,i} \end{bmatrix} + \begin{bmatrix} u_{1,i} \\ u_{2,i} \end{bmatrix}. \end{aligned} \tag{5.29}$$

The measurement (or level 1) model is used to handle the known sampling covariance matrix among the effect sizes. The *true* effect sizes are represented by the latent variables $f_{1,i}$ and $f_{2,i}$. The structural (or level 2) model is used to specify the relationship between the *true* effect sizes.

Figure 5.5 shows the SEM. The conditional sampling covariance matrix is known, that is, $\mathsf{Cov}\left(\begin{bmatrix} e_{1,i} \\ e_{2,i} \end{bmatrix}\right) = \begin{bmatrix} v_{1,1,i} & \\ v_{2,1,i} & v_{2,2,i} \end{bmatrix}$, whereas the covariance matrix between the *true* effect sizes is $\mathsf{Cov}\left(\begin{bmatrix} u_{1,i} \\ u_{2,i} \end{bmatrix}\right) = \begin{bmatrix} \tau^2_{1,1} & \\ 0 & \tau^2_{2,2} \end{bmatrix}$, which has to be estimated from the data. $\tau^2_{2,2}$ is the heterogeneity variance of $f_{2,i}$, while $\tau^2_{1,1}$ is the heterogeneity residual variance of $f_{1,i}$. An R^2 indice defined by

$$R^2 = \frac{\beta^2_{1,2}\tau^2_{2,2}}{\beta^2_{1,2}\tau^2_{2,2} + \tau^2_{1,1}} \tag{5.30}$$

can be used to indicate the percentage of the heterogeneity in $f_{1,i}$ that can be explained by $f_{2,i}$. It should be noted that $u_{1,i}$ and $u_{2,i}$ are conditionally independent because their dependence has already been accounted for by the regression coefficient $\beta_{1,2}$.

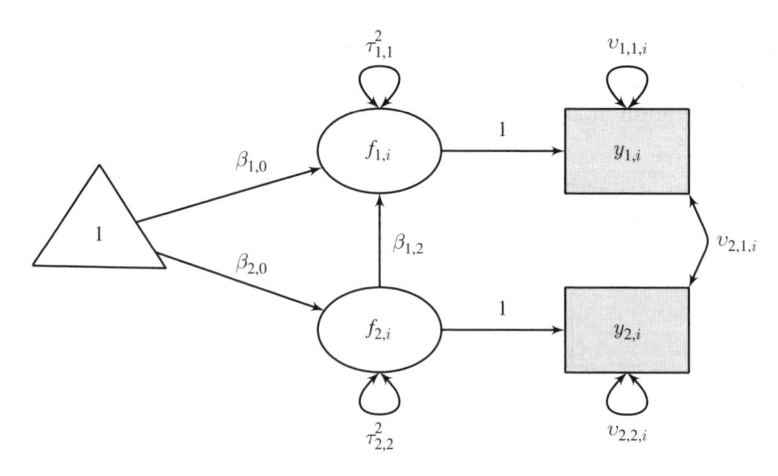

Figure 5.5 A regression model between two effect sizes.

Another point that should be noted is that this model is an equivalent model to the one displayed in Figure 5.2. This means that both models have the same model fit (−2LL). The parameter estimates can be directly transformed between the models in Figures 5.2 and 5.5. Let us denote the parameters in the multivariate meta-analysis in Figure 5.2 and in the mediating model in Figure 5.5 with subscripts (Mul) and (Med) in parentheses. The parameters are related by Equation 2.2:

$$\begin{aligned}
\beta_{1,0(\mathrm{Mul})} &= \beta_{1,0(\mathrm{Med})} + \beta_{1,2(\mathrm{Med})}\beta_{2,0(\mathrm{Med})}, \\
\beta_{2,0(\mathrm{Mul})} &= \beta_{2,0(\mathrm{Med})}, \\
\tau^2_{1,1(\mathrm{Mul})} &= \beta^2_{1,2(\mathrm{Med})}\tau^2_{2,2(\mathrm{Med})} + \tau^2_{1,1(\mathrm{Med})}, \\
\tau^2_{2,1(\mathrm{Mul})} &= \beta_{1,2(\mathrm{Med})}\tau^2_{2,2(\mathrm{Med})}, \quad \text{and} \\
\tau^2_{2,2(\mathrm{Mul})} &= \tau^2_{2,2(\mathrm{Med})}.
\end{aligned} \tag{5.31}$$

5.6.2 Mediating model

Suppose that there is a study characteristic x_i. On the basis of the theories, for example, researchers may want to test whether the effect from x_i to the *true* effect size f_1 is mediated by the *true* effect size f_2. As an example, Cheung (2009b) illustrated how to test a mediation model between two effect sizes. Data from 42 countries (World Values Study Group, 1994) were considered for the studies in a cross-cultural meta-analysis. The *true* effect size of the SMD on happiness between males and females was treated as the dependent variable, while the *true* effect size on the SMD on life control between males and females was considered as the mediator. The gross national product (GNP) of the country was treated as x_i.

The model in Equation 5.29 can be extended to handle the indirect effect between the *true* effect sizes. The measurement (or level 1) model and the structural (or level 2) model in the ith study are

$$\begin{aligned}
\text{Measurement model:} \quad \begin{bmatrix} y_{1,i} \\ y_{2,i} \end{bmatrix} &= \begin{bmatrix} f_{1,i} \\ f_{2,i} \end{bmatrix} + \begin{bmatrix} e_{1,i} \\ e_{2,i} \end{bmatrix} \quad \text{and} \\
\text{Structural model:} \quad \begin{bmatrix} f_{1,i} \\ f_{2,i} \end{bmatrix} &= \begin{bmatrix} \beta_{1,0} \\ \beta_{2,0} \end{bmatrix} + \begin{bmatrix} 0 & \beta_{1,2} \\ 0 & 0 \end{bmatrix} \begin{bmatrix} f_{1,i} \\ f_{2,i} \end{bmatrix} \\
&\quad + \begin{bmatrix} \gamma_1 \\ \gamma_2 \end{bmatrix} x_i + \begin{bmatrix} u_{1,i} \\ u_{2,i} \end{bmatrix}.
\end{aligned} \tag{5.32}$$

The measurement model is formulated to handle the conditional sampling covariance matrix between the effect sizes, while the structural model is used to model the indirect effect. Under the above model, the direct effects from x_i to the *true* effect size $f_{1,i}$ is γ_1, while the indirect effect via the *true* effect size $f_{2,i}$ is $\gamma_2\beta_{1,2}$. The total effect is $\gamma_2\beta_{1,2} + \gamma_1$. Standard procedures, such as the Sobel test, the bootstrap CI, or the likelihood-based confidence interval (LBCI), can be used to construct the CIs on the indirect effect (e.g., Cheung, 2007, 2009a).

Similar to the previous sections, there are two models to fit the mediation models. Figure 5.6 shows two models with the *true* effect size of $y_{2,i}$ as the mediator and

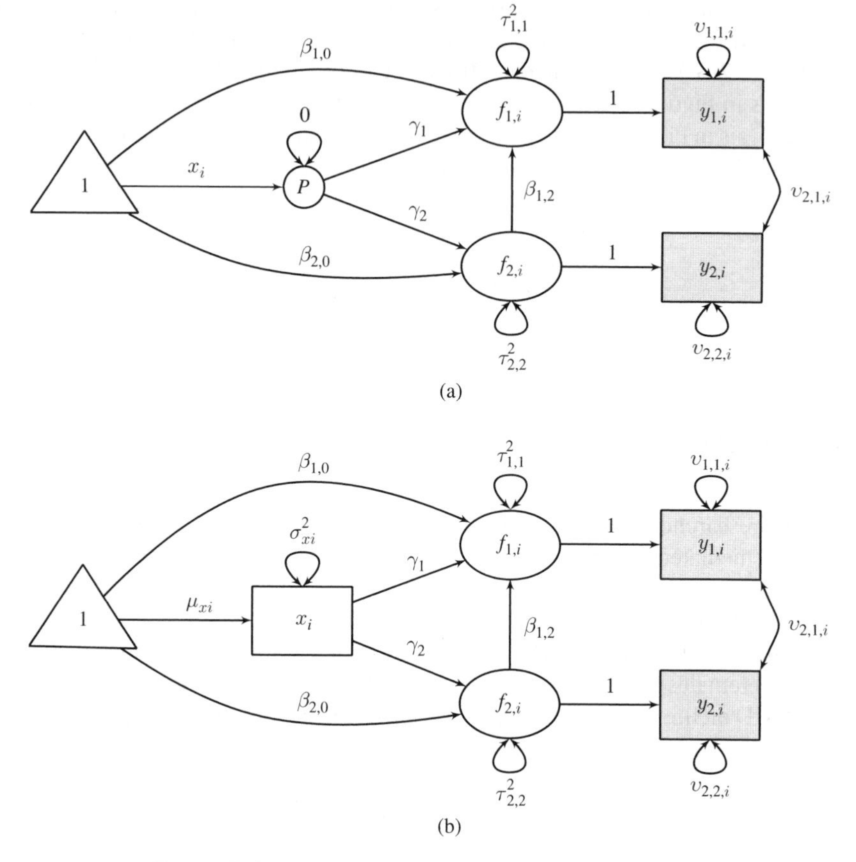

Figure 5.6 A mediation model with two effect sizes.

the *true* effect size of $y_{1,i}$ as the dependent variable. Figure 5.6a uses the phantom variable P to impose the effects from x_i to the dependent variable. Figure 5.6b fits the mediation model by introducing x_i as an observed variable. The main difference between these two models is that Figure 5.6a treats x_i as a design matrix, whereas Figure 5.6b considers x_i as a variable. Under both the models, the effects from x_i to the *true* effect sizes $f_{2,i}$ and $f_{1,i}$ are shown as γ_2 and γ_1, respectively. As a reminder, the mediation model is an equivalent model to the mixed-effects model discussed in Section 5.4. Thus, the mediation model should only be fitted based on theories.

5.6.3 Moderating model

Moderation is often used to test how the independent variables relate to the dependent variables where the relationship depends on a third variable called a moderator (Aguinis, 2004; Aiken et al., 1991). As an illustration, Cheung (2009b) showed how to test the moderation between two effect sizes. The author tested the effect of the *true* effect size of the SMD on life control to the *true* effect size of the SMD on happiness by considering GNP as the moderator. Statistically speaking, we cannot distinguish which variable is the independent variable and which variable is the moderator. As we are testing the model with one *true* effect size as the independent variable, we consider the study characteristic x_i as the moderator.

We may also extend the regression model in Equation 5.29 to test the moderation effect among the effect sizes. The measurement (or level 1) model and the structural (or level 2) model in the ith study are

$$\text{Measurement model:} \begin{bmatrix} y_{1,i} \\ y_{2,i} \end{bmatrix} = \begin{bmatrix} f_{1,i} \\ f_{2,i} \end{bmatrix} + \begin{bmatrix} e_{1,i} \\ e_{2,i} \end{bmatrix},$$

$$\text{Structural model:} \begin{bmatrix} f_{1,i} \\ f_{2,i} \end{bmatrix} = \begin{bmatrix} \beta_{1,0} \\ \beta_{2,0} \end{bmatrix} + \begin{bmatrix} 0 & \beta_{1,2} \\ 0 & 0 \end{bmatrix} \begin{bmatrix} f_{1,i} \\ f_{2,i} \end{bmatrix} + \begin{bmatrix} \gamma_1 \\ 0 \end{bmatrix} x_i \quad (5.33)$$
$$+ x_i \begin{bmatrix} 0 & \omega_{1,2} \\ 0 & 0 \end{bmatrix} \begin{bmatrix} f_{1,i} \\ f_{2,i} \end{bmatrix} + \begin{bmatrix} u_{1,i} \\ u_{2,i} \end{bmatrix}.$$

From the above equations, the equation for $f_{1,i}$ is

$$f_{1,i} = \beta_{1,0} + \gamma_1 x_i + (\beta_{1,2} + \omega_{1,2} x_i) f_{2,i} + u_{1,i}. \quad (5.34)$$

In the above equation, the effect from $f_{2,i}$ to $f_{1,i}$ is $\beta_{1,2} + \omega_{1,2} x_i$. Therefore, x_i moderates the effect from $f_{2,i}$ to $f_{1,i}$. There is a moderating effect when $\omega_{1,2}$ is nonzero. We usually center the moderator x_i to facilitate the interpretations. After centering x_i, $\beta_{1,2}$ represents the effect from $f_{2,i}$ to $f_{1,i}$ when x_i is at its average value, that is, zero. Figure 5.7 shows two models with the effect from the *true* effect size $f_{2,i}$ to the *true* effect size $f_{1,i}$ moderated by x_i. The main difference between these two models is that Figure 5.7a treats x_i as a design matrix, while Figure 5.7b considers x_i as a variable. Figure 5.7a tests the moderation effect by introducing two phantom variables, while Figure 5.7b displays the same model by treating x_i as an observed variable.

Before closing this section, we need to mention how the missing values in x_i are handled in the mixed-effects meta-analysis, the mediation, and the moderation models. Regardless of whether x_i is treated as a design matrix or a variable, the whole study will be deleted when there are missing values in x_i. If we want to keep the studies with missing values in x_i in the analysis, we may formulate the models so that x_i is treated as a dependent variable. FIML can then be used to handle the missing values in x_i (see Section 8.2 for details).

5.7 Illustrations using R

This section illustrates how to conduct a multivariate meta-analysis using the `metaSEM` package. The first data set tests the effectiveness of the Bacillus Calmette–Guerin (BCG) vaccine for preventing tuberculosis (Colditz et al., 1994).

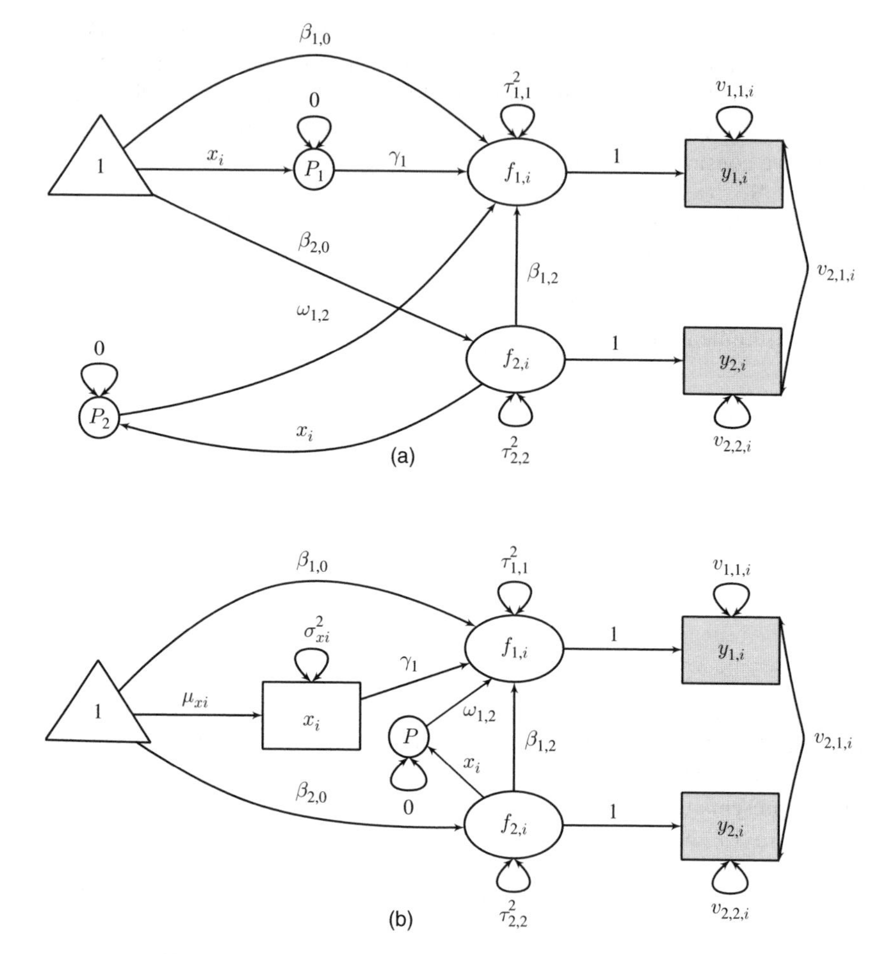

Figure 5.7 A moderation model with two effect sizes.

This data set has been used by several researchers (e.g., Berkey et al., 1998; Viechtbauer, 2010). It is used to illustrate the multivariate random-effects meta-analysis. The second data set was extracted from the World Values Survey II (World Values Study Group, 1994). It has been used to demonstrate the SEM-based multivariate meta-analysis in Cheung (2013b). This data set is also used to illustrate the mediation and the moderation analyses. The statistics reported in the illustrations were captured by using the `Sweave` function in R. The numbers of decimal places may be slightly different for those reported in the selected output and in the text.

5.7.1 BCG vaccine for preventing tuberculosis

When we conduct a meta-analysis on contingency tables, we usually calculate the logarithm of the odds ratio between the treatment group and the control group as the effect size. A univariate meta-analysis is applied to the data because there is only one effect size. van Houwelingen et al. (2002) argued that this practice may hide a lot of information, especially when there is a high degree of variation on the control and the intervention groups. They illustrated how a multivariate meta-analysis may be applied to the effect sizes of the control and the treatment groups.

The BCG data was stored in the object `BCG` in the `metaSEM` package. The effect sizes are `ln_Odd_V` (natural logarithm of the odds of the vaccinated group) and `ln_Odd_NV` (natural logarithm of the odds of the nonvaccinated group), while their sampling variances are `v_ln_Odd_V` (sampling variance of `ln_Odd_V`) and `v_ln_Odd_NV` (sampling variance of `ln_Odd_NV`). As the control group and the treatment group are independent, the sampling covariance between the effect size is 0 (`cov_V_NV`). Therefore, these two effect sizes are conditionally independent.

We display a few cases of the data by typing `head(BCG)`.

```
R> #### Load the metaSEM library
R> library("metaSEM")
R> ## Display the dataset
R> head(BCG)
```

```
  Trial                Author Year  VD   VWD NVD  NVWD Latitude
1     1               Aronson 1948   4   119  11   128       44
2     2       Ferguson & Simes 1949   6   300  29   274       55
3     3      Rosenthal et al 1960   3   228  11   209       42
4     4    Hart & Sutherland 1977  62 13536 248 12619       52
5     5 Frimodt-Moller et al 1973  33  5036  47  5761       13
6     6       Stein & Aronson 1953 180  1361 372  1079       44
  Allocation   ln_OR  v_ln_OR ln_Odd_V ln_Odd_NV v_ln_Odd_V cov_V_NV
1     random -0.9387 0.357125   -3.393    -2.454    0.25840        0
2     random -1.6662 0.208132   -3.912    -2.246    0.17000        0
3     random -1.3863 0.433413   -4.331    -2.944    0.33772        0
4     random -1.4564 0.020314   -5.386    -3.930    0.01620        0
5  alternate -0.2191 0.051952   -5.028    -4.809    0.03050        0
6  alternate -0.9581 0.009905   -2.023    -1.065    0.00629        0
  v_ln_Odd_NV
1    0.098722
2    0.038132
3    0.095694
4    0.004112
5    0.021450
6    0.003615
```

5.7.1.1 Random-effects model

As the effect sizes are conditionally independent, the covariance between these two effect sizes is zero. We conduct a random-effects meta-analysis using the `meta()` command with `y` and `v` as the arguments for the effect size and its sampling variance. We use the `cbind()` function to combine the effect sizes, as there is more than one effect size.

```
R> ## Covariance between the effect size is 0.
R> bcg1 <- meta(y=cbind(ln_Odd_V, ln_Odd_NV),
                v=cbind(v_ln_Odd_V, cov_V_NV, v_ln_Odd_NV),
                data=BCG, model.name="Random effects model")
R> summary(bcg1)
```

```
------------------------ Selected output ----------------------

95% confidence intervals: z statistic approximation
Coefficients:
           Estimate Std.Error lbound ubound z value Pr(>|z|)
Intercept1   -4.834     0.340 -5.501 -4.167  -14.21   <2e-16 ***
Intercept2   -4.096     0.435 -4.948 -3.244   -9.42   <2e-16 ***
Tau2_1_1      1.431     0.583  0.289  2.574    2.46    0.014 *
Tau2_2_1      1.757     0.724  0.338  3.177    2.43    0.015 *
Tau2_2_2      2.407     0.967  0.511  4.303    2.49    0.013 *
---
Signif. codes:  0 '***' 0.001 '**' 0.01 '*' 0.05 '.' 0.1 ' ' 1

Q statistic on the homogeneity of effect sizes: 5270
Degrees of freedom of the Q statistic: 24
P value of the Q statistic: 0

Heterogeneity indices (based on the estimated Tau2):
                             Estimate
Intercept1: I2 (Q statistic)     0.99
Intercept2: I2 (Q statistic)     1.00

Number of studies (or clusters): 13
Number of observed statistics: 26
Number of estimated parameters: 5
Degrees of freedom: 21
-2 log likelihood: 66.18
OpenMx status1: 0 ("0" or "1": The optimization is considered fine.
Other values indicate problems.)

------------------------ Selected output ----------------------
```

Before interpreting the results, we should check whether there are any estimation problems in the analysis. We check the `OpenMx status1` at the end of the output. If the status is either 0 or 1, the optimization is fine; otherwise, the results are not trustworthy. The test of homogeneity of effect sizes is $Q(\text{df} = 24) = 5270.3863, p < 0.001$, which is statistically significant. The estimated I^2 for the vaccinated and the nonvaccinated groups are 0.9887 and 0.9955, respectively. These indicate an extremely high degree of heterogeneity on the population effect sizes. We will discuss the estimated variance component soon.

5.7.1.2 Testing the average effects

The estimated average effect sizes for the vaccinated and the nonvaccinated groups (and their approximate 95% Wald CIs) are −4.8338 (−5.5005, −4.1670) and −4.0960 (−4.9481, −3.2439), respectively. We test whether the average population effect sizes are the same for these two groups by fitting a model with the equality constraint on the average population effect sizes and using the argument `intercept.constraints=c("0*Intercept","0*Intercept")`. As the label `(Intercept)` is the same for both intercepts, the intercepts are equally constrained.

```
R> bcg2 <- meta(y=cbind(ln_Odd_V, ln_Odd_NV), data=BCG,
               v=cbind(v_ln_Odd_V, cov_V_NV, v_ln_Odd_NV),
               intercept.constraints=c("0*Intercept",
               "0*Intercept"),
               model.name="Equal intercepts")
R> summary(bcg2)
```

```
----------------------- Selected output ----------------------

95% confidence intervals: z statistic approximation
Coefficients:
         Estimate Std.Error  lbound  ubound z value Pr(>|z|)
Intercept  -5.3750    0.4584 -6.2735 -4.4765  -11.72   <2e-16 ***
Tau2_1_1    1.7003    0.8551  0.0243  3.3763    1.99    0.047 *
Tau2_2_1    2.4556    1.3113 -0.1145  5.0257    1.87    0.061 .
Tau2_2_2    4.1087    1.9902  0.2079  8.0094    2.06    0.039 *
---
Signif. codes:  0 '***' 0.001 '**' 0.01 '*' 0.05 '.' 0.1 ' ' 1

Heterogeneity indices (based on the estimated Tau2):
                              Estimate
Intercept1: I2 (Q statistic)      0.99
Intercept2: I2 (Q statistic)      1.00

----------------------- Selected output ----------------------
```

This model is nested within the model without the constraint; we compare them with the LR statistic.

```
R> anova(bcg1, bcg2)
```

```
                  base       comparison ep minus2LL df   AIC diffLL
1 Random effects model             <NA>  5    66.18 21 24.18     NA
2 Random effects model Equal intercepts  4    77.06 22 33.06  10.88
  diffdf         p
1     NA        NA
2      1 0.0009708
```

The LR statistic is $\Delta\chi^2(\mathrm{df} = 1) = 10.8825, p = 0.001$. Therefore, the null hypothesis of equal average population effect sizes is rejected. The odds for the vaccinated and the nonvaccinated groups are different. As they are based on the logarithm, the results may be difficult to interpret. We convert the results back into odds. The vaccinated group has smaller odds than the nonvaccinated group. It should be noted, however, that this is a nonlinear transformation. The transformed average values and their CIs may not represent the *average* odds of the experimental and control groups very well (see Hafdahl (2009) for more details).

```
R> ## Extract the coefficient table from the summary
R> Est <- summary(bcg1)$coefficients
R> ## Only select the first 2 rows and the columns
R> ## related to estimate, lbound, and ubound
R> ## Convert them into odds
R> exp( Est[1:2, c("Estimate", "lbound", "ubound") ] )
```

```
            Estimate   lbound   ubound
Intercept1 0.007957 0.004085 0.01550
Intercept2 0.016639 0.007097 0.03901
```

Following van Houwelingen et al. (2002), we also calculate the logarithm of the odds ratio between the vaccinated and the nonvaccinated groups that is equivalent to `log_OR = ln_Odd_V - ln_Odd_NV`. The sampling variance of `log_OR` is

$$\begin{aligned}\mathrm{Var}(\log_{\mathrm{OR}}) =& \mathrm{Var}(\log\_\mathrm{Odd\_V}) + \mathrm{Var}(\log\_\mathrm{Odd\_NV}) \\ & - 2\mathrm{Cov}(\log\_\mathrm{Odd}\_V, \log\_\mathrm{Odd\_NV}),\end{aligned}$$

where $\mathrm{Var}(\log\_\mathrm{Odd\_V})$ and $\mathrm{Var}(\log\_\mathrm{Odd\_NV})$ are the sampling variances of `ln_Odd_V` and `ln_Odd_NV`, and $\mathrm{Cov}(\log\_\mathrm{Odd\_V}, \log\_\mathrm{Odd\_NV})$ is the sampling covariance between them. Therefore, we estimate the fixed-effects on the logarithm of the odds ratio as

```
R> ## Extract the fixed effects
R> ( fixed <- coef(bcg1, select="fixed") )
```

```
Intercept1 Intercept2
    -4.834     -4.096
```

```
R> ## Extract the sampling covariance matrix on the estimates
R> ( omega <- vcov(bcg1)[c("Intercept1","Intercept2"),
                          c("Intercept1","Intercept2")] )
```

```
           Intercept1 Intercept2
```

```
Intercept1      0.1157       0.136
Intercept2      0.1360       0.189
```

```
R> ## Calculate the logarithm on the odds ratio
R> ( log_OR <- fixed[1] - fixed[2] )
```

```
Intercept1
   -0.7378
```

```
R> ## Calculate the standard error on log_OR
R> ( se_log_OR <- sqrt(omega[1,1]+omega[2,2]-2*omega[2,1]) )
```

```
[1] 0.1811
```

The estimated effect size (with its SE) in terms of the logarithm of the odds ratio is −0.7378 (SE = 0.1811). The results are similar to those based on a random-effects meta-analysis on the logarithm of the odds ratio `ln_OR` as the effect size and `v_ln_OR` as the sampling variance.

```
R> summary( meta(y=ln_OR, v=v_ln_OR, data=BCG) )
```

```
------------------------ Selected output ----------------------

95% confidence intervals: z statistic approximation
Coefficients:
           Estimate Std.Error   lbound   ubound z value Pr(>|z|)
Intercept1 -0.74197   0.17863 -1.09208 -0.39185   -4.15  3.3e-05 ***
Tau2_1_1    0.30246   0.15663 -0.00453  0.60944    1.93    0.053 .
---
Signif. codes:  0 '***' 0.001 '**' 0.01 '*' 0.05 '.' 0.1 ' ' 1

Q statistic on the homogeneity of effect sizes: 163.2
Degrees of freedom of the Q statistic: 12
P value of the Q statistic: 0

Heterogeneity indices (based on the estimated Tau2):
                             Estimate
Intercept1: I2 (Q statistic)     0.91

Number of studies (or clusters): 13
Number of observed statistics: 13
Number of estimated parameters: 2
Degrees of freedom: 11
-2 log likelihood: 26.15
```

```
OpenMx status1: 0 ("0" or "1": The optimization is considered fine.
Other values indicate problems.)

------------------------ Selected output -----------------------
```

5.7.1.3 Testing the variance component of the random effects

$\hat{\tau}^2_{1,1}$ and $\hat{\tau}^2_{2,2}$ of the vaccinated group and the nonvaccinated group are 1.4314 and 2.4073, respectively. It seems that the nonvaccinated group has a higher degree of heterogeneity than the vaccinated group. We test the null hypothesis $H_0 : \tau^2_{1,1} = \tau^2_{2,2}$ by comparing the models with and without the equality constraint. We fit the model with the equality constraint on the variances with the following syntax.

```
R> bcg3 <- meta(y=cbind(ln_Odd_V, ln_Odd_NV), data=BCG,
                v=cbind(v_ln_Odd_V, cov_V_NV, v_ln_Odd_NV),
                RE.constraints=matrix(c("0.1*Tau2_Eq","0*Tau2_2_1",
                                        "0*Tau2_2_1","0.1*Tau2_Eq"),
                                      ncol=2, nrow=2),
                model.name="Equal variances")
R> summary(bcg3)
```

```
------------------------ Selected output -----------------------

95% confidence intervals: z statistic approximation
Coefficients:
           Estimate Std.Error lbound ubound z value Pr(>|z|)
Intercept1   -4.837     0.392 -5.606 -4.067  -12.33   <2e-16 ***
Intercept2   -4.083     0.389 -4.845 -3.320  -10.49   <2e-16 ***
Tau2_Eq       1.920     0.738  0.474  3.365    2.60   0.0092 **
Tau2_2_1      1.765     0.739  0.317  3.214    2.39   0.0169 *
---
Signif. codes:  0 '***' 0.001 '**' 0.01 '*' 0.05 '.' 0.1 ' ' 1

Heterogeneity indices (based on the estimated Tau2):
                              Estimate
Intercept1: I2 (Q statistic)      0.99
Intercept2: I2 (Q statistic)      0.99

------------------------ Selected output -----------------------
```

As this model is nested within the model without the constraint, we compare them with the LR statistic.

```
R> anova(bcg1, bcg3)
```

```
                 base       comparison ep minus2LL df    AIC diffLL
```

```
1 Random effects model                    <NA>  5     66.18 21 24.18       NA
2 Random effects model Equal variances    4     71.25 22 27.25    5.071
  diffdf       p
1     NA      NA
2      1 0.02433
```

The LR statistic is $\Delta\chi^2(\text{df} = 1) = 5.0712, p = 0.024$. Therefore, the null hypothesis of equal heterogeneity variances is rejected. The nonvaccinated group has a higher degree of heterogeneity than the vaccinated group. We extract the variance component by using the `coef()` function with the `select="random"` argument. We convert the values into a matrix using the `vec2symMat()` function. As it is difficult to inspect the covariance, we convert it into a correlation matrix.

```
R> ( T2 <- vec2symMat(coef(bcg1, select="random")) )
```

```
      [,1]  [,2]
[1,] 1.431 1.757
[2,] 1.757 2.407
```

```
R> cov2cor(T2)
```

```
       [,1]   [,2]
[1,] 1.0000 0.9467
[2,] 0.9467 1.0000
```

The estimated correlation between the random effects is 0.9467, which is extremely high. This indicates that studies with high odds in the nonvaccinated group also have higher odds in the vaccinated group. As special care has to be taken to interpret meta-analyses involving both baseline risk and between-treatment effect, readers are advised to refer to Arends et al. (2000), Ghidey et al. (2013), and van Houwelingen et al. (2002). The following plot (Figure 5.8) will graphically display the effect.

5.7.1.4 Plotting the figures

We may plot the effect sizes and their confidence ellipses with the `plot()` function for a multivariate meta-analysis (Cheung, 2013b). Figure 5.8 displays the average effect sizes and the individual effect sizes.

```
R> plot(bcg1, xlim=c(-8,0), ylim=c(-8,0))
```

Some explanations of the figure are required. The x- and the y-axes represent the first and the second effect sizes for the vaccinated group and the nonvaccinated group, respectively. The small circle dots are the observed effect sizes, whereas

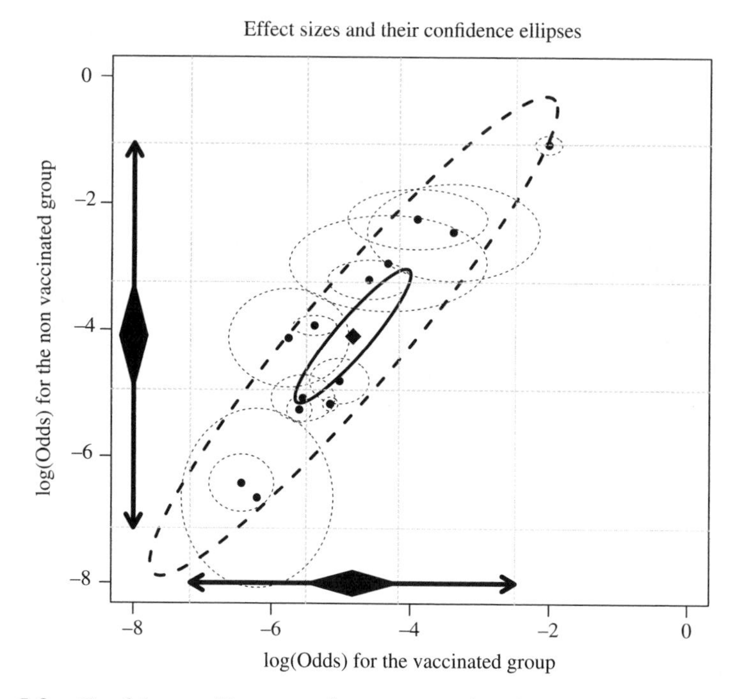

Figure 5.8 Confidence ellipses on the vaccinated and the nonvaccinated groups.

the dashed ellipses around them are the 95% confidence ellipses. A confidence ellipse is the bivariate generalization of the CI (see Friendly et al., 2013). If we were able to repeat Study i by collecting new data, 95% of such ellipses constructed in the replications will contain Study i's *true* bivariate effect sizes. The confidence ellipses around the studies are not tilted in the figure, showing that the effect sizes are conditionally independent.

The solid square in the location (−4.8338, −4.0960) represents the estimated average population effect sizes for the vaccinated and the nonvaccinated groups. The small ellipse in a solid line is the 95% confidence ellipse of the average effect sizes. It indicates the best estimates of the average population effect sizes for the vaccinated and the nonvaccinated groups in the long run. The large ellipse in a dashed line indicates the random effects for the 95% of studies that may fall inside this ellipse. It is constructed based on the estimated variance component of the random effects, which is a bivariate generalization of the 95% plausible value interval (Raudenbush, 2009). If we randomly select studies, 95% of the selected studies may fall inside the ellipse in long run. Therefore, the *true* population effect sizes of the studies vary greatly. Moreover, we also calculate the average effect size for the vaccinated group (−4.8338 in the x-axis) and the average effect size for the nonvaccinated group (−4.0960 in the y-axis) and their 95% CIs. They are shown by the diamonds near the x-axis and the y-axis. The arrows represent the 95% plausible value intervals.

We may combine the forest plots provided by the `metafor` package (Viechtbauer, 2010) with the confidence ellipses. We specify the `diag.panel=TRUE` argument in the `plot()` function to generate the diagonal panels for the forest plots. We use the `forest()` and the `rma()` functions in the `metafor` package to generate the forest plots.

```
R> ## Load the metafor package
R> library("metafor")
R> plot(bcg1, xlim=c(-8,0), ylim=c(-8,0), diag.panel=TRUE)
R> ## Forest plot for the vaccinated group
R> forest( rma(yi=ln_Odd_V, vi=v_ln_Odd_V, method="ML", data=BCG) )
R> title("Forest plot for the vaccinated group")
R> ## Forest plot for the non-vaccinated group
R> forest( rma(yi=ln_Odd_NV, vi=v_ln_Odd_NV, method="ML",
     data=BCG) )
R> title("Forest plot for the non-vaccinated group")
```

The forest plots visually display the strength of each effect size and its approximate 95% Wald CI and the average effect sizes. Figure 5.9 provides more

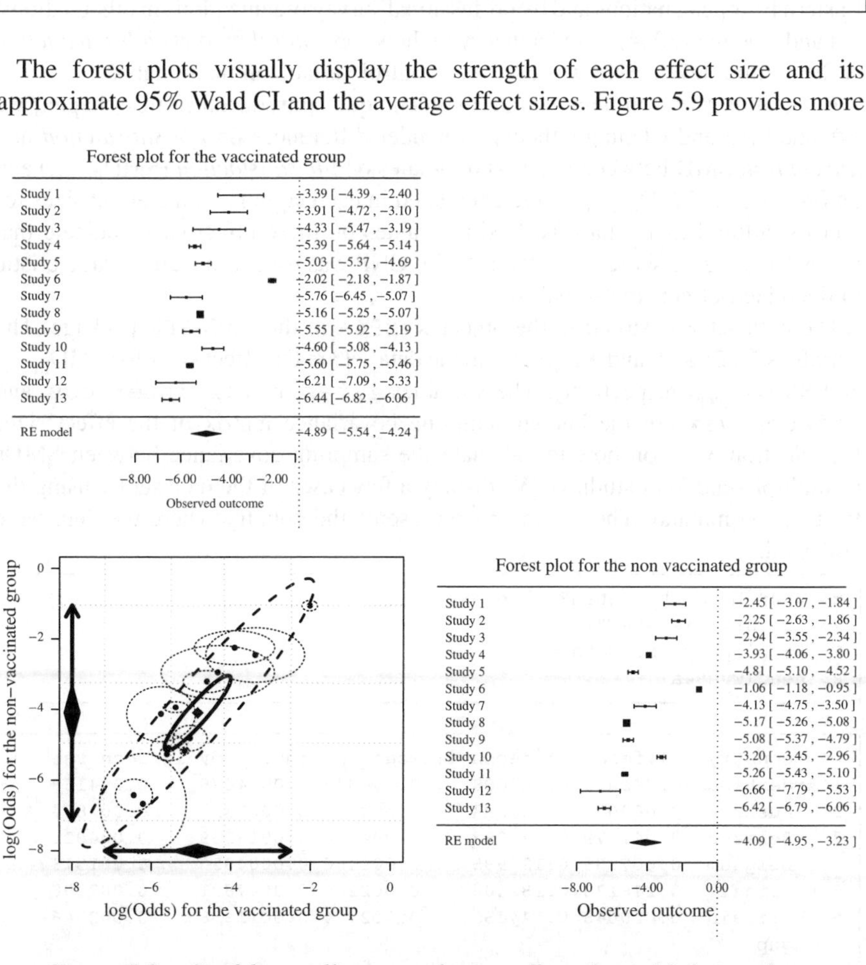

Figure 5.9 Confidence ellipses on the two effect sizes with forest plots.

information on the univariate and multivariate nature of the data. More specifically, researchers may get some insights on how the effect sizes are correlated when comparing the forest plots and the confidence ellipses. In this example, the high correlation (0.9467) between the random effects is mainly due to the base rate differences on the vaccinated and the nonvaccinated groups. It is not possible to get this information from two separate univariate meta-analyses.

5.7.2 Standardized mean differences between males and females on life satisfaction and life control

A data set from the World Values Survey II (World Values Study Group, 1994) was used to illustrate the procedures and analyses using the multivariate meta-analysis and the mediation and moderation analyses. Between 1990 and 1993, 57,561 adults aged 18 and above from 42 nations were interviewed by local academic institutes in Eastern European nations and by professional survey organizations in other nations. Au and Cheung (2004) tested a theory on how *job control* predicts *job satisfaction* at the cultural level. GNP was used as a control variable in their analyses.

As an illustration to demonstrate the techniques introduced in this chapter, we extended Au and Cheung's theory to gender differences on *life satisfaction* and *life control*. SMD between males and females on *life satisfaction* (SMD_{life_sat}) and on *life control* (SMD_{life_con}) were calculated in each country as the effect sizes for a cross-cultural meta-analysis. Positive values on these effect sizes indicate that males have higher scores than females do. GNP was used as a study characteristic in the mixed-effects meta-analysis.

The data set was stored in the object `wvs94a` in the `metaSEM` package. The variables `lifesat` and `lifecon` in the data set are the effect sizes for SMD_{life_sat} and SMD_{life_con}, respectively. The variables `lifesat_var`, `inter_cov`, and `lifecon_var` are the known sampling covariance matrix of the effect sizes (see Section 3.3.1 on how to calculate the sampling covariance between SMDs of multiple-endpoint studies). We display a few cases of the data set by using the `head()` command. The `country` represents the country where the data were collected.

```
R> #### Load the metaSEM library
R> library("metaSEM")
R> ## Display the dataset
R> head(wvs94a)
```

```
    country   lifesat  lifecon lifesat_var inter_cov lifecon_var
1 Argentina -0.032093 0.057608    0.004043 0.0014075    0.004158
2   Austria  0.080096 0.008893    0.002888 0.0009337    0.002894
3   Belarus  0.041979 0.074087    0.004010 0.0013359    0.004011
4   Belgium  0.007755 0.127995    0.001456 0.0004050    0.001513
5    Brazil  0.148138 0.182106    0.002266 0.0007891    0.002290
6   Britain  0.020048 0.044455    0.002724 0.0011858    0.002746
    gnp
```

```
1  2370
2  4900
3  3110
4 15540
5  2680
6 16100
```

5.7.2.1 Random-effects model

We employ a random-effects model with the following syntax.

```
R> ## Random-effects model
R> wvs1 <- meta(y=cbind(lifesat, lifecon),
             v=cbind(lifesat_var, inter_cov, lifecon_var),
             data=wvs94a, model.name="Random effects model")
R> summary(wvs1)
```

```
----------------------- Selected output ----------------------

95% confidence intervals: z statistic approximation
Coefficients:
            Estimate Std.Error    lbound    ubound z value Pr(>|z|)
Intercept1  0.001350  0.013856 -0.025808  0.028508    0.10  0.92239
Intercept2  0.068826  0.016820  0.035860  0.101792    4.09  4.3e-05
Tau2_1_1    0.004727  0.001762  0.001275  0.008180    2.68  0.00728
Tau2_2_1    0.003934  0.001687  0.000628  0.007241    2.33  0.01970
Tau2_2_2    0.008414  0.002537  0.003441  0.013387    3.32  0.00091

Intercept1
Intercept2 ***
Tau2_1_1   **
Tau2_2_1   *
Tau2_2_2   ***
---
Signif. codes:  0 '***' 0.001 '**' 0.01 '*' 0.05 '.' 0.1 ' ' 1

Q statistic on the homogeneity of effect sizes: 250
Degrees of freedom of the Q statistic: 82
P value of the Q statistic: 0

Heterogeneity indices (based on the estimated Tau2):
                            Estimate
Intercept1: I2 (Q statistic)    0.61
Intercept2: I2 (Q statistic)    0.73

Number of studies (or clusters): 42
Number of observed statistics: 84
Number of estimated parameters: 5
Degrees of freedom: 79
-2 log likelihood: -161.9
```

```
OpenMx status1: 0 ("0" or "1": The optimization is considered fine.
Other values indicate problems.)

------------------------ Selected output ------------------------
```

```
R> ## Extract the variance component of the random effects
R> ( T2 <- vec2symMat(coef(wvs1, select="random")) )
```

```
         [,1]     [,2]
[1,] 0.004727 0.003934
[2,] 0.003934 0.008414
```

```
R> ## Convert the covariance matrix to a correlation matrix
R> cov2cor(T2)
```

```
       [,1]   [,2]
[1,] 1.0000 0.6238
[2,] 0.6238 1.0000
```

First, we check whether there are any estimation problems in the analysis. The `OpenMx status1` is 0, indicating that everything is fine. The test of homogeneity of effect sizes is $Q(\text{df} = 82) = 250.0303, p < 0.001$, which is statistically significant. The I^2 for $\text{SMD}_{\text{life\_sat}}$ and $\text{SMD}_{\text{life\_con}}$ are 0.6129 and 0.7345, respectively. These indicate that the gender differences on *life satisfaction* and *life control* vary across different cultural groups. A random-effects model is more appropriately employed to describe the data. The estimated variance component is $\hat{T}^2 = \begin{bmatrix} 0.0047 & \\ 0.0039 & 0.0084 \end{bmatrix}$. As it is difficult to interpret the covariance, we convert it into a correlation matrix. The correlation between the random effects is 0.6238, which is moderate. This indicates that countries with higher $\text{SMD}_{\text{life\_sat}}$ tend to have higher $\text{SMD}_{\text{life\_con}}$.

The estimated average population effect sizes for $\text{SMD}_{\text{life\_sat}}$ and $\text{SMD}_{\text{life\_con}}$ (and their approximate 95% Wald CIs) are 0.0013 $(-0.0258, 0.0285)$ and 0.0688 $(0.0359, 0.1018)$, respectively. The results suggest that there is a gender difference on *life control* but not on *life satisfaction*. On an average, males and females have a similar perception of life satisfaction, whereas males perceive themselves as having more control over their life.

We plot the average effect sizes with their confidence ellipses and the individual effect sizes with the `plot()` function. As there are too many studies involved, we skip the confidence ellipses on the individual effect sizes by specifying the `study.ellipse.plot=FALSE` argument.

```
R> plot(wvs1, axis.labels=c("SMD on life satisfaction",
```

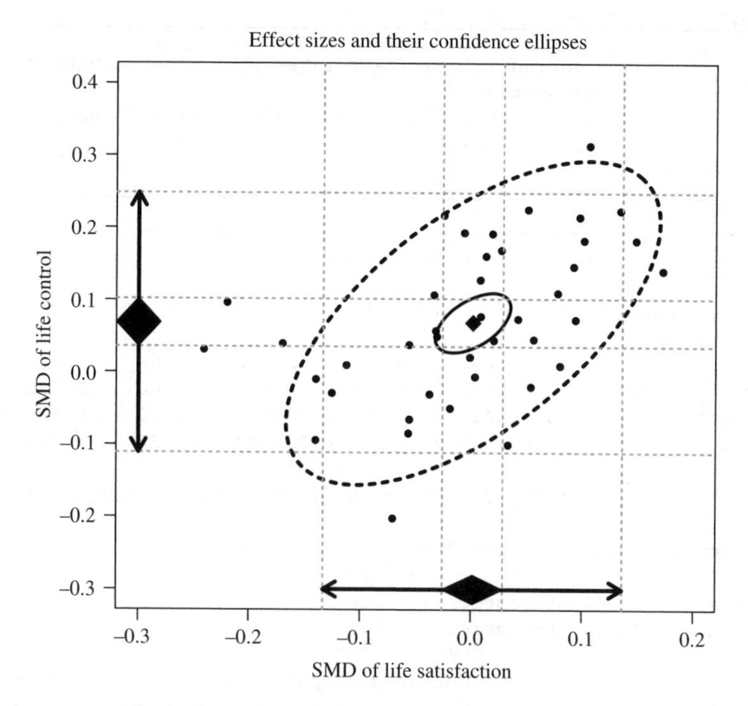

Figure 5.10 Confidence ellipses on the SMDs of life satisfaction and life control.

```
                         "SMD on life control"),
       study.ellipse.plot=FALSE,
       xlim=c(-0.3, 0.2), ylim=c(-0.3,0.4))
```

Figure 5.10 displays the plot. The figure shows that the average $\mathrm{SMD}_{\mathrm{life\_sat}}$ (the x-axis) is centered around zero, whereas the average $SMD_{\mathrm{life\_con}}$ (the y-axis) is above zero. The confidence ellipse for the random effects (the dashed ellipse) is quite large, showing that there is a high degree of heterogeneity. Most of the studies are located within this confidence ellipse.

When there are multiple dependent variables, it is preferable to simultaneously test both effect sizes. The multivariate test controls the overall Type I error better in testing the means. If the overall test is significant, we may test the individual effect sizes. We fit a model by fixing 0 at both average effect sizes with the `intercept.constraints` argument. As there are two effect sizes per study, the `intercept.constraints` argument expects a 1×2 matrix as input. This model is nested within the model without the constraints; we compare the models with an LR statistic.

```
R> ## Random-effects model with both effect sizes fixed at 0
R> wvs2 <- meta(y=cbind(lifesat, lifecon),
```

```
                v=cbind(lifesat_var, inter_cov, lifecon_var),
                data=wvs94a,
                intercept.constraints=matrix(0, nrow=1, ncol=2),
                model.name="Both effect sizes fixed at 0")
R> summary(wvs2)
```

```
------------------------ Selected output -----------------------

95% confidence intervals: z statistic approximation
Coefficients:
         Estimate Std.Error   lbound   ubound z value Pr(>|z|)
Tau2_1_1 0.004632  0.001744 0.001214 0.008050    2.66  0.00790 **
Tau2_2_1 0.004131  0.001922 0.000363 0.007898    2.15  0.03165 *
Tau2_2_2 0.013083  0.003554 0.006116 0.020049    3.68  0.00023 ***
---
Signif. codes:  0 '***' 0.001 '**' 0.01 '*' 0.05 '.' 0.1 ' ' 1

Heterogeneity indices (based on the estimated Tau2):
                             Estimate
Intercept1: I2 (Q statistic)     0.61
Intercept2: I2 (Q statistic)     0.81

------------------------ Selected output -----------------------
```

```
R> ## Compare the nested models
R> anova(wvs1, wvs2)
```

```
                   base                    comparison ep minus2LL df
1 Random effects model                           <NA>  5   -161.9 79
2 Random effects model Both effect sizes fixed at 0  3   -143.5 81
     AIC diffLL diffdf        p
1 -319.9     NA     NA       NA
2 -305.5  18.45      2 9.86e-05
```

The LR statistic is $\Delta\chi^2(\text{df} = 2) = 18.4488, p < 0.001$. Therefore, the null hypothesis that both effect sizes are zero is rejected.

5.7.2.2 Mixed-effects model

We test the mixed-effects model by using GNP as a moderator. To improve the numerical stability of the results, GNP was centered and divided by 10,000 in the analyses.

```
R> ## Mixed-effects model
R> ## gnp is divided by 10000 and centered by using
R> ## scale(gnp/10000, scale=FALSE)
```

```
R> wvs3 <- meta(y=cbind(lifesat, lifecon),
                v=cbind(lifesat_var, inter_cov, lifecon_var),
                x=scale(gnp/10000, scale=FALSE), data=wvs94a,
                model.name="GNP as a predictor")
R> summary(wvs3)
```

```
------------------------ Selected output ------------------------

95% confidence intervals: z statistic approximation
Coefficients:
             Estimate Std.Error    lbound    ubound z value Pr(>|z|)
Intercept1   0.001300  0.014575 -0.027266  0.029866    0.09   0.9289
Intercept2   0.070591  0.017136  0.037004  0.104178    4.12  3.8e-05
Slope1_1    -0.024051  0.015305 -0.054049  0.005947   -1.57   0.1161
Slope2_1    -0.037204  0.017948 -0.072382 -0.002027   -2.07   0.0382
Tau2_1_1     0.004600  0.001798  0.001076  0.008123    2.56   0.0105
Tau2_2_1     0.003592  0.001681  0.000297  0.006886    2.14   0.0326
Tau2_2_2     0.007479  0.002471  0.002635  0.012323    3.03   0.0025
---
Signif. codes:  0 '***' 0.001 '**' 0.01 '*' 0.05 '.' 0.1 ' ' 1

Explained variances (R2):
                             y1   y2
Tau2 (no predictor)     0.00473 0.01
Tau2 (with predictors)  0.00460 0.01
R2                      0.02697 0.11

------------------------ Selected output ------------------------
```

From the above output, we may write down the equations on predicting the average *true* effect sizes:

$$\hat{f}_{\text{life\_sat}} = 0.0013 - 0.0241\text{GNP} \quad \text{and}$$
$$\hat{f}_{\text{life\_con}} = 0.0706 - 0.0372\text{GNP}.$$

GNP was significant in predicting $f_{\text{life\_con}}$ but not in $f_{\text{life\_sat}}$. The estimated residual variance component is $\hat{T}^2 = \begin{bmatrix} 0.0046 & \\ 0.0036 & 0.0075 \end{bmatrix}$ after controlling for GNP. The R^2 on $f_{\text{life\_sat}}$ and $f_{\text{life\_con}}$ are 0.0270 and 0.1111, respectively. The gender difference on *life control* is larger for countries with lower GNP. In other words, females tend to perceive lower *life control* in countries with lower GNP.

5.7.3 Mediation and moderation models

Following the illustrations by Cheung (2009b), we formulate a mediation model and a moderation model using the $SMD_{\text{life\_sat}}$ as the dependent variable. As functions to fit mediation and moderation models among the *true* effect sizes have not been implemented in the metaSEM package, we use the OpenMx package (Boker et al., 2011) to fit these models. In the following sections, we first illustrate the steps involved in fitting the regression model between two effect sizes. We then extend the steps to fit the mediation and moderation models.

5.7.3.1 Regression between two effect sizes

We fit a regression model by regressing the *true* effect size $f_{\text{life\_sat}}$ on the *true* effect size $f_{\text{life\_con}}$. This is equivalent to the model in Figure 5.5 by using $\text{SMD}_{\text{life\_sat}}$ and $\text{SMD}_{\text{life\_con}}$ as $y_{1,i}$ and $y_{2,i}$, respectively. We delete the missing values in GNP before the analysis.

We specify the structural equation models in OpenMx via the reticular action model (RAM) specification (McArdle and McDonald, 1984; see also Section 2.2.3). Four matrices are required to specify the model. All of the latent and observed variables are combined together: f_lifesat, f_lifecon, lifesat, and lifecon. The $\boldsymbol{A}$ matrix is used to specify the regression coefficients or factor loadings among the variables. The $\boldsymbol{S}$ matrix is used to specify the variance–covariances of the variables or the residuals. The $\boldsymbol{M}$ matrix is used to represent the means or the intercepts of the variables. The $\boldsymbol{F}$ matrix is used to select the observed variables (see Section 2.2.3 for details). The followings are the complete syntax for the analysis.

```
R> ## Remove the missing values in gnp
R> ##  and exclude the first column "country"
R> my.df <- wvs94a[!is.na(wvs94a$gnp), -1]
R> ## Center gnp and divide it by 10000 to
R> ## improve numerical stability
R> my.df$gnp <- scale(my.df$gnp, scale=FALSE)/10000
R> head(my.df)
```

```
    lifesat  lifecon lifesat_var inter_cov lifecon_var     gnp
1 -0.032093 0.057608    0.004043 0.0014075    0.004158 -0.8528
2  0.080096 0.008893    0.002888 0.0009337    0.002894 -0.5998
3  0.041979 0.074087    0.004010 0.0013359    0.004011 -0.7788
4  0.007755 0.127995    0.001456 0.0004050    0.001513  0.4642
5  0.148138 0.182106    0.002266 0.0007891    0.002290 -0.8218
6  0.020048 0.044455    0.002724 0.0011858    0.002746  0.5202
```

The observed effect sizes are lifesat and lifecon. We create two latent variables, f_lifesat and f_lifecon, to represent the *true* effect sizes. We may use the matrix() function to create a character matrix and convert this matrix into the mxMatrix class using the as.mxMatrix() function. If the elements are numeric, they are treated as fixed parameters fixed with the specific values as the numeric inputs. If the elements are characters, they are treated as free parameters. For example, "0.1*beta_1_2" in the $\boldsymbol{A}$ matrix means that the regression coefficient from $f_{\text{life\_con}}$ to $f_{\text{life\_sat}}$ is a free parameter with the label beta_1_2 and 0.1

as the starting value. If we want to impose an equality constraint on some parameters, we use the same label for both parameters. The factor loadings from the latent variables to their observed variables are fixed at 1.

```
R> ## A: asymmetric paths for regression coefficients
R> ##    and factor loadings
R> A <- matrix(c(0, "0.1*beta1_2", 0, 0,
                 0, 0, 0, 0,
                 1, 0, 0, 0,
                 0, 1, 0, 0),
               ncol=4, nrow=4, byrow=TRUE)
R> dimnames(A) <- list(c("f_lifesat","f_lifecon",
                         "lifesat","lifecon"),
                       c("f_lifesat","f_lifecon","lifesat",
                         "lifecon"))
R> ## Show the content of A
R> A
```

```
          f_lifesat f_lifecon     lifesat lifecon
f_lifesat "0"       "0.1*beta1_2" "0"     "0"
f_lifecon "0"       "0"           "0"     "0"
lifesat   "1"       "0"           "0"     "0"
lifecon   "0"       "1"           "0"     "0"
```

```
R> ## Convert it into OpenMx matrix
R> A <- as.mxMatrix(A)
```

We may also directly create the matrices using the `mxMatrix()` function. As the sampling covariance matrix on `lifesat` and `lifecon` are known, we need to fix these values by using the definition variables. If the label begins with "`data.`," this parameter will be fixed with the observed variable from the data set. For example, the label `data.lifesat_var` means that the error variance on `lifesat` will be fixed using the values of `lifesat_var` in the data set. As the values of `lifesat_var` may vary in the data set, the error variance of `lifesat` may also vary across subjects (or studies).

```
R> ## S: symmetric covariances and variances
R> S <- mxMatrix(type="Symm", nrow=4, ncol=4, byrow=TRUE,
                 free=c(TRUE,
                        FALSE,TRUE,
                        FALSE,FALSE,FALSE,
                        FALSE,FALSE,FALSE,FALSE),
                 values=c(0.1,
                          0,0.1,
                          0,0,0,
                          0,0,0,0),
                 labels=c("tau2_1_1",
                          NA,"tau2_2_2",
```

```
                              NA,NA,"data.lifesat_var",
                              NA,NA,"data.inter_cov","data.lifecon_var"),
                      name = "S")
R> ## Show the labels of S
R> S@labels
```

```
     [,1]       [,2]       [,3]               [,4]
[1,] "tau2_1_1" NA         NA                 NA
[2,] NA         "tau2_2_2" NA                 NA
[3,] NA         NA         "data.lifesat_var" "data.inter_cov"
[4,] NA         NA         "data.inter_cov"   "data.lifecon_var"
```

We need to create a selection matrix ***F*** to select the observed variables `lifesat` and `lifecon`.

```
R> ## F: select observed variables
R> F <- matrix(c(0, 0, 1, 0,
                 0, 0, 0, 1), nrow = 2, ncol = 4, byrow = TRUE)
R> dimnames(F) <- list(c("lifesat","lifecon"),
                       c("f_lifesat","f_lifecon","lifesat",
                         "lifecon"))
R> ## Show the content of F
R> F
```

```
        f_lifesat f_lifecon lifesat lifecon
lifesat         0         0       1       0
lifecon         0         0       0       1
```

```
R> F <- as.mxMatrix(F)
```

We also need to create a matrix to represent the means (and the intercepts) of the variables. As shown in Figure 5.4, only the mean (and the intercept) of the latent variables are estimated, while the mean (and the intercept) of the observed variables are fixed at 0.

```
R> ## M: intercepts or means
R> M <- matrix(c("0*beta1_0","0*beta2_0",0,0), nrow=1, ncol=4)
R> dimnames(M)[[2]] <- c("f_lifesat","f_lifecon",
                         "lifesat","lifecon")
R> M
```

```
     f_lifesat   f_lifecon   lifesat lifecon
[1,] "0*beta1_0" "0*beta2_0" "0"     "0"
```

```
R> M <- as.mxMatrix(M)
```

When fitting a regression analysis, we may want to compute the R^2 (see Equation 5.30), the percentage of variance that can be explained by the predictor. We create an object called R2 for the calculations using the mxAlgebra() function.

```
R> ## Formula for R2
R> R2 <- mxAlgebra(beta1_2^2*tau2_2_2/(beta1_2^2*tau2_2_2 + tau2_1_1),
                   name="R2")
```

Now, we build a model called reg by putting all of the matrices together. The function mxExpectationRAM() converts the RAM formulation into the model-implied matrices with Equation 2.6, while the function mxFitFunctionML() fits the data against the model-implied matrices. We conduct the analysis by calling the mxRun() function. As we are going to calculate the R^2, we also request the LBCI on the R^2 by calling the mxCI() function in mxModel(). We also need to specify the intervals=TRUE argument in the mxRun() function. When we check the OpenMx status[[1]], it is "6," indicating that the optimizer could find no way to improve the estimate. We rerun the analysis by calling up the mxRun() function again to see if we can get rid of this error; otherwise, we cannot trust the results. In this example, the OpenMx status[[1]] is fine after rerunning the model.

```
R> ## Build the model
R> reg <- mxModel("Regression",
                  mxData(observed=my.df, type="raw"),
                  A, S, F, M, R2, mxCI("R2"),
                  mxExpectationRAM(A="A", S="S",
                                   F="F", M="M",
                  dimnames = c("f_lifesat","f_lifecon",
                               "lifesat","lifecon")),
                  mxFitFunctionML())
```

```
R> ## Run the analysis
R> reg.fit <- mxRun(reg, intervals=TRUE, silent=TRUE)
R> ## Check the status of the results
R> reg.fit@output$status[[1]]
```

```
[1] 6
```

```
R> ## Rerun the analysis
R> reg.fit <- mxRun(reg.fit, intervals=TRUE, silent=TRUE)
R> ## Check the status of the results
R> reg.fit@output$status[[1]]
```

```
[1] 0
```

```
R> summary(reg.fit)
```

```
------------------------ Selected output ----------------------

free parameters:
      name matrix row       col  Estimate Std.Error lbound ubound
1  beta1_2      A   1         2  0.498821  0.152612
2 tau2_1_1      S   1         1  0.002902  0.001341
3 tau2_2_2      S   2         2  0.008731  0.002771
4  beta1_0      M   1 f_lifesat -0.033950  0.016493
5  beta2_0      M   1 f_lifecon  0.070645  0.018110

confidence intervals:
                   lbound estimate ubound
Regression.R2[1,1] 0.0657   0.4281 0.7727

observed statistics:  74
estimated parameters:  5
degrees of freedom:  69
-2 log likelihood:  -142.5
number of observations:  37

------------------------ Selected output ----------------------
```

The estimated regression coefficient (`beta1_2` or `A[1,2]` in the output) from the *true* effect size $f_{\text{life\_con}}$ to the *true* effect size $f_{\text{life\_sat}}$ is 0.4988, SE = 0.1526, $z = 0.4988/0.1526 = 3.2685$, and $p = 0.0011$. The equation on predicting the average *true* effect size $f_{\text{life\_sat}}$ is

$$\hat{f}_{\text{life\_sat}} = -0.0340 + 0.4988 f_{\text{life\_con}}.$$

The estimated R^2 and its 95% LBCI is 0.4281 (0.0657, 0.7727). This indicates that $f_{\text{life\_con}}$ is quite strong in predicting $f_{\text{life\_sat}}$.

5.7.3.2 Mediation model

We fit a mediation model similar to the one in Figure 5.6 by considering GNP as x_i and the *true* effect size $f_{\text{life\_con}}$ as the mediator $f_{2,i}$. As shown in the previous sections, we may either treat x_i as a design matrix or a variable. We consider GNP as a variable in this illustration. The variables are arranged as `gnp`, `f_lifesat`, `f_lifecon`, `lifesat`, and `lifecon`.

```
R> A <- matrix(c(0,0,0,0,0,
                 "0*gamma1",0,"0*beta1_2",0,0,
                 "0*gamma2",0,0,0,0,
                 0,1,0,0,0,
                 0,0,1,0,0), ncol=5, nrow=5, byrow=TRUE)
```

```
R> dimnames(A) <- list(c("gnp","f_lifesat","f_lifecon",
                         "lifesat","lifecon"),
                       c("gnp","f_lifesat","f_lifecon",
                         "lifesat","lifecon"))
R> A
```

```
          gnp         f_lifesat f_lifecon   lifesat lifecon
gnp       "0"         "0"       "0"         "0"     "0"
f_lifesat "0*gamma1"  "0"       "0*beta1_2" "0"     "0"
f_lifecon "0*gamma2"  "0"       "0"         "0"     "0"
lifesat   "0"         "1"       "0"         "0"     "0"
lifecon   "0"         "0"       "1"         "0"     "0"
```

```
R> A <- as.mxMatrix(A)
```

```
R> S <- mxMatrix(type="Symm", nrow=5, ncol=5, byrow=TRUE,
              free=c(TRUE,
                     FALSE,TRUE,
                     FALSE,FALSE,TRUE,
                     FALSE,FALSE,FALSE,FALSE,
                     FALSE,FALSE,FALSE,FALSE,FALSE),
              values=c(1,
                       0,0.01,
                       0,0,0.1,
                       0,0,0,0,
                       0,0,0,0,0),
              labels=c("sigma2_x",
                       NA,"tau2_1_1",
                       NA,NA,"tau2_2_2",
                       NA,NA,NA,"data.lifesat_var",
                       NA,NA,NA,"data.inter_cov","data.lifecon_var"),
              name="S")
R> S@labels
```

```
     [,1]       [,2]       [,3]       [,4]
[1,] "sigma2_x" NA         NA         NA
[2,] NA         "tau2_1_1" NA         NA
[3,] NA         NA         "tau2_2_2" NA
[4,] NA         NA         NA         "data.lifesat_var"
[5,] NA         NA         NA         "data.inter_cov"
     [,5]
[1,] NA
[2,] NA
[3,] NA
[4,] "data.inter_cov"
[5,] "data.lifecon_var"
```

```
R> F <- matrix(c(1,0,0,0,0,
```

```
                  0,0,0,1,0,
                  0,0,0,0,1), nrow=3, ncol=5, byrow=TRUE)
R> dimnames(F) <- list(c("gnp","lifesat","lifecon"),
                       c("gnp","f_lifesat","f_lifecon",
                         "lifesat","lifecon"))
R> F
```

```
        gnp f_lifesat f_lifecon lifesat lifecon
gnp       1         0         0       0       0
lifesat   0         0         0       1       0
lifecon   0         0         0       0       1
```

```
R> F <- as.mxMatrix(F)
```

```
R> M <- matrix(c("0*mu_x","0*beta1_0","0*beta2_0",0,0),
               nrow=1, ncol=5)
R> dimnames(M)[[2]] <- c("gnp", "f_lifesat","f_lifecon",
                         "lifesat","lifecon")
R> M
```

```
     gnp      f_lifesat   f_lifecon   lifesat lifecon
[1,] "0*mu_x" "0*beta1_0" "0*beta2_0" "0"     "0"
```

```
R> M <- as.mxMatrix(M)
```

We define the indirect, direct, and total effects using the `mxAlgebra()` function. We also obtain the 95% LBCI on these estimates by specifying the `intervals=TRUE` argument in calling up the `mxRun()` function.

```
R> ## Define the direct effect
R> direct <- mxAlgebra(gamma1, name="direct")
R> ## Define the indirect effect
R> indirect <- mxAlgebra(gamma2*beta1_2, name="indirect")
R> ## Define the total effect
R> total <- mxAlgebra(gamma1+gamma2*beta1_2, name="total")
R> med <- mxModel("Mediation",
                  mxData(observed=my.df, type="raw"),
                  A, S, F, M, direct, indirect, total,
                  mxCI(c("direct","indirect","total")),
                  mxExpectationRAM(A="A", S="S", F="F", M="M",
                  dimnames=c("gnp","f_lifesat","f_lifecon",
                             "lifesat","lifecon")),
                  mxFitFunctionML())
```

```
R> med.fit <- mxRun(med, intervals=TRUE)
```

```
R> med.fit@output$status[[1]]
```

```
[1] 0
```

```
R> summary(med.fit)
```

```
------------------------ Selected output -----------------------

free parameters:
      name matrix row       col   Estimate Std.Error lbound ubound
1   gamma1      A   2         1 -6.185e-03  0.014694
2   gamma2      A   3         1 -3.720e-02  0.017948
3  beta1_2      A   2         3  4.802e-01  0.167966
4 sigma2_x      S   1         1  9.195e-01  0.213771
5 tau2_1_1      S   2         2  2.875e-03  0.001324
6 tau2_2_2      S   3         3  7.479e-03  0.002471
7     mu_x      M   1       gnp -8.795e-10  0.157640
8  beta1_0      M   1 f_lifesat -3.260e-02  0.017158
9  beta2_0      M   1 f_lifecon  7.059e-02  0.017136

confidence intervals:
                             lbound  estimate     ubound
Mediation.direct[1,1]      -0.03609 -0.006185  0.0232042
Mediation.indirect[1,1] -0.04482 -0.017867 -0.0004997
Mediation.total[1,1]       -0.05506 -0.024051  0.0065157

observed statistics:  111
estimated parameters:  9
degrees of freedom:  102
-2 log likelihood:  -45.01
number of observations:  37

------------------------ Selected output -----------------------
```

The structural equation for the predicted average effect sizes with the parameter estimates as shown in Equation 5.32 is

$$\begin{bmatrix} \hat{f}_{\text{life\_sat}} \\ \hat{f}_{\text{life\_con}} \end{bmatrix} = \begin{bmatrix} -0.0326 \\ 0.0706 \end{bmatrix} + \begin{bmatrix} 0 & 0.4802 \\ 0 & 0 \end{bmatrix} \begin{bmatrix} f_{\text{life\_sat}} \\ f_{\text{life\_con}} \end{bmatrix} + \begin{bmatrix} -0.0062 \\ -0.0372 \end{bmatrix} \text{GNP}.$$

The estimated indirect, direct, and total effects (with their 95% LBCI) are -0.0179 $(-0.0448, -0.0005)$, -0.0062 $(-0.0361, 0.0232)$, and -0.0241 $(-0.0551, 0.0065)$, respectively. Although the estimated indirect effect is statistically significant, the effect size is very small. Both the direct effect and the total effect are not significant.

5.7.3.3 Moderation model

In this section, we illustrate how to analyze the moderating effect between two *true* effect sizes by treating the GNP as the moderator. We treat the GNP as a variable in the analysis (see Figure 5.7b). The variables are arranged as `gnp`, `f_lifesat`, `f_lifecon`, `lifesat`, `lifecon`, and `P`. As shown in the figure and Equation 5.33, the regression coefficient from $f_{\text{life\_con}}$ to $f_{\text{life\_sat}}$ depends on the value of x_i. We create a phantom variable P to account for this moderation. The regression coefficient from $f_{\text{life\_con}}$ to P is fixed by the value x_i via a definition variable. GNP is used twice—one as the predictor and the other in the definition variable. We may specify the $\boldsymbol{A}$ matrix using the `mxMatrix()` function. Alternatively, we may prepare the $\boldsymbol{A}$ matrix as usual. Then we may change the elements related to the definition variable in an ad hoc way, for example,

```
R> A <- matrix(c(0,0,0,0,0,0,
                 "0*gamma1",0,"0*beta1_2",0,0,"0*omega1_2",
                 0,0,0,0,0,0,
                 0,1,0,0,0,0,
                 0,0,1,0,0,0,
                 0,0,"0*data_gnp",0,0,0),
               ncol=6,nrow=6,byrow=TRUE)
R> dimnames(A) <- list(c("gnp","f_lifesat","f_lifecon",
                         "lifesat","lifecon","P"),
                       c("gnp","f_lifesat","f_lifecon",
                         "lifesat","lifecon","P"))
R> A
```

```
          gnp        f_lifesat f_lifecon    lifesat lifecon
gnp       "0"        "0"       "0"          "0"     "0"
f_lifesat "0*gamma1" "0"       "0*beta1_2"  "0"     "0"
f_lifecon "0"        "0"       "0"          "0"     "0"
lifesat   "0"        "1"       "0"          "0"     "0"
lifecon   "0"        "0"       "1"          "0"     "0"
P         "0"        "0"       "0*data_gnp" "0"     "0"
          P
gnp       "0"
f_lifesat "0*omega1_2"
f_lifecon "0"
lifesat   "0"
lifecon   "0"
P         "0"
```

```
R> A <- as.mxMatrix(A)
R> ## Change the elements related to the definition variable
R> ## A[6,3] is fixed by the variable gnp in the data
R> A@labels[6,3] <- "data.gnp"
R> ## A[6,3] is a fixed parameter
R> A@free[6,3] <- FALSE
R> A
```

```
FullMatrix 'A'

$labels
     [,1]     [,2] [,3]       [,4] [,5] [,6]
[1,] NA       NA   NA         NA   NA   NA
[2,] "gamma1" NA   "beta1_2"  NA   NA   "omega1_2"
[3,] NA       NA   NA         NA   NA   NA
[4,] NA       NA   NA         NA   NA   NA
[5,] NA       NA   NA         NA   NA   NA
[6,] NA       NA   "data.gnp" NA   NA   NA

$values
     [,1] [,2] [,3] [,4] [,5] [,6]
[1,]    0    0    0    0    0    0
[2,]    0    0    0    0    0    0
[3,]    0    0    0    0    0    0
[4,]    0    1    0    0    0    0
[5,]    0    0    1    0    0    0
[6,]    0    0    0    0    0    0

$free
      [,1]  [,2]  [,3]  [,4]  [,5]  [,6]
[1,] FALSE FALSE FALSE FALSE FALSE FALSE
[2,]  TRUE FALSE  TRUE FALSE FALSE  TRUE
[3,] FALSE FALSE FALSE FALSE FALSE FALSE
[4,] FALSE FALSE FALSE FALSE FALSE FALSE
[5,] FALSE FALSE FALSE FALSE FALSE FALSE
[6,] FALSE FALSE FALSE FALSE FALSE FALSE

$lbound: No lower bounds assigned.

$ubound: No upper bounds assigned.
```

As the phantom variable P is only introduced to impose the moderating effect, its variance is fixed at zero.

```
R> S <- mxMatrix(type="Symm", nrow=6, ncol=6, byrow=TRUE,
               free=c(TRUE,
                      FALSE,TRUE,
                      TRUE,FALSE,TRUE,
                      FALSE,FALSE,FALSE,FALSE,
                      FALSE,FALSE,FALSE,FALSE,FALSE,
                      FALSE,FALSE,FALSE,FALSE,FALSE,FALSE),
               values=c(1,
                        0,0.01,
                        0,0,0.1,
                        0,0,0,0,
                        0,0,0,0,0,
                        0,0,0,0,0,0),
               labels=c("sigma2_x",
                        NA,"tau2_1_1",
```

```
                                "cov_x_lifecon",NA,"tau2_2_2",
                                NA,NA,NA,"data.lifesat_var",
                                NA,NA,NA,"data.inter_cov","data.lifecon_var",
                                NA,NA,NA,NA,NA,NA),
                      name="S")
R> S@labels
```

```
     [,1]            [,2]       [,3]            [,4]
[1,] "sigma2_x"      NA         "cov_x_lifecon" NA
[2,] NA              "tau2_1_1" NA              NA
[3,] "cov_x_lifecon" NA         "tau2_2_2"      NA
[4,] NA              NA         NA              "data.lifesat_var"
[5,] NA              NA         NA              "data.inter_cov"
[6,] NA              NA         NA              NA
     [,5]               [,6]
[1,] NA                 NA
[2,] NA                 NA
[3,] NA                 NA
[4,] "data.inter_cov"   NA
[5,] "data.lifecon_var" NA
[6,] NA                 NA
```

```
R> F <- matrix(c(1,0,0,0,0,0,
                 0,0,0,1,0,0,
                 0,0,0,0,1,0), nrow=3, ncol=6, byrow=TRUE)
R> dimnames(F) <- list(c("gnp","lifesat","lifecon"),
                       c("gnp","f_lifesat","f_lifecon","lifesat",
                         "lifecon","P"))
R> F
```

```
        gnp f_lifesat f_lifecon lifesat lifecon P
gnp       1         0         0       0       0 0
lifesat   0         0         0       1       0 0
lifecon   0         0         0       0       1 0
```

```
R> F <- as.mxMatrix(F)
```

```
R> M <- matrix(c("0*mu_x","0*beta1_0","0*beta2_0",0,0,0),
               nrow=1, ncol=6)
R> dimnames(M)[[2]] <- c("gnp","f_lifesat","f_lifecon",
                         "lifesat","lifecon","P")
R> M
```

```
     gnp      f_lifesat   f_lifecon   lifesat lifecon P
[1,] "0*mu_x" "0*beta1_0" "0*beta2_0" "0"     "0"     "0"
```

```
R> M <- as.mxMatrix(M)
```

We build the model by putting all matrices together. We also request the 95% LBCI on the parameter omega1_2 that indicates the moderating effect.

```
R> mod <- mxModel("Moderator",
                  mxData(observed=my.df, type="raw"),
                  A, S, F, M,
                  mxCI("omega1_2"),
                  mxExpectationRAM(A="A", S="S", F="F", M="M",
                      dimnames=c("gnp","f_lifesat","f_lifecon",
                                 "lifesat","lifecon","P")),
                  mxFitFunctionML())
```

```
R> mod.fit <- mxRun(mod, intervals=TRUE, silent=TRUE)
R> ## Check the optimization status
R> mod.fit@output$status[[1]]
```

```
[1] 0
```

```
R> summary(mod.fit)
```

```
----------------------  Selected output  ----------------------

free parameters:
          name matrix row       col  Estimate Std.Error lbound
1       gamma1      A   2         1 -1.021e-02  0.017521
2      beta1_2      A   2         3  4.707e-01  0.169707
3     omega1_2      A   2         6  7.284e-02  0.171198
4     sigma2_x      S   1         1  9.195e-01  0.213771
5     tau2_1_1      S   2         2  2.803e-03  0.001322
6  cov_x_lifecon    S   1         3 -3.463e-02  0.018333
7     tau2_2_2      S   3         3  8.720e-03  0.002764
8         mu_x      M   1       gnp -2.082e-10  0.157640
9      beta1_0      M   1 f_lifesat -2.930e-02  0.018795
10     beta2_0      M   1 f_lifecon  7.044e-02  0.018089

confidence intervals:
         lbound estimate ubound
omega1_2 -0.2734  0.07284 0.4315

observed statistics:  111
estimated parameters:  10
degrees of freedom:  101
```

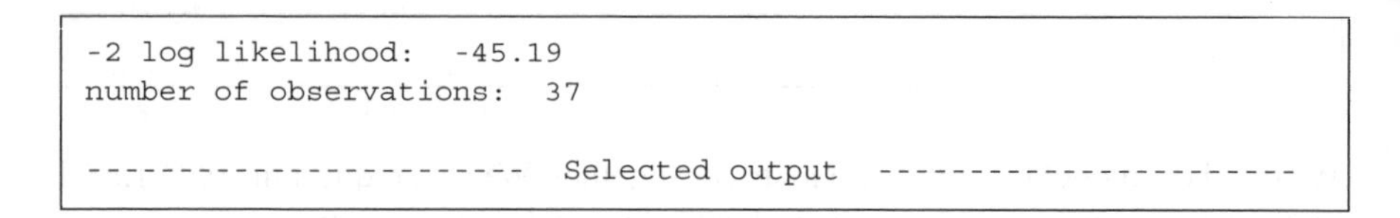

```
-2 log likelihood:   -45.19
number of observations:  37

------------------------  Selected output  ----------------------
```

The optimization status code is “0,” indicating that the optimization is fine. From the above outputs, the equation on predicting the average $f_{\text{life\_sat}}$ is

$$\hat{f}_{\text{life\_sat}} = -0.0293 - 0.0102\text{GNP} + (0.4707 + 0.0728\text{GNP})f_{\text{life\_con}}.$$

The estimated moderating effect (with its 95% LBCI) is 0.0728 (−0.2734, 0.4315), which is not significant at $\alpha = 0.05$. Thus, the analysis does not support the claim that GNP is a moderator of the relationship from $f_{\text{life\_con}}$ to $f_{\text{life\_con}}$.

5.8 Concluding remarks and further readings

This chapter introduced why multivariate meta-analysis was preferred to handle multiple effect sizes. We briefly discussed the benefits of multivariate meta-analysis to several univariate meta-analyses from a missing data perspective. Readers may refer to Jackson et al. (2011) and the commentaries for more advantages and issues of the multivariate meta-analysis. This chapter also introduced the basic models of multivariate meta-analysis and how the multivariate meta-analysis could be formulated under the SEM-based meta-analysis (Cheung, 2013b). One common issue when applying multivariate meta-analysis is missing information on the correlation of the multiple effect sizes. This chapter briefly discussed some of the common methods to handle the missing correlation. As there is still a lack of clear consensus on what the best method is to handle the missing correlation, future studies should explore this issue in more details. We also provided statistical models to test mediation and moderator models among the effect sizes. Readers may refer to the work by Shadish (1992, 1996) and Shadish and Sweeney (1991) for the conceptual theories behind this approach.

References

Aguinis H 2004. *Regression analysis for categorical moderators*. Guilford Press, New York.

Aiken LS, West SG and Reno RR 1991. *Multiple regression: testing and interpreting interactions*. Sage Publications, Inc., Newbury Park, CA.

Arends LR, Hoes AW, Lubsen J, Grobbee DE and Stijnen T 2000. Baseline risk as predictor of treatment benefit: three clinical meta-re-analyses. *Statistics in Medicine* **19**(24), 3497–3518.

Arends LR, Vokó Z and Stijnen T 2003. Combining multiple outcome measures in a meta-analysis: an application. *Statistics in Medicine* **22**(8), 1335–1353.

Au K and Cheung MWL 2004. Intra-cultural variation and job autonomy in 42 countries. *Organization Studies* **25**(8), 1339–1362.

Baron RM and Kenny DA 1986. The moderator-mediator variable distinction in social psychological research: conceptual, strategic, and statistical considerations. *Journal of Personality and Social Psychology* **51**(6), 1173–1182.

Becker BJ 1992. Using results from replicated studies to estimate linear models. *Journal of Educational Statistics* **17**(4), 341–362.

Becker BJ 1995. Corrections to "using results from replicated studies to estimate linear models". *Journal of Educational and Behavioral Statistics* **20**(1), 100–102.

Becker BJ 2007. Multivariate meta-analysis: contributions of Ingram Olkin. *Statistical Science* **22**(3), 401–406.

Berkey CS, Hoaglin DC, Antczak-Bouckoms A, Mosteller F and Colditz GA 1998. Meta-analysis of multiple outcomes by regression with random effects. *Statistics in Medicine* **17**(22), 2537–2550.

Boker S, Neale M, Maes H, Wilde M, Spiegel M, Brick T, Spies J, Estabrook R, Kenny S, Bates T, Mehta P and Fox J 2011. OpenMx: an open source extended structural equation modeling framework. *Psychometrika* **76**(2), 306–317.

Bollen KA 1989. *Structural equations with latent variables*. John Wiley & Sons, Inc., New York.

Bujkiewicz S, Thompson JR, Sutton AJ, Cooper NJ, Harrison MJ, Symmons DP and Abrams KR 2013. Multivariate meta-analysis of mixed outcomes: a Bayesian approach. *Statistics in Medicine* **32**(22), 3926–3943.

Buonaccorsi JP 2010. *Measurement error: models, methods, and applications*. CRC Press, Boca Raton, FL.

Chen H, Manning AK and Dupuis J 2012. A method of moments estimator for random effect multivariate meta-analysis. *Biometrics* **68**(4), 1278–1284.

Cheung MWL 2007. Comparison of approaches to constructing confidence intervals for mediating effects using structural equation models. *Structural Equation Modeling: A Multidisciplinary Journal* **14**(2), 227–246.

Cheung MWL 2009a. Comparison of methods for constructing confidence intervals of standardized indirect effects. *Behavior Research Methods* **41**(2), 425–438.

Cheung MWL 2009b. Modeling multivariate effect sizes with structural equation models. Paper presented at the Association for Psychological Science 21st Annual Convention, San Francisco, CA, USA.

Cheung MWL 2013a. Implementing restricted maximum likelihood estimation in structural equation models. *Structural Equation Modeling: A Multidisciplinary Journal* **20**(1), 157–167.

Cheung MWL 2013b. Multivariate meta-analysis as structural equation models. *Structural Equation Modeling: A Multidisciplinary Journal* **20**(3), 429–454.

Cheung MWL 2014a. Fixed- and random-effects meta-analytic structural equation modeling: examples and analyses in R. *Behavior Research Methods* **46**(1), 29–40.

Cheung MWL 2014b. metaSEM: An R package for meta-analysis using structural equation modeling. Frontiers in Psychology 5(1521).

Cheung MWL 2014c. Modeling dependent effect sizes with three-level meta-analyses: a structural equation modeling approach. *Psychological Methods* **19**(2), 211–229.

Cheung MWL and Chan W 2004. Testing dependent correlation coefficients via structural equation modeling. *Organizational Research Methods* **7**(2), 206–223.

Cochran W 1954. The combination of estimates from different experiments. *Biometrics* **10**(1), 101–129.

Colditz GA, Brewer TF, Berkey CS, Wilson ME, Burdick E, Fineberg HV and Mosteller F 1994. Efficacy of BCG vaccine in the prevention of tuberculosis. Meta-analysis of the published literature. *JAMA: The journal of the American Medical Association* **271**(9), 698–702.

Demidenko E 2013. *Mixed models: theory and applications with R*, 2nd edn. Wiley-Interscience, Hoboken, NJ.

DerSimonian R and Laird N 1986. Meta-analysis in clinical trials. *Controlled Clinical Trials* **7**(3), 177–188.

Digman JM 1997. Higher-order factors of the Big Five. *Journal of Personality and Social Psychology* **73**(6), 1246–1256.

Enders CK 2010. *Applied missing data analysis*. Guilford Press, New York.

Friendly M, Monette G and Fox J 2013. Elliptical insights: understanding statistical methods through elliptical geometry. *Statistical Science* **28**(1), 1–39.

Fuller WA 1987. *Measurement error models, Wiley series in probability and statistics*. John Wiley & Sons, Inc., New York.

Ghidey W, Stijnen T and Houwelingen HCv 2013. Modelling the effect of baseline risk in meta-analysis: a review from the perspective of errors-in-variables regression. *Statistical Methods in Medical Research* **22**(3), 307–323.

Hafdahl AR 2007. Combining correlation matrices: simulation analysis of improved fixed-effects methods. *Journal of Educational and Behavioral Statistics* **32**(2), 180–205.

Hafdahl AR 2008. Combining heterogeneous correlation matrices: simulation analysis of fixed-effects methods. *Journal of Educational and Behavioral Statistics* **33**(4), 507–533.

Hafdahl AR 2009. *Meta-analysis for functions of heterogeneous multivariate effect sizes* Master of arts. Washington University in St. Louis United States, Washington, DC.

Hayduk LA 1987. *Structural equation modeling with LISREL: essentials and advances*. Johns Hopkins University Press, Baltimore, MD.

Hedges LV and Olkin I 1985. *Statistical methods for meta-analysis*. Academic Press, Orlando, FL.

Hedges LV, Tipton E and Johnson MC 2010. Robust variance estimation in meta-regression with dependent effect size estimates. *Research Synthesis Methods* **1**(1), 39–65.

Higgins JPT and Thompson SG 2002. Quantifying heterogeneity in a meta-analysis. *Statistics in Medicine* **21**(11), 1539–1558.

Hox JJ 2010. *Multilevel analysis: techniques and applications*, 2nd edn. Routledge, New York.

Ishak KJ, Platt RW, Joseph L and Hanley JA 2008. Impact of approximating or ignoring within-study covariances in multivariate meta-analyses. *Statistics in Medicine* **27**(5), 670–686.

Jackson D, Riley R and White IR 2011. Multivariate meta-analysis: potential and promise. *Statistics in Medicine* **30**(20), 2481–2498.

Jackson D, White IR and Riley RD 2012. Quantifying the impact of between-study heterogeneity in multivariate meta-analyses. *Statistics in Medicine* **31**(29), 3805–3820.

Jackson D, White IR and Thompson SG 2010. Extending DerSimonian and laird's methodology to perform multivariate random effects meta-analyses. *Statistics in Medicine* **29**(12), 1282–1297.

Jamshidian M and Bentler PM 1999. ML estimation of mean and covariance structures with missing data using complete data routines. *Journal of Educational and Behavioral Statistics* **24**(1), 21–24.

Kalaian HA and Raudenbush SW 1996. A multivariate mixed linear model for meta-analysis. *Psychological Methods* **1**(3), 227–235.

MacKinnon D 2008. *Introduction to statistical mediation analysis*. Lawrence Erlbaum Associates, New York, NY.

McArdle JJ and McDonald RP 1984. Some algebraic properties of the reticular action model for moment structures. *British Journal of Mathematical and Statistical Psychology* **37**(2), 234–251.

Mulaik SA 2009. *Linear causal modeling with structural equations*. CRC Press, Boca Raton, FL.

Muthén B, Kaplan D and Hollis M 1987. On structural equation modeling with data that are not missing completely at random. *Psychometrika* **52**(3), 431–462.

Nam I, Mengersen K and Garthwaite P 2003. Multivariate meta-analysis. *Statistics in Medicine* **22**(14), 2309–2333.

Oberski DL and Satorra A 2013. Measurement error models with uncertainty about the error variance. *Structural Equation Modeling: A Multidisciplinary Journal* **20**(3), 409–428.

Pinheiro JC and Bates DM 1996. Unconstrained parameterizations for variance-covariance matrices. *Statistics and Computing* **6**(3), 289–296.

Raudenbush SW 2009. Analyzing effect sizes: random effects models In *The handbook of research synthesis and meta-analysis* (ed. Cooper HM, Hedges LV and Valentine JC), 2nd edn. Russell Sage Foundation, New York, pp. 295–315.

Raudenbush SW, Becker BJ and Kalaian H 1988. Modeling multivariate effect sizes. *Psychological Bulletin* **103**(1), 111–120.

Riley RD 2009. Multivariate meta-analysis: the effect of ignoring within-study correlation. *Journal of the Royal Statistical Society: Series A (Statistics in Society)* **172**(4), 789–811.

Riley RD, Thompson JR and Abrams KR 2008. An alternative model for bivariate random-effects meta-analysis when the within-study correlations are unknown. *Biostatistics* **9**(1), 172–186.

Schafer JL 1997. *Analysis of incomplete multivariate data*. Chapman & Hall, London.

Schafer JL and Graham JW 2002. Missing data: our view of the state of the art. *Psychological Methods* **7**(2), 147–177.

Shadish WR 1992. Do family and marital psychotherapies change what people do? A meta-analysis of behavioral outcomes In *Meta-analysis for explanation: a casebook* (ed. Cook TD, Cooper HM, Cordray DS, Hartmann H, Hedges LV, Light RJ, Louis TA and Mosteller F). Russell Sage Foundation, New York, pp. 129–208.

Shadish WR 1996. Meta-analysis and the exploration of causal mediating processes: a primer of examples, methods, and issues. *Psychological Methods* **1**(1), 47–65.

Shadish WR and Sweeney RB 1991. Mediators and moderators in meta-analysis: there's a reason we don't let dodo birds tell us which psychotherapies should have prizes. *Journal of Consulting and Clinical Psychology* **59**(6), 883–893.

Stoel RD, Garre FG, Dolan C and van den Wittenboer G 2006. On the likelihood ratio test in structural equation modeling when parameters are subject to boundary constraints. *Psychological Methods* **11**(4), 439–455.

van Houwelingen HC, Arends LR and Stijnen T 2002. Advanced methods in meta-analysis: multivariate approach and meta-regression. *Statistics in Medicine* **21**(4), 589–624.

Viechtbauer W 2010. Conducting meta-analyses in R with the metafor package. *Journal of Statistical Software* **36**(3), 1–48.

Wei Y and Higgins JP 2013. Bayesian multivariate meta-analysis with multiple outcomes. *Statistics in Medicine* **32**(17), 2911–2934.

World Values Study Group 1994. *World Values Survey, 1981–1984 and 1990–1993 [Computer file]*. Inter-university Consortium for Political and Social Research, Ann Arbor, MI.

Wothke W 1993. Nonpositive definite matrices in structural modeling In *Testing structural equation models* (ed. Bollen KA and Long JS). Lawrence Erlbaum Associates, Newbury Park, CA, pp. 256–293.

6

Three-level meta-analysis

This chapter reviews issues of nonindependent effect sizes in a meta-analysis and some of the conventional approaches used to address these issues. A three-level meta-analysis is then put forward to address the problem of dependence in the effect sizes. A model and analyses of three-level meta-analysis are introduced. This chapter also seeks to extend the key concepts of Q statistics, I^2, and R^2 from a two-level meta-analysis to a three-level meta-analysis. A structural equation modeling (SEM) approach to conducting a three-level meta-analysis is introduced. The relationships between a three-level meta-analysis and a multivariate meta-analysis are described. An example is used to illustrate the procedures in the R statistical environment.

6.1 Introduction

Most statistical methods used in meta-analyses assume that the effect sizes are independent. The assumption of independence among the effect sizes does not seem plausible in many research settings. When the effect sizes are nonindependent, the results of conventional meta-analyses conducted on the assumption that effect sizes are independent are no longer correct. Broadly speaking, there are two types of dependence—either the conditional sampling covariance matrices of the studies are known or it is unknown. When the conditional sampling covariance matrices can be estimated from the studies, the multivariate meta-analysis introduced in Chapter 5 may be used to model the dependence. This chapter focuses on situations where the conditional sampling covariances matrices cannot be estimated from the studies.

6.1.1 Examples of dependent effect sizes with unknown degree of dependence

In many research settings, the degree of dependence is unknown. For example, in a cross-cultural meta-analysis, participants from the same ethnic group may share similar values, beliefs, and psychological attributes (Fischer and Boer, 2011; Fischer et al., 2012; Hanke and Fischer, 2013). The effect sizes reported by the same research teams or authors may be more similar to each other than those reported by other research teams or authors (Cooper, 2010; Shin, 2009). Dependence on the effect sizes may also be introduced by using the same data sets in different publications (Shin, 2009). Single-case studies may report multiple effect sizes on the same subjects (Owens and Ferron, 2012). Studies may also report multiple effect sizes on similar constructs, over multiple time points or conditions while the correlations among the effect sizes remain unknown (Van den Bussche et al., 2009).

Let us consider Nguyen and Benet-Martínez (2013) as an example. These authors extracted 935 correlation coefficients between biculturalism (e.g., behavior, values, and identity) and adjustment (e.g., life satisfaction and grades) from a total of 141 studies. Schmidt and Hunter (2015) called the multiple effect sizes reported in the same study as *conceptual replication* because these effect sizes are also qualified as the effect sizes in a meta-analysis. As multiple effect sizes have been reported, it is not reasonable to assume that the effect sizes are independent. On the other hand, the multivariate meta-analysis introduced in Chapter 5 is not applicable because there is not enough information to estimate the sampling covariances among the effect sizes.

Dependence on the data may be introduced by either the researchers conducting the primary studies or the reviewers doing the meta-analysis. For example, the researchers conducting the primary studies may collect data from a multisite setting; they may compare different treatment groups against the same control group or use multiple measures on the same construct. When the summary statistics are used in a meta-analysis, these effect sizes are not independent.

On the other hand, dependence may also be introduced by the reviewers who conceptualize the meta-analysis. For example, reviewers conducting a cross-cultural meta-analysis may hypothesize that culture plays an important role on psychological processes. It is expected that the effect sizes reported by participants from the same cultural group may be more similar than the effect sizes from other cultural groups. When conducting a meta-analysis, studies will be naturally grouped under cultural groups although the primary studies will have been independently conducted by different researchers (e.g., Fischer and Boer, 2011; Fischer et al., 2012; Hanke and Fischer, 2013).

6.1.2 Common methods to handling dependent effect sizes

Several strategies have been used to handle the dependent effect sizes (Borenstein et al., 2010; Cooper, 2010; Schmidt and Hunter, 2015). They are ignoring the dependence, averaging the dependent effect sizes within studies, selecting one effect size per study, and shifting the unit of analysis. This section reviews some of these approaches and their associated limitations.

6.1.2.1 Ignoring the Dependence

One *incorrect* strategy is simply ignoring the dependence and analyzing the data as if they were independent. A great deal of work has been done on the effects of ignoring the dependence in multilevel modeling (e.g., Hox, 2010; Snijders and Bosker, 2012). The general conclusion is that the fixed-effects parameter estimates, such as the intercept and the regression coefficients, can be biased. The associated standard errors (SEs) of the parameter estimates are also likely underestimated. Thus, the empirical Type I error rates are usually inflated. Researchers may incorrectly conclude that there is an effect but in fact the significance is due to ignoring the dependence of the effect sizes. Ignoring the dependence is generally not acceptable in a meta-analysis (Becker et al., 2004).

6.1.2.2 Averaging the Dependent Effect Sizes within Studies

One approach to addressing the dependence is to average the dependent effect sizes into a single effect size and to use the average effect sizes in the subsequent analyses. This approach is also known as *aggregation*. Several procedures have been suggested on how to average the dependent effect sizes (e.g., Cheung and Chan, 2004, 2008, 2014; Marín-Martínez and Sánchez-Meca, 1999; Rosenthal and Rubin, 1986). One simple method is to weigh the dependent effect sizes equally within a study. Suppose that there are two dependent effect sizes y_{1j} and y_{2j} with their sampling variances (v_{1j} and v_{2j}) and their covariance (v_{12j}) in the jth Study. If v_{12j} is unknown, an ad hoc estimate based on expert knowledge or other sources may be used. We may compute an average effect size $y_j = 0.5(y_{1j} + y_{2j})$ with its sampling variance $v_j = 0.25(v_{1j} + v_{2j} + 2v_{12j})$ (e.g., Borenstein et al., 2010; Gleser and Olkin, 1994). There are other more accurate weighting approaches that take the dependence into account (e.g., Cheung and Chan, 2004, 2008, 2014).

The main advantage of these approaches is that they are easy to apply. After the averaging, there is only one effect size per study. Conventional meta-analytic techniques can be directly applied. Although these approaches can effectively remove the dependence among the effect sizes, there are several limitations. First, the statistical power of the tests may be affected because some information on the data will have been lost. Second, these approaches may limit what research questions can be addressed in the meta-analysis. For example, the reviewers may want to investigate the effectiveness of a teaching program on students. While it is common for studies to report teaching effectiveness based on student reports y_{1j} as well as teacher evaluations y_{2j}, if reviewers average the effect sizes reported by students and teachers, it would not be possible to test the differences between the student reports and the teacher evaluations.

When averaging the effect sizes within studies, researchers usually calculate an average effect size for each study by using either a simple average or a weighted average. This means that the effect sizes are assumed to be homogeneous within studies. The differences among the effect sizes within a study are merely due to the sampling error. This assumption of homogeneity of effect sizes within a study may be questionable in a meta-analysis. Even though the same samples are used within the same study, the *true* population effect sizes within the same study may still vary because different measures are used.

6.1.2.3 Selecting One Effect Size Per Study

Another approach is to select only one effect size per study. This is also known as *elimination*. As there is only one effect size per study, the effect sizes are independent. Similar to the problems listed in the averaging approach, the elimination approach affects the statistical power of the meta-analysis because some effect sizes have been excluded (Hedges and Pigott, 2001, 2004). Using the previous example on the effectiveness of a teaching program as an illustration, the reviewers may need to choose between the teaching effectiveness based on the students' own reports or on the teachers' evaluations as the effect sizes. Moreover, the ability to test study characteristics using the mixed-effects model is also limited, because not all effect sizes can be included in the meta-analysis.

6.1.2.4 Shifting the Unit of Analysis

Cooper (2010) proposed an approach called *shifting the unit of analysis*. The basic idea is to select the unit of analysis and then apply averaging on the effect sizes within the units. For example, when an overall estimate of the pooled effect size is required, the unit of analysis is a study. The dependent effect sizes are first averaged within studies before the analysis is conducted. As each study contributes only one effect size, the average effect is calculated based on independent samples. When study characteristics are being examined, for example, comparing the effects of different gender groups, the dependent effect sizes are averaged within gender groups. This approach can minimize the effect of dependence.

It is not without limitations, however. Under this approach, issues such as a loss of information and the assumption of homogeneity within studies in the averaging process have not been resolved. As different data sets are used to address different research questions, there is no guarantee that the findings will be consistent. As noted by Cooper (2010), the process of managing data also becomes complicated because the data structures have to be changed according to the research questions.

Ahn et al. (2012) reviewed 56 meta-analyses published in eight journals in education between 2000 and 2010. Among other findings, one was that in 28 of these meta-analyses issues relating to dependent effect sizes were encountered because of multiple measures of the same construct, multiple outcomes or interventions, multiple time points, or multiple comparison groups. The reported methods used to address the issue of dependence were (i) averaging the dependent effect sizes within studies (18 studies), (ii) shifting the unit of analysis (8 studies), and (iii) selecting one effect size per study and combining this approach with other methods

(5 studies). For those studies that did not mention issues of dependence, it is unclear whether there was no such issue or whether the issues were ignored.

Scammacca et al. (2013) compared several approaches to handle dependent effect sizes by using a case study. They found that the estimates of the overall effect sizes were similar, whereas the estimated indices on heterogeneity varied. As their study was based on a single data set, further simulation studies may be needed to address the generalizability of their findings. From these reviews, it is clear that it is quite often to encounter dependent effect sizes when conducting meta-analyses.

6.2 Three-level model

This section reviews how three-level multilevel models (e.g., Goldstein, 2011; Raudenbush and Bryk, 2002; Snijders and Bosker, 2012) can be applied to meta-analysis to address the dependence of the effect sizes. The issues of quantifying the degree of heterogeneity and the explained variance in a mixed-effects meta-analysis are also discussed.

6.2.1 Random-effects model

The fixed-effects model is a special case of the conventional two-level random-effects model, while the conventional two-level random-effects model is a special case of the three-level meta-analysis. To fix the notations, y_{ij} be the ith effect size in the jth cluster. The definition of cluster depends on the data structure and research questions. For example, y_{ij} represents one of the multiple effect sizes in the jth study (e.g., Nguyen and Benet-Martínez, 2013); it represents one of the studies in the jth cultural group in a cross-cultural meta-analysis (e.g., Fischer and Boer, 2011; Fischer et al., 2012; Hanke and Fischer, 2013). In single-case studies, y_{ij} represents one of the measures in the jth subject (Owens and Ferron, 2012). As the most common type of dependence is multiple effect sizes nested within studies, in this chapter, we often denote the ith effect size in the jth study for ease of discussion.

The three-level random-effects meta-analysis is depicted as follows:

$$\begin{aligned} \text{Level 1:} \quad & y_{ij} = \lambda_{ij} + e_{ij}, \\ \text{Level 2:} \quad & \lambda_{ij} = f_j + u_{(2)ij}, \\ \text{Level 3:} \quad & f_j = \beta_0 + u_{(3)j}, \end{aligned} \tag{6.1}$$

where λ_{ij} is the *true* effect size and e_{ij} is the known sampling variance in the ith effect size in the jth cluster, f_j is the *true* effect size in the jth cluster, β_0 is the average population effect, and $\mathsf{Var}(u_{(2)ij}) = \tau^2_{(2)}$ and $\mathsf{Var}(u_{(3)j}) = \tau^2_{(3)}$ are the level-2 and level-3 heterogeneity variances, respectively. Therefore, β_0 represents the average effect across all effect sizes, while the study-specific level-2 and level-3 random effects allow each effect size y_{ij} has its own *true* population effect size.

Similar to the two-level model, the equations can be combined into a single equation:

$$y_{ij} = \beta_0 + u_{(2)ij} + u_{(3)j} + e_{ij}. \tag{6.2}$$

Standard assumptions in multilevel modeling are applied. Random effects at different levels and the sampling error are assumed to be independent:

$$\mathsf{Cov}(u_{(2)ij}, u_{(3)j}) = \mathsf{Cov}(u_{(2)ij}, e_{ij}) = \mathsf{Cov}(u_{(3)j}, e_{ij}) = 0. \tag{6.3}$$

Moreover, the level-2 random effects within the same cluster and the level-3 between-cluster random effects are independent:

$$\begin{aligned} \mathsf{Cov}(u_{(2)ij}, u_{(2)kj}) &= 0 \quad \text{and} \\ \mathsf{Cov}(u_{(3)m}, u_{(3)n}) &= 0. \end{aligned} \tag{6.4}$$

On the basis of the above assumptions, it follows that

$$\begin{aligned} \mathsf{E}(y_{ij}) &= \beta_0 \quad \text{and} \\ \mathsf{Var}(y_{ij}) &= \mathsf{Var}(u_{(2)ij}) + \mathsf{Var}(u_{(3)j}) + \mathsf{Var}(e_{ij}) \\ &= \tau^2_{(2)} + \tau^2_{(3)} + v_{ij}. \end{aligned} \tag{6.5}$$

The covariance of two effect sizes within the same jth cluster is

$$\begin{aligned} \mathsf{Cov}(y_{ij}, y_{kj}) &= \mathsf{Cov}(u_{(2)ij} + u_{(3)j} + e_{ij}, u_{(2)kj} + u_{(3)j} + e_{kj}) \\ &= \mathsf{Cov}(u_{(3)j}, u_{(3)j}) \quad \text{as the other terms are zero,} \\ &= \tau^2_{(3)}. \end{aligned} \tag{6.6}$$

The covariance of two effect sizes from different clusters is

$$\mathsf{Cov}(y_{ij}, y_{mn}) = 0. \tag{6.7}$$

In plain language, the unconditional sampling variance of the effect size equals the sum of the level-2 and level-3 heterogeneity and the known sampling variance. Effect sizes in the same clusters share the same covariance $\tau^2_{(3)}$, whereas effect sizes in different clusters are independent. Readers may refer to Konstantopoulos (2011) for details.

6.2.1.1 Testing the homogeneity of effect sizes

As with a conventional (two-level) meta-analysis, reviewers may want to test the hypothesis of the homogeneity of effect sizes. Suppose that there are k clusters with a maximum of p effect sizes per study; the null hypothesis of the homogeneity of effect sizes is $H_0 : \mu_{11} = \mu_{21} = \cdots = \mu_{1k} = \cdots = \mu_{pk}$. When all of the population effect sizes are the same under the null hypothesis, there is no cluster effect. Thus, the conventional Q statistic proposed by Cochran (1954) can be directly applied. It is defined by using Equation 4.14 as follows:

$$Q = \sum_{i=1}^{n} w_i (y_i - \hat{\beta}_F)^2, \tag{6.8}$$

where $w = 1/v_i$, $\hat{\beta}_F = \sum_{i=1}^{n} w_i y_i / \sum_{i=1}^{n} w_i$, and n is the total number of effect sizes. Under the null hypothesis, the Q statistic has an approximate chi-square distribution with $(n - 1)$ degree of freedoms (dfs). It should be noted that this test assumes that the effect sizes are conditionally independent. The test statistic may not be accurate when the observed effect sizes are not independent even the null hypothesis is correct. If the degree of dependence of the effect sizes can be estimated, the modified Q statistic (Equation 5.6) that takes the dependence into account may be used.

6.2.1.2 Testing $H_0 : \tau^2_{(3)} = 0$

It is clear that the two-level meta-analysis is a special case of the three-level meta-analysis by fixing $\tau^2_{(3)} = 0$. As these two models are nested, we may test the null hypothesis $H_0 : \tau^2_{(3)} = 0$ by using a likelihood ratio (LR) statistic. If the test statistic is nonsignificant, we cannot reject the null hypothesis of $\tau^2_{(3)} = 0$. In order words, the three-level model is not statistically better than the two-level model. Alternatively, a likelihood-based confidence interval (LBCI) can be constructed on $\hat{\tau}^2$. If a 95% LBCI does not include 0, the null hypothesis $H_0 : \tau^2_{(3)} = 0$ is rejected at $\alpha = 0.05$.

Two points of caution should be raised. First, even if the null hypothesis is true, the test statistic is not distributed as a chi-square distribution with 1 df. As $H_0 : \tau^2_{(3)} = 0$ is tested on the boundary, $\tau^2_{(3)}$ cannot be negative in theory. The test statistic is distributed as a 50:50 mixture of a degenerate random variable with all of its probability mass concentrated at zero and a chi-square random variable with 1 df. One simple strategy to correct for this bias is to use 2α instead of α as the alpha level (Pinheiro and Bates, 2000). That is, we may reject the null hypothesis when the observed p value is larger than 0.10 for $\alpha = 0.05$ (see Section 4.3.2 for details).

Second, it is not advisable to decide between a two-level or a three-level model by testing $H_0 : \tau^2_{(3)} = 0$. Similar to the choice between a fixed-effects or a random-effects model in a conventional meta-analysis, the decision should be based on whether a conditional or an unconditional inference is required (e.g., Hedges and Vevea, 1998). If reviewers want to generalize the findings to both level 2 and level 3, a three-level model should be used even though the test of $H_0 : \tau^2_{(3)} = 0$ is nonsignificant. If the $\hat{\tau}^2_{(3)} = 0$, the model becomes a conventional two-level meta-analysis.

6.2.1.3 Testing $H_0 : \tau^2_{(2)} = 0$

We may compare the conceptual differences between the three-level meta-analysis and conventional approaches, such as aggregation and elimination, in handling dependent effect sizes. The conventional approaches assume that $\tau^2_{(2)} = 0$. With this assumption, the multiple effect sizes within the same study are direct replication of each other. The observed differences are only due to the sampling error. As they are direct replication of each other, the elimination approach chooses only one effect size per study. The aggregation approach combines these multiple effect sizes with the aim of minimizing the sampling error. Under the three-level meta-analysis, whether $H_0 : \tau^2_{(2)} = 0$ is an empirical question that can be tested based on the data. We can test the null hypothesis $H_0 : \tau^2_{(2)} = 0$. $\tau^2_{(2)}$ indicates the heterogeneity of *true* effect sizes within a cluster. Similar to the case in testing $H_0 : \tau^2_{(3)} = 0$, it is testing on the boundary. Moreover, it is not advisable to fix $\tau^2_{(2)} = 0$ even the test statistic is nonsignificant.

6.2.1.4 Testing $H_0 : \tau^2_{(2)} = \tau^2_{(3)}$

Reviewers may sometimes want to compare the magnitude of the heterogeneity variances between level 2 and level 3. Let us consider cross-cultural research as an example. Traditionally, researchers have mainly been interested in the between-cultural variation, for example, the differences between Americans, Chinese, and Japanese. Some researchers, however, argue that the variations within a culture, which is known as intracultural variation, is also theoretically meaningful (e.g., Au and Cheung, 2004). If a cross-cultural meta-analysis is conducted by treating studies as level 2 and cultural groups as level 3, reviewers may want to test $H_0 : \tau^2_{(2)} = \tau^2_{(3)}$. If the null hypothesis is not rejected, the magnitude of the intracultural variation is similar to that of the between-cultural variation; otherwise, one is greater than the other.

It should be emphasized that this research hypothesis cannot be tested by using common methods for handling dependent effect sizes, such as averaging the dependent effect sizes, selecting one effect size per study, and shifting the unit of analysis. The above null hypothesis can easily be tested in the SEM approach, which is introduced in next section, by imposing an equality constraint on $\tau^2_{(2)}$ and $\tau^2_{(3)}$. Under the null hypothesis, the difference in the LRs between the models with and without the constraint has a chi-square distribution with 1 df.

6.2.1.5 Quantifying the degree of the heterogeneity of the effect sizes

Besides the statistical tests, reviewers may want to quantify the heterogeneity of the effect sizes at both levels. To quantify the heterogeneity of the effect sizes, the I^2 in Section 4.3.3 proposed by Higgins and Thompson (2002) can be extended to a three-level meta-analysis. We define the I^2 at level 2 and level 3 as

$$\begin{aligned} I^2_{(2)} &= \frac{\hat{\tau}^2_{(2)}}{\hat{\tau}^2_{(2)} + \hat{\tau}^2_{(3)} + \tilde{v}} \quad \text{and} \\ I^2_{(3)} &= \frac{\hat{\tau}^2_{(3)}}{\hat{\tau}^2_{(2)} + \hat{\tau}^2_{(3)} + \tilde{v}}, \end{aligned} \tag{6.9}$$

where $\tilde{v}$ is a *typical* within-study sampling variance, as discussed in Section 4.3.3.

$I^2_{(2)}$ and $I^2_{(3)}$ can be interpreted as proportions of the total variation of the effect size due to the level-2 and level-3 between-study heterogeneity, respectively. For instance, suppose that level 2 refers to multiple measures while level 3 refers to studies. $I^2_{(2)}$ and $I^2_{(3)}$ can be interpreted as proportions of the total variation due to the use of multiple measures within a study and to the between-study effects, respectively. In a cross-cultural meta-analysis by using cultural groups as the cluster effect, $I^2_{(2)}$ and $I^2_{(3)}$ can be interpreted as the proportions of the intracultural variation and the between-cultural variation, respectively. As the sample sizes ($\tilde{v}$) are involved in the calculations of $I^2_{(2)}$ and $I^2_{(3)}$, these indices are not estimating any population parameters. Rather, they are mainly serving as descriptive indices for the studies included in the meta-analysis.

We may also define two intraclass correlations (ICCs) that involve only the level-2 and level-3 quantities (e.g., Snijders and Bosker, 2012). It can easily be shown that these ICCs are simply one-to-one transformations of $I^2_{(2)}$ and $I^2_{(3)}$,

$$\begin{aligned} ICC_{(2)} &= \frac{\hat{\tau}^2_{(2)}}{\hat{\tau}^2_{(2)} + \hat{\tau}^2_{(3)}} = \frac{I^2_{(2)}}{I^2_{(2)} + I^2_{(3)}} \quad \text{and} \\ ICC_{(3)} &= \frac{\hat{\tau}^2_{(3)}}{\hat{\tau}^2_{(2)} + \hat{\tau}^2_{(3)}} = \frac{I^2_{(3)}}{I^2_{(2)} + I^2_{(3)}}. \end{aligned} \tag{6.10}$$

These two indices can be interpreted as the proportions of the total between-study heterogeneity of the effect size that are due to the level 2 and the level 3 between studies. One advantage of the ICCs over I^2 is that they are sample size free. Therefore, they are estimating the population quantities $\tau^2_{(2)}/(\tau^2_{(2)} + \tau^2_{(3)})$ and $\tau^2_{(3)}/(\tau^2_{(2)} + \tau^2_{(3)})$.

Although the above indices seem intuitive, there is a lack of studies on the empirical performance of these indices. This is because quantifying the degree of heterogeneity is still a new topic in three-level meta-analysis (Cheung, 2014b). There are a couple of issues that should be addressed in the future. First, there are several definitions of what constitutes a *typical* within-study sampling variance (see Section 4.3.3). It is not clear which definition fits best in a three-level meta-analysis. Second, v_i and thus $\tilde{v}$ are treated as constants without considering the cluster effect or the dependency of the data. Calculations of $\tilde{v}$ in a conventional two-level meta-analysis featured in Section 4.3.3 are directly applied here. Further studies may investigate whether it is necessary to adjust for the cluster effect when calculating indices of heterogeneity. Third, ICCs are seldom used in meta-analyses because they are not defined in a conventional two-level meta-analysis. Whether ICCs or I^2 are better at quantifying the degree of heterogeneity in a three-level meta-analysis needs to be determined.

6.2.2 Mixed-effects model

The random-effects model can be extended to a mixed-effects model by including study-specific characteristics as moderators. Let x be a moderator that can be either x_{ij} for a level-2 moderator or x_j for a level-3 moderator. x_j (without the i subscript) indicates that the value is the same in the jth cluster, whereas x_{ij} indicates that the value may vary across the effect sizes in the jth cluster. Without the loss of generality, we use x_{ij} in the model notation. The mixed-effect model with one moderator is

$$y_{ij} = \beta_0 + \beta_1 x_{ij} + u_{(2)ij} + u_{(3)j} + e_{ij}. \tag{6.11}$$

The values conditional at x_{ij} are

$$\begin{aligned} \mathsf{E}(y_{ij}|x_{ij}) &= \beta_0 + \beta_1 x_{ij}, \\ \mathsf{Var}(y_{ij}|x_{ij}) &= \tau^2_{(2)} + \tau^2_{(3)} + v_{ij}, \\ \mathsf{Cov}(y_{ij}, y_{kj}|x_{ij}) &= \tau^2_{(3)}, \quad \text{and} \\ \mathsf{Cov}(y_{ij}, y_{mn}|x_{ij}) &= 0. \end{aligned} \tag{6.12}$$

The interpretations of these terms are similar to those in Equation 6.2 except that $\tau^2_{(2)}$ and $\tau^2_{(3)}$ are now the level-2 and level-3 residual heterogeneity variances after controlling for the moderator x_{ij}.

6.2.2.1 Explained variance

The R^2 discussed in Section 4.5.2 can be extended to three-level meta-analysis. We may define the level-2 $R^2_{(2)}$ and level-3 $R^2_{(3)}$ as

$$\begin{aligned} R^2_{(2)} &= 1 - \frac{\hat{\tau}^2_{(2)1}}{\hat{\tau}^2_{(2)0}} \quad \text{and} \\ R^2_{(3)} &= 1 - \frac{\hat{\tau}^2_{(3)1}}{\hat{\tau}^2_{(3)0}}, \end{aligned} \tag{6.13}$$

where $\hat{\tau}^2_{(2)1}$ and $\hat{\tau}^2_{(2)0}$ are the estimated heterogeneity with predictors and without predictors at level 2, respectively, and $\hat{\tau}^2_{(3)1}$ and $\hat{\tau}^2_{(3)0}$ are the estimated heterogeneity with predictors and without predictors at level 3, respectively. $R^2_{(2)}$ and $R^2_{(3)}$ can be interpreted as the percentage of the variance of the heterogeneity that can be explained by the moderators at level 2 and level 3, respectively. Similar to the conventional three-level model, adding predictors at level 2 (or level 3) may affect the R^2 at the other level. When the estimated $R^2_{(2)}$ and $R^2_{(3)}$ are negative, they are truncated to zero.

Compared to other approaches to handling dependent effect sizes, such as aggregation and elimination, the three-level meta-analysis disentangles the effects (heterogeneity and explained variances) to their corresponding levels. This allows researchers to investigate which level contributes to the heterogeneity (or explained variance) of the effect sizes.

6.3 Structural equation modeling approach

In this section, we first review how SEM can be used to model nested data. The link between the dependence in a multilevel model and the covariance in a structural equation model is highlighted. Then, the SEM approach proposed by Cheung (2014b) to model the three-level meta-analysis is introduced.

6.3.1 Two representations of the same model

There are two equivalent model representations for a conventional two-level meta-analysis. Table 6.1 shows the data structure arranged in the so-called *long* format. One row represents one study. Figure 6.1a shows a structural equation model of the random-effects meta-analysis. As the rows are independent, only one variable y_i is required to represent the effect sizes. The model-implied mean and variance are

$$\begin{aligned}\underset{1\times1}{\mu_i(\theta)} &= \beta_0 \quad \text{and} \\ \underset{1\times1}{\Sigma_i(\theta)} &= \tau^2 + v_i.\end{aligned} \tag{6.14}$$

It should be noted that there is a subscript i in $\Sigma_i(\theta)$ indicating that the variance may vary across studies.

Another representation of the same model is to treat subjects (or studies in a meta-analysis) as variables (Mehta and Neale, 2005). Table 6.2 shows the data structure arranged in the so-called *wide* format. Each study is represented by one variable. When there are k studies, k variables are created to represent the studies. Figure 6.1b shows the model with three studies. The effect sizes of these studies are represented by three variables y_1, y_2, and y_3. There are three key features in this model representation.

First, the expected means labeled β_0 are the same for all studies. Second, the subjects (or studies in the meta-analysis) are independent. Therefore, the covariances among y_1, y_2, and y_3 are all zero. Third, only one row is used to represent the whole data set as the number of subjects (or studies in the meta-analysis) is represented by the variables. The model-implied mean vector and the model-implied covariance matrix are

$$\begin{aligned}\underset{k\times1}{\boldsymbol{\mu}(\boldsymbol{\theta})} &= \underset{k\times1}{\mathbf{1}}\,\beta_0 = \begin{bmatrix}\beta_0\\ \vdots \\ \beta_0\end{bmatrix} \quad \text{and} \\ \underset{k\times k}{\boldsymbol{\Sigma}(\boldsymbol{\theta})} &= \underset{k\times k}{\boldsymbol{I}}\,\tau^2 + \text{Diag}(v_1, v_2, \cdots, v_k) \\ &= \begin{bmatrix}\tau^2 & & & \\ 0 & \tau^2 & & \\ \vdots & \ddots & \ddots & \\ 0 & \cdots & 0 & \tau^2\end{bmatrix} + \begin{bmatrix}v_1 & & & \\ 0 & v_2 & & \\ \vdots & \ddots & \ddots & \\ 0 & \cdots & 0 & v_k\end{bmatrix},\end{aligned} \tag{6.15}$$

where $\underset{k\times1}{\mathbf{1}}$ is a $k \times 1$ vector of ones and $\underset{k\times k}{\boldsymbol{I}}$ is a $k \times k$ identity matrix. From the SEM perspective, there is only *one* subject in this analysis. The model-implied means and covariance matrix are structured in such a way that there are only two parameters—β_0 and τ^2. As it is difficult and inefficient to maintain such a large

Table 6.1 Long format data for a two-level meta-analysis.

Study	Effect size
1	y_1
2	y_2
$\vdots$	$\vdots$
k	y_k

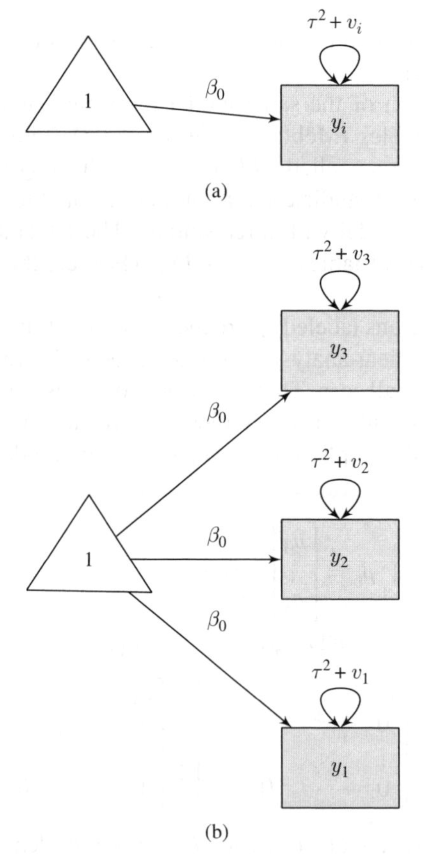

Figure 6.1 Two representations of a random-effects model.

Table 6.2 Wide format data for a two-level meta-analysis.

Study	Effect size 1	Effect size 2	...	Effect size k
1	y_1	y_2	...	y_k

model, this approach is rarely used in SEM except when handling multilevel data (see Mehta and Neale, 2005; Preacher, 2011). By using variables as subjects, we may convert a two-level model into a single-level model. We will apply this technique to model a three-level meta-analysis.

6.3.2 Random-effects model

Now, we treat each cluster as a *subject* and the effect sizes within clusters as *variables* to model the three-level effect sizes in SEM. Suppose that the maximum number of effect sizes per cluster is p; we create p variables to represent the effect sizes nested within studies. For example, if the maximum number of effect sizes per cluster is five, we create five variables. Table 6.3 shows a sample data structure for this example. There are two effect sizes in Study 1, whereas there are five effect sizes in Study 2 and only one effect size in Study k.

Table 6.3 Wide format data for a three-level meta-analysis.

Cluster	y_1	y_2	y_3	y_4	y_5
1	$y_{1,1}$	$y_{2,1}$	NA	NA	NA
2	$y_{1,2}$	$y_{2,2}$	$y_{3,2}$	$y_{4,2}$	$y_{5,2}$
⋮	⋮	⋮	⋮	⋮	
k	$y_{1,k}$	NA	NA	NA	NA

Abbreviation: NA, not available. $y_{i,j}$ represents the ith effect size in the jth cluster.

There are a few points worth mentioning. First, the variables y_1 to y_5 are exchangeable. It does not matter whether a particular effect size, say $y_{1,1}$, is placed under y_1 or y_5 as long as it is placed under Study 1. Second, the incomplete effect sizes in clusters with fewer than five effect sizes are treated as missing values and handled using the full information maximum likelihood (FIML) estimation method. Third, the clusters are independent. Thus, SEM can be used to directly model the data.

Figure 6.2 shows two equivalent SEM models for a three-level meta-analysis with two effect sizes per cluster. Figure 6.2a displays the random-effects model for Equation 6.1 with latent variables. The level-3 study effect in the jth cluster is represented by the latent variable f_j, whereas the ith level-2 random effects in the j cluster are represented by $u_{(2)ij}$. The ith known sampling variance in the jth cluster is represented by v_{ij}. This model is similar to the conventional CFA model where

f_j represents a general factor and $u_{(2)ij}$ and e_{ij} are the specific factor and the measurement error, respectively. The advantage of this model representation is that the level-2 and level-3 random effects are conceptualized as latent variables in SEM. f_j is the *true* effect size in the jth cluster, whereas $u_{(2)ij}$ is the random effect because of using different measures within the jth cluster. The *true* effect size f_j can be used as an independent or dependent variable in more sophisticated modeling.

Figure 6.2b displays an equivalent model without the latent variables. It mainly shows the expected means and the covariance structure. One advantage of this

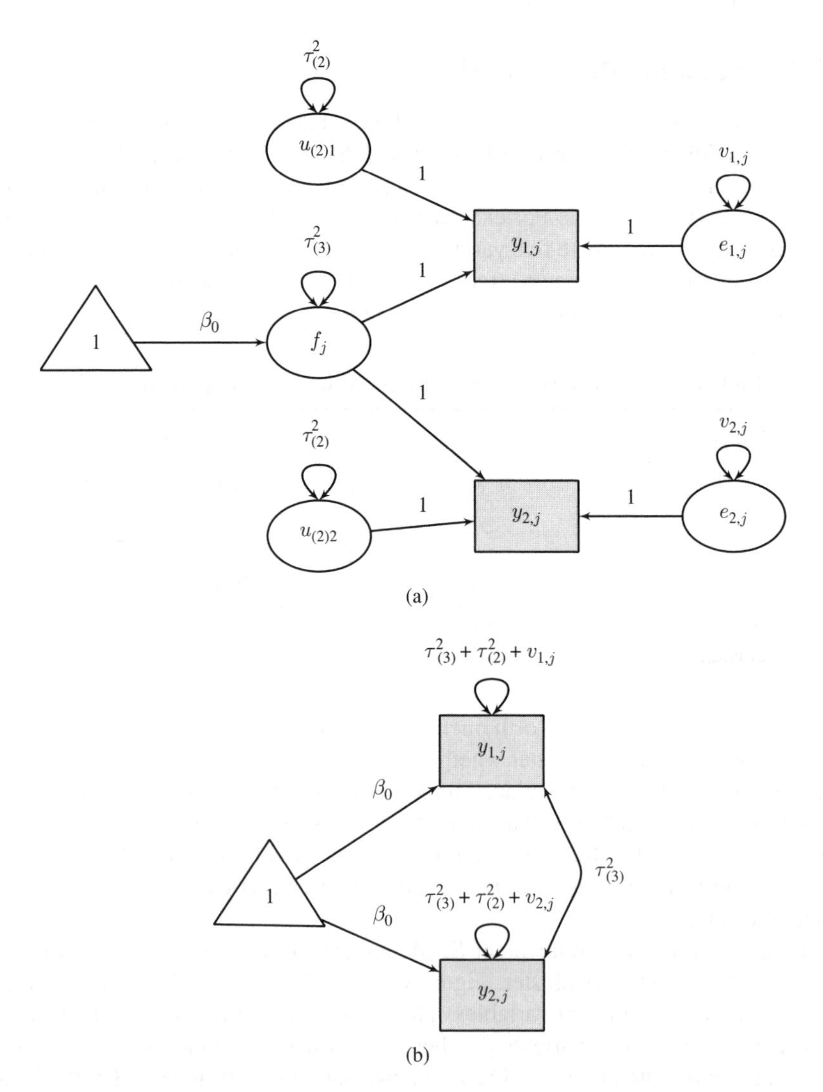

Figure 6.2 A three-level random-effects model with two studies in the j*th cluster.*

approach is that the model can be handled in some multilevel modeling software packages, for example, SAS (Konstantopoulos, 2011).

Because the variables are exchangeable, the model-implied means for the jth cluster is a $p \times 1$ vector of β_0,

$$\underset{p\times 1}{\boldsymbol{\mu}_j(\boldsymbol{\theta})} = \underset{p\times 1}{\boldsymbol{1}}\beta_0 = \begin{bmatrix} \beta_0 \\ \vdots \\ \beta_0 \end{bmatrix}. \tag{6.16}$$

The model-implied covariance matrix for the jth cluster is a $p \times p$ matrix of the sum of three matrices: a $p \times p$ matrix with all elements of $\tau^2_{(3)}$ representing the level-3 heterogeneity, a $p \times p$ diagonal matrix with elements of $\tau^2_{(2)}$ representing the level-2 heterogeneity, and a $p \times p$ diagonal matrix of known sampling variances with elements v_{ij}. That is,

$$\begin{aligned}
\underset{p\times p}{\boldsymbol{\Sigma}_j(\boldsymbol{\theta})} &= \underset{p\times p}{\boldsymbol{1}}\,\tau^2_{(3)} + \underset{p\times p}{\boldsymbol{I}}\,\tau^2_{(2)} + \underset{p\times p}{\boldsymbol{V}_j} \\
&= \begin{bmatrix} \tau^2_{(3)} & & & \\ \tau^2_{(3)} & \tau^2_{(3)} & & \\ \vdots & \ddots & \ddots & \\ \tau^2_{(3)} & \cdots & \tau^2_{(3)} & \tau^2_{(3)} \end{bmatrix} + \begin{bmatrix} \tau^2_{(2)} & & & \\ 0 & \tau^2_{(2)} & & \\ \vdots & \ddots & \ddots & \\ 0 & \cdots & 0 & \tau^2_{(2)} \end{bmatrix} \\
&\quad + \begin{bmatrix} v_{1j} & & & \\ 0 & v_{2j} & & \\ \vdots & \ddots & \ddots & \\ 0 & \cdots & 0 & v_{kj} \end{bmatrix}.
\end{aligned} \tag{6.17}$$

It is of interest to review two special cases under the SEM framework. The first case is when $\tau^2_{(3)} = 0$. The model becomes the conventional two-level meta-analysis using the *wide* format representation in SEM. The second case is when $\tau^2_{(2)} = 0$. This model assumes that all effect sizes within the study are direct replication of each other. There is no level-2 heterogeneity but a sampling error.

6.3.3 Mixed-effects model

When there is a high degree of heterogeneity at level 2 and (or) level 3, we may want to explain the heterogeneity by conducting a mixed-effects meta-analysis. Figure 6.3 shows two equivalent model representations for a model with a moderator x_{ij} in the jth cluster. Figure 6.3a uses a latent variable approach by using f_j and $u_{(2)}$ as the *true* effect size for the jth cluster and the level-2 study-specific effect, respectively. Figure 6.3b excludes the latent variables and focuses on the model-implied mean vectors and the covariance matrix. A phantom variable P with zero variance is introduced for the purpose of fixing the values of the covariate. The specific values of x_{ij} are imposed via definition variables, while β_1 is the regression coefficient (e.g., Cheung, 2010). If the moderators are at the cluster level, x_{ij} will be the same at the jth cluster.

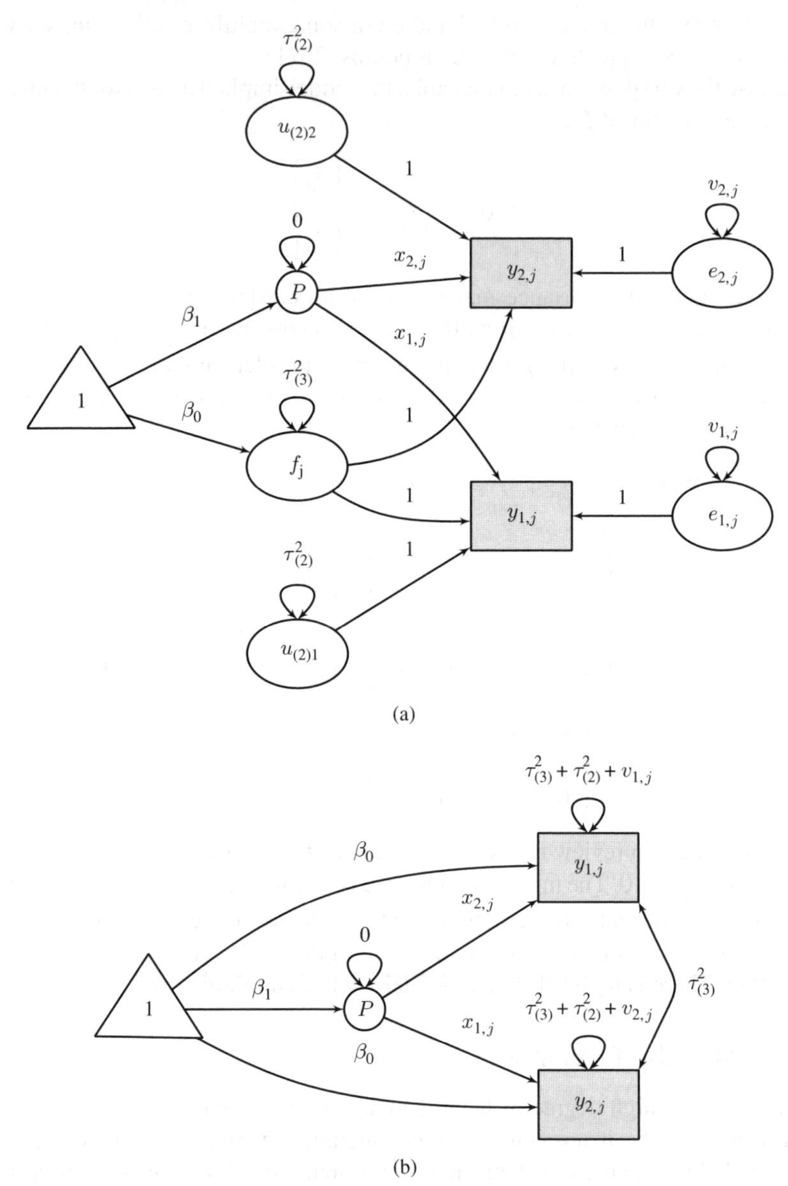

Figure 6.3 A three-level mixed-effects model with two studies and one moderator in the jth cluster.

Suppose that there is one moderator $\boldsymbol{x}$; the conditional model-implied means and the conditional model-implied covariance matrix for the jth cluster are

$$\underset{p\times 1}{\boldsymbol{\mu}_j(\boldsymbol{\theta}|\boldsymbol{x}_j)} = \underset{p\times 1}{\boldsymbol{1}}\beta_0 + \underset{p\times 1}{\boldsymbol{x}}\beta_1 = \begin{bmatrix} \beta_0 + \beta_1 x_{1j} \\ \vdots \\ \beta_0 + \beta_1 x_{pj} \end{bmatrix} \quad \text{and}$$
$$\underset{p\times p}{\boldsymbol{\Sigma}_j(\boldsymbol{\theta}|\boldsymbol{x}_j)} = \underset{p\times p}{\boldsymbol{1}}\tau^2_{(3)} + \underset{p\times p}{\boldsymbol{I}}\tau^2_{(2)} + \underset{p\times p}{\boldsymbol{V}_j}. \tag{6.18}$$

It should be noted that the form of the conditional model-implied covariance matrix is the same as that of the random-effects model. The only difference is that they are now the residuals of the heterogeneity after controlling for the moderator $\boldsymbol{x}$.

6.4 Relationship between the multivariate and the three-level meta-analyses

Thus far, we have discussed the multivariate meta-analysis and the three-level meta-analysis as two separate topics. In this section, we are going to compare and contrast these models. In Section 6.4.1, we show that the three-level meta-analysis is indeed a special case of the multivariate meta-analysis. In Section 6.4.2, we discuss under what conditions the multivariate meta-analysis can be fitted using the three-level meta-analysis when the degree of dependence is unknown. In Section 6.4.3, we combine the multivariate meta-analysis with the three-level model for the purpose of addressing more complicated research questions involving multiple effect sizes where the effect sizes are nested within clusters.

6.4.1 Three-level meta-analysis as a special case of the multivariate meta-analysis

For ease of discussion, we duplicate the random-effects multivariate meta-analysis in Equation 5.27 as follows:

$$\boldsymbol{\mu}_i(\boldsymbol{\theta}) = \boldsymbol{\beta}_{\mathrm{R}} \quad \text{and}$$
$$\boldsymbol{\Sigma}_i(\boldsymbol{\theta}) = \boldsymbol{T}^2 + \boldsymbol{V}_i. \tag{6.19}$$

There are three key features in the above multivariate meta-analysis. First, no structure has been imposed on $\boldsymbol{\mu}_i(\boldsymbol{\theta})$, meaning that the means of different effects can be different. Second, the off-diagonals of $\boldsymbol{V}_i$ are usually nonzero, meaning that the observed effect sizes are conditionally dependent. Third, $\boldsymbol{T}^2$ is an unstructured nonnegative matrix, meaning that the degree of heterogeneity can vary in different effect sizes with different degree of correlation (covariance).

When we compare the multivariate meta-analysis shown in Figure 5.2b and the three-level meta-analysis shown in of Figure 6.2b, it becomes apparent that the three-level meta-analysis is indeed a special case of the multivariate meta-analysis. As shown in Table 6.2, the effect sizes within a cluster are exchangeable in a three-level meta-analysis. That is, it does not matter whether these effect sizes are placed under y_1 or y_p within the same cluster. This imposes a particular structure on the model-implied means and the model-implied covariance matrix.

Specifically, the three-level meta-analysis can be formulated as a special case of the multivariate meta-analysis by imposing three constraints:

(i) $\underset{p\times 1}{\boldsymbol{\mu}_i(\boldsymbol{\theta})} = \underset{p\times 1}{\boldsymbol{1}}\beta_0$;

(ii) $\underset{p\times p}{\boldsymbol{V}_j}$ is a diagonal matrix; and

(iii) $\underset{p\times p}{\boldsymbol{T}^2} = \underset{p\times p}{\boldsymbol{1}}\tau^2_{(3)} + \underset{p\times p}{\boldsymbol{I}}\tau^2_{(2)}$ is a compound symmetry matrix.

The first constraint indicates that all of the p population effect sizes are equal under the three-level model because the effect sizes are exchangeable within a cluster. The second constraint shows that the p effect sizes are conditionally independent within a cluster. The third constraint indicates that the variances and the covariances of the *true* effect sizes are $\tau^2_{(2)} + \tau^2_{(3)}$ and $\tau^2_{(3)}$, respectively.

The second and the third constraints also highlight the main differences between a multivariate meta-analysis and a three-level meta-analysis. In the multivariate meta-analysis, the assumption is that the observed effect sizes are dependent because of the conditional sampling covariance matrix $\boldsymbol{V}_i$, whereas the *true* effect sizes are also correlated under the structure of $\boldsymbol{T}^2$. On the other hand, in the three-level meta-analysis, the assumption is that the observed effect sizes are conditionally independent. Thus, $\boldsymbol{V}_i$ is a diagonal matrix. The so-called dependence in the three-level meta-analysis is due to the *true* level-2 and level-3 random effects.

Practically speaking, software packages for the multivariate meta-analysis that allow for the imposition of the above constraints, for example, the `meta()` function in the `metaSEM` package (Cheung, 2014a), may be used to fit the three-level meta-analysis. After fitting the multivariate meta-analysis with the above constraints, the estimated heterogeneity covariance matrix is $\hat{\boldsymbol{T}}^2 = \begin{bmatrix} \hat{\tau}^2_V & & & \\ \hat{\tau}^2_C & \hat{\tau}^2_V & & \\ \vdots & \ddots & \ddots & \\ \hat{\tau}^2_C & \cdots & \hat{\tau}^2_C & \hat{\tau}^2_V \end{bmatrix}$. We may convert the parameter estimates under the multivariate meta-analysis into the parameter estimates of the three-level meta-analysis using $\hat{\tau}^2_{(3)} = \hat{\tau}^2_C$ and $\hat{\tau}^2_{(2)} = \hat{\tau}^2_V - \hat{\tau}^2_C$. Researchers may then calculate the $I^2_{(2)}$ and $I^2_{(3)}$ and $\text{ICC}^2_{(2)}$ and $\text{ICC}^2_{(3)}$. The `meta3()` function in the `metaSEM` package automatically calculates these indices.

6.4.2 Approximating a multivariate meta-analysis with a three-level meta-analysis

As discussed in Chapter 5, one critical assumption in multivariate meta-analysis is that the conditional sampling covariances of the multivariate effect sizes are known. Sometimes, the information required to calculate the sampling covariances is not available. Because of the similarity between a multivariate meta-analysis and a three-level meta-analysis, some researchers (e.g., Cheung and Chan, 2014; Van den Noortgate et al., 2013) have suggested conducting a multivariate meta-analysis with a three-level meta-analysis when the degree of dependence is unknown. In this section, we discuss how and when a three-level meta-analysis may be used to analyze the multivariate effect sizes even when the degree of dependence is unknown.

The three-level model at first appears to be quite restrictive because it assumes that the p effect sizes have to be exchangeable. Therefore, it is not appropriate to model distinct multivariate effect sizes such as the standardized mean differences on SAT-math and SAT-verbal. This is not entirely true. We may include level-2 dummy variables to indicate different types of effect sizes that are being modeled, that is,

$$\underset{p\times 1}{\boldsymbol{\mu}_j(\boldsymbol{\theta})} = \underset{p\times p}{\begin{bmatrix} 1 & 0 & \cdots & 0 \\ 0 & 1 & \ddots & \vdots \\ \vdots & \ddots & \ddots & 0 \\ 0 & \cdots & 0 & 1 \end{bmatrix}} \underset{p\times 1}{\begin{bmatrix} \beta_1 \\ \beta_2 \\ \vdots \\ \beta_p \end{bmatrix}}, \tag{6.20}$$

where β_1 to β_p are the average population effect sizes for different types of effect sizes. If the effect sizes are missing, the correspondent diagonals in the identity matrix $\underset{p\times p}{\boldsymbol{I}}$ are replaced by 0.

The main challenge is how to model the covariance structure. As we have discussed earlier, *dependence* means something slightly different in the context of a multivariate meta-analysis than it does in a three-level meta-analysis. In a multivariate meta-analysis, there are a conditional sampling covariance matrix $\boldsymbol{V}_i$ that varies across studies and a covariance matrix of the *true* effect sizes $\boldsymbol{T}^2$ that applies to all studies. In a three-level meta-analysis, the effect sizes are conditionally independent, that is, $\boldsymbol{V}_i = \text{Diag}(v_1, v_2, \cdots, v_k)$, while the dependence is due to the $\tau^2_{(3)}$ of the *true* effect sizes in the same cluster.

To model a multivariate meta-analysis using a three-level model, we have to make a few assumptions. The first one is that the heterogeneity variances are the same for all effect sizes. In our example, we assume that the degree of heterogeneity is similar for the SAT-math and SAT-verbal. We may then impose the following assumptions on the multivariate effect sizes:

(i) Common degree of heterogeneity on the *true* effect sizes (diagonals of $\boldsymbol{T}^2$): $\tau^2_{\text{V}} = \tau^2_{11} = \tau^2_{22} = \cdots = \tau^2_{pp}$;

(ii) Common degree of covariance among the *true* effect sizes (off-diagonals of $\boldsymbol{T}^2$): $\tau^2_{\text{C}} = \tau^2_{21} = \tau^2_{31} = \cdots = \tau^2_{p(p-1)}$, and

(iii) Common degree of conditional covariance among the observed effect sizes (off-diagonals of all V_i) for all studies: $\breve{v} = v_{21} = v_{31} = \cdots = v_{p(p-1)}$.

It is instructive to show the model-implied covariance matrices for the multivariate meta-analysis $\boldsymbol{\Sigma}_{\text{Mul}j}(\boldsymbol{\theta})$ under the above assumptions and the model-implied covariance matrix for the three-level meta-analysis $\boldsymbol{\Sigma}_{3\text{L}j}(\boldsymbol{\theta})$ in the j unit together:

$$\boldsymbol{\Sigma}_{\text{Mul}j}(\boldsymbol{\theta}) = \begin{bmatrix} \tau^2_{\text{V}} + v_{11.j} & & & \\ \tau^2_{\text{C}} + \breve{v} & \tau^2_{\text{V}} + v_{22.j} & & \\ \vdots & \ddots & \ddots & \\ \tau^2_{\text{C}} + \breve{v} & \cdots & \tau^2_{\text{C}} + \breve{v} & \tau^2_{\text{V}} + v_{pp.j} \end{bmatrix} \text{ and}$$
$$\boldsymbol{\Sigma}_{3\text{L}j}(\boldsymbol{\theta}) = \begin{bmatrix} \tau^2_{(3)} + \tau^2_{(2)} + v_{1.j} & & & \\ \tau^2_{(3)} & \tau^2_{(3)} + \tau^2_{(2)} + v_{2.j} & & \\ \vdots & \ddots & \ddots & \\ \tau^2_{(3)} & \cdots & \tau^2_{(3)} & \tau^2_{(3)} + \tau^2_{(2)} + v_{p.j} \end{bmatrix}, \quad (6.21)$$

where the subscript j on $v_{i.j}$ indicates that $v_{i.j}$ varies across the cluster.

The parameters in the multivariate meta-analysis and the three-level meta-analysis are now related using the following equations:

(i) $\tau^2_{(3)} = \tau^2_{\text{C}} + \breve{v}$; and

(ii) $\tau^2_{(2)} = \tau^2_{\text{V}} - \tau^2_{\text{C}} - \breve{v}$.

As there are three parameters under a multivariate meta-analysis and there are only two parameters under a three-level meta-analysis, it is not possible to recover the original parameters in a multivariate meta-analysis. If we are willing to assume that $\breve{v} \approx 0$, we may compute τ^2_{C} and τ^2_{V} from the above equations.

In practice, some or all of the above assumptions may be wrong. The empirical performance of the three-level meta-analysis on the multivariate meta-analysis depends on the degree of violation and sensitivity of these assumptions. Van den Noortgate et al. (2013) showed that the above approach worked reasonably well under their settings in a simulation study.

On the other hand, Cheung and Chan (2014) used a different model to analyze multivariate effect sizes within a three-level meta-analysis. These authors assumed that the level-2 heterogeneity variance was 0, meaning that the effect sizes within the cluster were perfect replication of each other. The observed differences on the effect sizes within the study were mainly due to a sampling error. Their preliminary simulation results showed that this approach works reasonably well. These two simulation studies seem to suggest that the above approximation works reasonably well. Further studies may address the empirical performance of these approaches in various realistic conditions.

6.4.3 Three-level multivariate meta-analysis

When the research questions and data are complicated, it may not be sufficient to conduct a three-level meta-analysis with only one effect size. For example, multiple effect sizes may be reported in a cross-cultural meta-analysis where the researchers want to separate the between-culture variation (level 3) and the within-culture variation (level 2). Another example was given by Wilson (2013). Wilson was interested in conducting a meta-analytic structural equation modeling (MASEM; see Chapter 7) on a pool of correlation coefficients. As each study may contribute more than one effect size (correlation coefficient), the effect sizes are nested within studies. This is similar to the hierarchical dependence discussed by Stevens and Taylor (2009). To address these questions properly, we may combine the multivariate meta-analysis (Cheung, 2013) and the three-level meta-analysis (Cheung, 2014b).

Without loss of generality, we assume that there are two effect sizes x_{ij} and y_{ij} with their known sampling covariance matrix $\boldsymbol{V}_{ij}$ in the ith set of the effect size in the jth cluster. We further suppose that there are k clusters with p sets of effect sizes nested within the studies. We assume that the sets of effect sizes are conditionally independent within a cluster. That is, the set of (x_{ij} and y_{ij}) and (x_{kj} and y_{kj}) are conditionally independent. The dependence between them is due to the level-2 variance component of the random effects. Table 6.4 illustrates a sample data set with a maximum of three sets of effect sizes per cluster. There is only one set of effect sizes in cluster 1, while there are three sets of effect sizes in cluster 2. There are incomplete effect sizes in cluster k.

Table 6.4 Two effect sizes nested within k clusters

Cluster	y_1	x_1	y_2	x_2	y_3	x_3
1	$y_{1,1}$	$x_{1,1}$	NA	NA	NA	NA
2	$y_{1,2}$	$x_{1,2}$	$y_{2,2}$	$x_{2,2}$	$y_{3,2}$	$x_{3,2}$
⋮	⋮	⋮	⋮	⋮	⋮	⋮
k	$y_{1,k}$	NA	NA	$x_{2,k}$	NA	NA

Abbreviation: NA, not available.

The model for the ith set of effect sizes x_{ij} and y_{ij} in the j cluster is

$$x_{ij} = \mu_x + u_{x(2)ij} + u_{x(3)j} + e_{x.ij} \quad \text{and}$$
$$y_{ij} = \mu_y + u_{y(2)ij} + u_{y(3)j} + e_{y.ij}, \tag{6.22}$$

where μ_x is the average population effect of x_{ij}, $u_{x(2)ij}$ and $u_{x(3)j}$ are the level-2 and level-3 random effects for x_{ij}, $e_{x.ij}$ is the sampling error of x_{ij}, and the other quantities are similarly defined for y_{ij}.

To conduct a multivariate three-level meta-analysis, we need to specify both the mean structure and the covariance structures. The mean structure $\boldsymbol{\mu}(\boldsymbol{\theta})$ consists of the fixed effects, such as intercepts and the regression coefficients. Thus, there is only one mean structure regardless of how many levels there are. In contrast, we need to specify the level-2 covariance structure $\boldsymbol{\Sigma}_{(2)}(\boldsymbol{\theta})$ and the level-3 covariance structure $\boldsymbol{\Sigma}_{(3)}(\boldsymbol{\theta})$ for the variables.

When there is no direction of prediction imposed between x_{ij} and y_{ij}, the model-implied mean structure and the model-implied covariance structures are

$$\underset{2\times 1}{\boldsymbol{\mu}_{ij}}(\boldsymbol{\theta}) = \mathsf{E}\left(\begin{bmatrix} x_{ij} \\ y_{ij} \end{bmatrix}\right) = \begin{bmatrix} \mu_x \\ \mu_y \end{bmatrix},$$
$$\underset{2\times 2}{\boldsymbol{\Sigma}_{(2)ij}}(\boldsymbol{\theta}) = \mathsf{Cov}\left(\begin{bmatrix} u_{x(2)ij} \\ u_{y(2)ij} \end{bmatrix}\right) = \boldsymbol{T}^2_{(2)} = \begin{bmatrix} \tau^2_{(2)xx} & \\ \tau^2_{(2)xy} & \tau^2_{(2)yy} \end{bmatrix}, \quad \text{and} \tag{6.23}$$
$$\underset{2\times 2}{\boldsymbol{\Sigma}_{(3)j}}(\boldsymbol{\theta}) = \mathsf{Cov}\left(\begin{bmatrix} u_{x(3)j} \\ u_{y(3)j} \end{bmatrix}\right) = \boldsymbol{T}^2_{(3)} = \begin{bmatrix} \tau^2_{(3)xx} & \\ \tau^2_{(3)xy} & \tau^2_{(3)yy} \end{bmatrix},$$

where the subscripts (2) and (3) represent the level-2 and level-3 heterogeneity variance components, respectively. The above model can be extended to models with predictions or mediators (see Section 5.6).

Once we have specified the model-implied mean structure and the model-implied covariance structures, we may convert them into a model that takes the nested structure into account similar to that in Equations 6.16 and 6.17. In this example, there are p sets of effect sizes and two effect sizes per each set. Thus, the model for the jth cluster is

$$\underset{2p\times 1}{\boldsymbol{\mu}_{j}}(\boldsymbol{\theta}) = \underset{p\times 1}{\boldsymbol{1}} \otimes \underset{2\times 1}{\boldsymbol{\mu}_{ij}}(\boldsymbol{\theta}) \quad \text{and}$$
$$\underset{2p\times 2p}{\boldsymbol{\Sigma}_{j}}(\boldsymbol{\theta}) = \underset{p\times p}{\boldsymbol{1}} \otimes \underset{2\times 2}{\boldsymbol{\Sigma}_{(3)j}}(\boldsymbol{\theta}) + \underset{p\times p}{\boldsymbol{I}} \otimes \underset{2\times 2}{\boldsymbol{\Sigma}_{(2)ij}}(\boldsymbol{\theta}) \tag{6.24}$$
$$+ \underset{2p\times 2p}{\mathrm{Diag}(\boldsymbol{V}_{1j}, \boldsymbol{V}_{2j}, \cdots, \boldsymbol{V}_{pj})},$$

where $\boldsymbol{A} \otimes \boldsymbol{B}$ is the Kronecker product between matrices $\boldsymbol{A}$ and $\boldsymbol{B}$. Suppose that $\underset{m\times n}{\boldsymbol{A}}$ and $\underset{r\times s}{\boldsymbol{B}}$; $\boldsymbol{A} \otimes \boldsymbol{B} = \underset{mr\times ns}{\begin{bmatrix} a_{11}\boldsymbol{B} & \cdots & a_{1n}\boldsymbol{B} \\ \vdots & \ddots & \vdots \\ a_{m1}\boldsymbol{B} & \cdots & a_{mn}\boldsymbol{B} \end{bmatrix}}$.

As far as the author knows, the only SEM package that can fit this model is the `OpenMx` package (Boker et al., 2011). Users may have to specify the above model-implied mean vectors and covariance matrices in `OpenMx`. Another challenge is that we still need to know the conditional sampling covariance matrix $\boldsymbol{V}_{ij}$. However, studies may not provide enough information to calculate $\boldsymbol{V}_{ij}$. The correlations among the effect sizes have to be estimated from other sources or imputed. The empirical performance of this approach remains unclear. Future studies are needed to determine how useful this model is and what its empirical performance is.

6.5 Illustrations using R

This section demonstrates how to conduct a three-level meta-analysis using the `metaSEM` package. How to make comparisons between the conventional two-level meta-analysis and the three-level meta-analysis is also illustrated. This section also demonstrates how to quantify the heterogeneity and explained variance at level 2 and level 3. It should be noted that the statistics reported in the illustrations were captured by using the `Sweave` function in R. The numbers of decimal places may be slightly different for those reported in the selected output and in the text.

The illustrations in this section were based on a data set from Bornmann et al. (2007). Marsh et al. (2009) reanalyzed the same data set in more detail. Bornmann et al. (2007) extracted 66 effect sizes from 21 studies of gender differences in the peer reviews of grant and fellowship applications. The effect size was an odds ratio that measured the odds of being approved among the female applicants divided by the odds of being approved among the male applicants. To improve the normality assumption with regard to the effect sizes, a logarithm was applied to the odds ratio. If the effect size was positive, female applicants were favored to receive the grant or fellowship, whereas if it was negative, male applicants were favored to obtain grant or fellowship.

Bornmann et al. (2007) and Marsh et al. (2009) used a three-level meta-analysis to handle dependent effect sizes. Level 1 was the authors, whereas levels 2 and 3 were the proposals and studies, respectively. Cheung (2014b) provided the R code to replicate the results of Table 3 of Marsh et al. (2009). The focus of this section is how to conduct the analyses in R. Readers should refer to Bornmann et al. (2007) and Marsh et al. (2009) for substantive interpretations of the results.

6.5.1 Inspecting the data

The data set was stored as `Bornmann07` in the `metaSEM` package. The effect size and its sampling variance are `logOR` and `v`, respectively, while `Cluster` is the cluster effect at level 3. The covariates are `Year` (Year of publication), `Type` (`Fellowship` vs `Grants`), `Discipline` (`Physical sciences`, `Life sciences/biology`, `Social sciences/humanities`, or `Multi-disciplinary`), and `Country` (`United States`, `Canada`, `Australia`, `United Kingdom`, or `Europe`). We inspect the data set by the following commands. The numbers of effect sizes per cluster varies from 1 to 9.

```
R> library("metaSEM")
R> ## Show the first few cases in the data set
R> head(Bornmann07)
```

```
  Id                        Study Cluster    logOR       v Year
1  1 Ackers (2000a; Marie Curie)        1 -0.40108 0.01392 1996
2  2 Ackers (2000b; Marie Curie)        1 -0.05727 0.03429 1996
3  3 Ackers (2000c; Marie Curie)        1 -0.29852 0.03391 1996
4  4 Ackers (2000d; Marie Curie)        1  0.36094 0.03404 1996
5  5 Ackers (2000e; Marie Curie)        1 -0.33336 0.01282 1996
6  6 Ackers (2000f; Marie Curie)        1 -0.07173 0.01361 1996
        Type               Discipline Country
1 Fellowship        Physical sciences  Europe
2 Fellowship        Physical sciences  Europe
```

```
3 Fellowship          Physical sciences  Europe
4 Fellowship          Physical sciences  Europe
5 Fellowship Social sciences/humanities  Europe
6 Fellowship          Physical sciences  Europe
```

```
R>
R> ## Show the last few cases in the data set
R> ## tail(Bornmann07)
R> ## Display the no. of effect sizes per cluster
R> t(aggregate(logOR~ Cluster, data=Bornmann07, FUN=length))
```

```
        [,1] [,2] [,3] [,4] [,5] [,6] [,7] [,8] [,9] [,10] [,11]
Cluster    1    2    3    4    5    6    7    8    9    10    11
logOR      7    7    1    1    5    3    1    3    2     4     1
        [,12] [,13] [,14] [,15] [,16] [,17] [,18] [,19] [,20] [,21]
Cluster    12    13    14    15    16    17    18    19    20    21
logOR       8     1     1     2     3     1     1     1     4     9
```

6.5.2 Fitting a random-effects model

The syntax for the `meta3()` function is similar to that of the `meta()` function. The arguments `y` and `v` are used to specify the effect size and its conditional sampling variance, respectively. The `cluster` argument is used to specify how the effect sizes are clustered.

```
R> ## Model 0: Random-effects model
R> summary( Model0 <- meta3(y=logOR, v=v, cluster=Cluster,
                            data=Bornmann07, model.name="3 level") )
```

```
----------------------- Selected output -----------------------

95% confidence intervals: z statistic approximation
Coefficients:
           Estimate Std.Error   lbound   ubound z value Pr(>|z|)
Intercept -0.10078   0.04013 -0.17944 -0.02212   -2.51    0.012 *
Tau2_2     0.00380   0.00272 -0.00154  0.00913    1.40    0.163
Tau2_3     0.01414   0.00914 -0.00379  0.03206    1.55    0.122
---
Signif. codes:  0 '***' 0.001 '**' 0.01 '*' 0.05 '.' 0.1 ' ' 1

Q statistic on the homogeneity of effect sizes: 221.3
Degrees of freedom of the Q statistic: 65
P value of the Q statistic: 0

Heterogeneity indices (based on the estimated Tau2):
```

```
                                    Estimate
I2_2 (Typical v: Q statistic)          0.16
I2_3 (Typical v: Q statistic)          0.58

Number of studies (or clusters): 21
Number of observed statistics: 66
Number of estimated parameters: 3
Degrees of freedom: 63
-2 log likelihood: 25.8
OpenMx status1: 0 ("0" or "1": The optimization is considered fine.
Other values indicate problems.)

----------------------  Selected output  ----------------------
```

Before interpreting the results, we should check whether there are any estimation problems in the analysis. We may check the `OpenMx status1` at the end of the output. If the status is either 0 or 1, the optimization is fine; otherwise, the results are not trustworthy. If there are problems in the optimization, we may try to rerun the analysis with the `rerun()` function (see Section 7.6.3 for an example). From the above output, there are a total of 66 effect sizes (level 2) nested within 21 clusters (level 3). The Q statistic on the homogeneity of the effect sizes is $Q(df = 65) = 221.2809, p < 0.001$. The estimated average population effect size (with its approximate 95% Wald confidence interval (CI)) is -0.1008 $(-0.1794, -0.0221)$. As the 95% Wald CI does not include 0, this result indicates that male applicants have a slightly higher chance of obtaining the grants.

The estimated level-2 and level-3 heterogeneity variances are 0.0038 and 0.0141, respectively. It should be noted that the approximate CIs are based on the Wald test, while the variances are not likely to be distributed normally unless the sample sizes are huge. We should avoid interpreting the Wald CIs of the heterogeneity variances. As it is difficult to interpret how much (or small) the heterogeneity variances are, we may inspect the I^2. Level 2 and level 3 explain 15.68% and 58.39% of the total variation, respectively, while the remaining 26% is due to the within-study known sampling variance.

6.5.3 Obtaining the likelihood-based confidence interval

We can obtain an LBCI on the parameter estimates. The LBCI is more accurate than the CI based on the Wald statistic (see Section 2.4.5). This is particularly useful for quantifying the uncertainty in τ^2 and I^2. We may obtain the LBCI by specifying the `intervals.type="LB"` argument in the call.

```
R> ## Model 0: Random-effects model with LBCI
R> summary( meta3(y=logOR, v=v, cluster=Cluster,
                  data=Bornmann07, model.name="3 level",
                  intervals.type="LB") )
```

```
----------------------  Selected output  ----------------------

95% confidence intervals: Likelihood-based statistic
Coefficients:
           Estimate Std.Error     lbound     ubound z value Pr(>|z|)
Intercept -0.100778        NA -0.186035 -0.016481      NA       NA
Tau2_2     0.003796        NA  0.000821  0.014756      NA       NA
Tau2_3     0.014135        NA  0.003397  0.048409      NA       NA

Heterogeneity indices (I2) and their 95% likelihood-based CIs:
                              lbound Estimate ubound
I2_2 (Typical v: Q statistic) 0.0288   0.1568   0.54
I2_3 (Typical v: Q statistic) 0.1795   0.5839   0.84

----------------------  Selected output  ----------------------
```

The estimated level-2 and level-3 heterogeneity variances (with their 95% LBCI) are 0.0038 (0.0008, 0.0148) and 0.0141 (0.0034, 0.0484), respectively. The level-2 $I^2_{(2)}$ and level-3 $I^2_{(3)}$ with their 95% LBCI are 0.1568 (0.0288, 0.5429) and 0.5839 (0.1795, 0.8436), respectively. The 95% LBCIs are very wide, indicating that the estimates may vary from sample to sample if we were able to collect new studies with similar settings for a new meta-analysis.

6.5.4 Testing $\tau^2_{(3)} = 0$

As an illustration, we fit a conventional two-level model and compare it against the three-level model. We fit the two-level model either with the `meta()` function or with the `meta3()` function by applying the constraint of $\tau^2_{(3)} = 0$. After fitting the two-level model (`Model1` in this example), we compare it against the three-level model (`Model0` in this example) with the `anova()` function.

```
R> ## Model 1: Testing tau^2_3 = 0
R> Model1 <- meta3(logOR, v, cluster=Cluster, data=Bornmann07,
                   RE3.constraints=0, model.name="2 level")
R> ## Alternative approach
R> ## Model1 <- meta(logOR, v, data=Bornmann07, model.name="2 level")
R>
R> anova(Model0, Model1)
```

```
     base comparison ep minus2LL df     AIC diffLL diffdf        p
1 3 level       <NA>  3    25.80 63 -100.20     NA     NA       NA
2 3 level    2 level  2    36.02 64  -91.98  10.22      1 0.001389
```

The LR statistic on comparing the level-3 and level-2 models is $\chi^2(\text{df} = 1) = 10.2202, p = 0.0014$. It indicates that the three-level model is statistically better than the two-level model. As a reminder, we should still prefer to use the three-level model even when the test is not statistically significant.

6.5.5 Testing $\tau^2_{(2)} = 0$

We can also test the null hypothesis $H_0 : \tau^2_{(2)} = 0$ by specifying the `RE2.constraints=0` argument in calling the `meta3()` function.

```
R> ## Model 2: Testing tau^2_2 = 0
R> Model2 <- meta3(logOR, v, cluster=Cluster, data=Bornmann07,
                   RE2.constraints=0, model.name="tau2_2 EQ 0")
R> anova(Model0, Model2)
```

```
     base   comparison ep minus2LL df      AIC diffLL diffdf        p
1 3 level         <NA>  3    25.80 63 -100.20     NA     NA       NA
2 3 level tau2_2 EQ 0  2    38.52 64  -89.48  12.72      1 0.000362
```

The LR statistic on comparing the level-3 model and model with $\tau^2_{(2)} = 0$ is χ^2 (df = 1) = 12.7187, $p = 0.0004$. It indicates that the three-level model is statistically better than the model with $\tau^2_{(2)} = 0$. Therefore, the effect sizes within a study are not merely direct replication of each other—there are true differences among them.

6.5.6 Testing $\tau^2_{(2)} = \tau^2_{(3)}$

We test whether the level-2 and the level-3 heterogeneity variances are the same. We fit a model (`Model3` in this example) with the constraint $\tau^2_{(2)} = \tau^2_{(3)}$ by using the same label (e.g., `Eq_tau2` in the example) in the `RE2.constraints` and `RE3.constraints` arguments. The value of `0.1` is the starting value in the example.

```
R> ## Model 3: Testing tau^2_2 = tau^2_3
R> Model3 <- meta3(logOR, v, cluster=Cluster, data=Bornmann07,
                   RE2.constraints="0.1*Eq_tau2",
                   RE3.constraints="0.1*Eq_tau2",
                   model.name="Eq tau2")
```

```
----------------------- Selected output -----------------------

95% confidence intervals: z statistic approximation
Coefficients:
           Estimate Std.Error    lbound    ubound z value Pr(>|z|)
Intercept -0.09602   0.03461 -0.16385 -0.02819   -2.77   0.0055 **
Eq_tau2    0.00738   0.00301  0.00149  0.01328    2.45   0.0141 *
---
Signif. codes:  0 '***' 0.001 '**' 0.01 '*' 0.05 '.' 0.1 ' ' 1

Heterogeneity indices (based on the estimated Tau2):
                              Estimate
```

```
I2_2 (Typical v: Q statistic)     0.35
I2_3 (Typical v: Q statistic)     0.35

----------------------  Selected output  ----------------------
```

By imposing the equality constraint $\tau^2_{(2)} = \tau^2_{(3)}$, the common heterogeneity variance is 0.0074. As $\tau^2_{(2)} = \tau^2_{(3)}$, the level-2 $I^2_{(2)}$ and the level-3 $I^2_{(3)}$ are also the same. We test $H_0 : \tau^2_{(2)} = \tau^2_{(3)}$ by comparing the models with and without the constraint.

```
R> anova(Model0, Model3)
```

```
     base comparison ep minus2LL df    AIC diffLL diffdf      p
1 3 level       <NA>  3    25.80 63 -100.2     NA     NA     NA
2 3 level    Eq tau2  2    27.16 64 -100.8  1.359      1 0.2437
```

The LR statistic is $\chi^2(\mathrm{df} = 1) = 1.3591, p = 0.2437$. Therefore, we cannot reject $H_0 : \tau^2_{(2)} = \tau^2_{(3)}$. It seems that the heterogeneity variances are similar at both levels.

6.5.7 Testing types of proposals (grant versus fellowship)

There are two types of studies in the data set. They are the applications for either a *grant* or a *fellowship*. We may test whether the effects are the same for them. First, we show the first 10 cases of `Type` in the data set. We then create a dummy variable `Type2` with the `ifelse()` function, where `0` and `1` represent *grant* and *fellowship*, respectively. We conduct a mixed-effects three-level meta-analysis by specifying the `x=Type2` argument.

```
R> ## Show the first 10 cases of Type
R> Bornmann07$Type[1:10]
```

```
 [1] Fellowship Fellowship Fellowship Fellowship Fellowship
    Fellowship
 [7] Fellowship Grant      Grant      Grant
Levels: Grant Fellowship
```

```
R> ## Convert characters into a dummy variable
R> ## Type2=0 (Grant); Type2=1 (Fellowship)
R> Type2 <- ifelse(Bornmann07$Type=="Fellowship", yes=1, no=0)
R> ## Show the first 10 cases of Type2
R> Type2[1:10]
```

```
 [1] 1 1 1 1 1 1 1 0 0 0
```

```
R> ## Model 4: Type2 as the covariate
R> summary( meta3(y=logOR, v=v, x=Type2, cluster=Cluster,
                 data=Bornmann07) )
```

```
----------------------- Selected output -----------------------

95% confidence intervals: z statistic approximation
Coefficients:
           Estimate Std.Error    lbound    ubound z value Pr(>|z|)
Intercept -0.00661   0.03711 -0.07935  0.06613   -0.18  0.85870
Slope_1   -0.19559   0.05416 -0.30175 -0.08943   -3.61  0.00031 ***
Tau2_2     0.00353   0.00243 -0.00123  0.00830    1.45  0.14601
Tau2_3     0.00291   0.00312 -0.00320  0.00902    0.93  0.35107
---
Signif. codes:  0 '***' 0.001 '**' 0.01 '*' 0.05 '.' 0.1 ' ' 1

Explained variances (R2):
                        Level 2 Level 3
Tau2 (no predictor)     0.00380    0.01
Tau2 (with predictors) 0.00353    0.00
R2                      0.06926    0.79

----------------------- Selected output -----------------------
```

As `Type2=0` represents *grant*, the estimated `Intercept` is the average effect on *grant*. The estimated average effect (with its 95% Wald CI) on *grant* is -0.0066 $(-0.0793, 0.0661)$. The estimated average difference between *fellowship* and *grant* (`Slope_1` in the output) (with its 95% Wald CI) is -0.1956 $(-0.3017, -0.0894)$. As the 95% Wald CI does not include 0, it is statistically significant at $\alpha = 0.05$. When compared to *grant*, male applicants are more likely to get a *fellowship* than female participants. The level-2 $R^2_{(2)}$ and level-3 $R^2_{(3)}$ are 0.0693 and 0.7943, respectively. The result indicates that the predictor `Type2` is mainly useful in explaining the level-3 heterogeneity.

6.5.8 Testing the effect of the year of application

Marsh et al. (2009) tested the effect of publication year. These authors hypothesized that *year* had a quadratic effect on the effect size. Continuous covariates are usually centered in the analysis to improve the numerical stability of the results and to facilitate interpretations of the intercept. We may use `scale(Year, center=TRUE, scale=FALSE)` to center *year* in R. After the centering, the intercept represents the predicted effect size when the moderators are at their mean values.

As Marsh et al. (2009) standardized *year* in their analyses, we follow their practice in this illustration. We use `scale(Year, center=TRUE, scale=TRUE)` to standardize *year* in R. When there are more than one moderators, we combine them with the `cbind()` function. In the output, `Slope_1` and `Slope_2` refer to the regression coefficients for *year* and its quadratic term, respectively.

After fitting this model (`Model5` in this example), we compare it against the model without the moderators. Under the null hypothesis that both regression coefficients are zero, the LR statistic has a chi-square distribution with 2 dfs.

```
R> ## Model 5: Year and Year^2 as covariates
R> summary( Model5 <- meta3(y=logOR, v=v,
                          x=cbind(scale(Year), scale(Year)^2),
                          cluster=Cluster, data=Bornmann07,
                          model.name="Model 5") )
```

```
----------------------- Selected output -----------------------

95% confidence intervals: z statistic approximation
Coefficients:
            Estimate Std.Error    lbound    ubound z value Pr(>|z|)
Intercept -0.086273  0.041256 -0.167133 -0.005413   -2.09    0.037 *
Slope_1   -0.000953  0.023652 -0.047310  0.045405   -0.04    0.968
Slope_2   -0.011768  0.006600 -0.024704  0.001167   -1.78    0.075 .
Tau2_2     0.002874  0.002068 -0.001180  0.006927    1.39    0.165
Tau2_3     0.014794  0.009261 -0.003357  0.032946    1.60    0.110
---
Signif. codes:  0 '***' 0.001 '**' 0.01 '*' 0.05 '.' 0.1 ' ' 1

Explained variances (R2):
                        Level 2 Level 3
Tau2 (no predictor)     0.00380    0.01
Tau2 (with predictors)  0.00287    0.01
R2                      0.24301    0.00

----------------------- Selected output -----------------------
```

```
R> ## Testing H_0: beta_{Year} = beta_{Year^2}=0
R> anova(Model5, Model0)
```

```
     base comparison ep minus2LL df     AIC diffLL diffdf     p
1 Model 5       <NA>  5    22.38 61  -99.62     NA     NA    NA
2 Model 5    3 level  3    25.80 63 -100.20  3.419      2 0.181
```

The estimated regression coefficients (and their 95% CIs) for *year* and its quadratic term are -0.0010 $(-0.0473, 0.0454)$ and -0.0118 $(-0.0247, 0.0012)$, respectively. The level-2 $R^2_{(2)}$ and level-3 $R^2_{(3)}$ are 0.2430 and 0.0000, respectively. When comparing this model to the model without any moderator, the LR test statistic is $\chi^2(\text{df} = 2) = 3.4190, p = 0.1810$. Therefore, we cannot reject $H_0 : \beta_{\text{year}} = \beta_{\text{year}^2} = 0$. There is not enough evidence to support the hypothesis that *year* explains the effect size.

6.5.9 Testing the country effect

We test whether the effect sizes are different across countries or regions. There are five countries or regions in the data set. As there are five categories, we may create four dummy variables to represent them. Alternatively, we may create five indicators (D_{USA} to D_{UK}) to represent the countries or regions. The value for that particular country is 1; otherwise, it is 0. As the model with the intercept and the five indicators is unidentified, we need to fix the intercept at zero. This constraint can be imposed by using the `intercept.constraint=0` argument. The advantage of this parameterization is that we can estimate the average effect sizes for all countries or regions. The model is

$$y_{ij} = D_{\mathrm{USA}}\beta_1 + D_{\mathrm{Aus}}\beta_2 + D_{\mathrm{Can}}\beta_3 + D_{\mathrm{Eur}}\beta_4 + D_{\mathrm{UK}}\beta_5 + u_{(2)ij} + u_{(3)j} + e_{ij}. \quad (6.25)$$

```
R> ## Create indicators for countries
R> USA <- ifelse(Bornmann07$Country=="United States", yes=1, no=0)
R> Aus <- ifelse(Bornmann07$Country=="Australia", yes=1, no=0)
R> Can <- ifelse(Bornmann07$Country=="Canada", yes=1, no=0)
R> Eur <- ifelse(Bornmann07$Country=="Europe", yes=1, no=0)
R> UK  <- ifelse(Bornmann07$Country=="United Kingdom", yes=1, no=0)
```

```
R> ## Model 6: indicators for country as moderators
R> summary( Model6 <- meta3(y=logOR, v=v, intercept.constraint=0,
                             x=cbind(USA, Aus, Can, Eur, UK),
                             cluster=Cluster, data=Bornmann07,
                             model.name="Model 6") )
```

```
---------------------- Selected output ----------------------

95% confidence intervals: z statistic approximation
Coefficients:
         Estimate Std.Error   lbound   ubound z value Pr(>|z|)
Slope_1   0.00257   0.05978 -0.11459  0.11973    0.04     0.97
Slope_2  -0.02144   0.09266 -0.20305  0.16017   -0.23     0.82
Slope_3  -0.13151   0.10265 -0.33270  0.06968   -1.28     0.20
Slope_4  -0.21851   0.05008 -0.31667 -0.12035   -4.36  1.3e-05 ***
Slope_5   0.05629   0.07905 -0.09864  0.21122    0.71     0.48
Tau2_2    0.00334   0.00235 -0.00127  0.00794    1.42     0.16
Tau2_3    0.00480   0.00448 -0.00399  0.01358    1.07     0.28
---
Signif. codes:  0 '***' 0.001 '**' 0.01 '*' 0.05 '.' 0.1 ' ' 1

Explained variances (R2):
                        Level 2 Level 3
Tau2 (no predictor)     0.00380    0.01
Tau2 (with predictors)  0.00334    0.00
R2                      0.12086    0.66

---------------------- Selected output ----------------------
```

```
R> ## Testing H_0: all countries have the average effect sizes
R> anova(Model6, Model0)
```

```
     base comparison ep minus2LL df     AIC diffLL diffdf       p
1 Model 6       <NA>  7    14.18 59 -103.8     NA     NA      NA
2 Model 6    3 level  3    25.80 63 -100.2  11.62      4 0.02041
```

The estimated average effect sizes (and their 95% Wald CIs) for *United States*, *Australia*, *Canada*, *Europe*, and *United Kingdom* were 0.0026 (−0.1146, 0.1197), −0.0214 (−0.2031, 0.1602), −0.1315 (−0.3327, 0.0697), −0.2185 (−0.3167, −0.1204), and 0.0563 (−0.0986, 0.2112), respectively. When comparing this model to the model without any moderator, the LR test statistic is $\chi^2(\text{df} = 4) = 11.6200, p = 0.0204$. Therefore, the null hypothesis on the equality of population effect sizes is rejected. All of the average effect sizes are nonsignificant except that for *Europe* (−0.2185). The level-2 $R^2_{(2)}$ and level-3 $R^2_{(3)}$ are 0.1209 and 0.6606, respectively. The results suggest that *country* explains more heterogeneity at level 3 than at level 2 in terms of percentage.

6.6 Concluding remarks and further readings

This chapter reviewed several common approaches to handling nonindependent effect sizes in a meta-analysis. A three-level meta-analysis was introduced to model the dependence by considering both the level-2 and level-3 data structure. Reviewers are advised to consider the dependence in the effect sizes as a research opportunity rather than as a statistical problem. When the reviewers include all effect sizes in the analysis, the number of effect sizes will be larger than the number of studies. Moreover, both level-2 and level-3 moderators can be entered into the meta-analysis. In theory, a three-level meta-analysis is more powerful than conventional approaches to handling the dependence. Reviewers may also explore new research questions that cannot be addressed in conventional approaches. Researchers may explore how the heterogeneity be decomposed into different levels. Readers may refer to Cheung (2014b), Konstantopoulos (2011), and Van den Noortgate et al. (2013) for more details on the three-level meta-analysis. As there are only limited simulation studies on the empirical performance of three-level meta-analysis, it remains unclear how robust it is when the numbers of level 2 or (and) level 3 units are small or when the *true* effect sizes are not normally distributed. More studies should investigate its robustness to violation of some of these assumptions.

Another promising approach to addressing the issue of dependence effect size is the robust variance estimation (Hedges et al., 2010a,b; Raudenbush, 2009; Tanner-Smith and Tipton, 2014; Tipton, 2013). The robust variance estimation approach corrects the SEs owing to the dependence. As we have shown earlier, the conventional two-level meta-analysis can be considered as a three-level meta-analysis with misspecified covariance structure where $\tau^2_{(3)} = 0$. The use of robust statistics to address the covariance structure misspecification is popular in the SEM literature (e.g., Yuan and Bentler, 2007). Future research may compare the strengths and limitations of these two approaches to addressing the dependence in the effect sizes.

References

Ahn S, Ames AJ and Myers ND 2012. A review of meta-analyses in education: methodological strengths and weaknesses. *Review of Educational Research* **82**(4), 436–476.

Au K and Cheung MWL 2004. Intra-cultural variation and job autonomy in 42 countries. *Organization Studies* **25**(8), 1339–1362.

Becker BJ, Hedges LV and Pigott TD 2004. Campbell collaboration statistical analysis policy brief. Technical report, Campbell Collaboration, Oslo.

Boker S, Neale M, Maes H, Wilde M, Spiegel M, Brick T, Spies J, Estabrook R, Kenny S, Bates T, Mehta P and Fox J 2011. OpenMx: an open source extended structural equation modeling framework. *Psychometrika* **76**(2), 306–317.

Borenstein M, Hedges LV, Higgins JP and Rothstein HR 2010. A basic introduction to fixed-effect and random-effects models for meta-analysis. *Research Synthesis Methods* **1**(2), 97–111.

Bornmann L, Mutz R and Daniel HD 2007. Gender differences in grant peer review: a meta-analysis. *Journal of Informetrics* **1**(3), 226–238.

Cheung MWL 2010. Fixed-effects meta-analyses as multiple-group structural equation models. *Structural Equation Modeling: A Multidisciplinary Journal* **17**(3), 481–509.

Cheung MWL 2013. Multivariate meta-analysis as structural equation models. *Structural Equation Modeling: A Multidisciplinary Journal* **20**(3), 429–454.

Cheung MWL 2014a. metaSEM: An R package for meta-analysis using structural equation modeling. Frontiers in Psychology 5(1521).

Cheung MWL 2014b. Modeling dependent effect sizes with three-level meta-analyses: a structural equation modeling approach. *Psychological Methods* **19**(2), 211–229.

Cheung SF and Chan DKS 2004. Dependent effect sizes in meta-analysis: incorporating the degree of interdependence. *Journal of Applied Psychology* **89**(5), 780–791.

Cheung SF and Chan DKS 2008. Dependent correlations in meta-analysis the case of heterogeneous dependence. *Educational and Psychological Measurement* **68**(5), 760–777.

Cheung SF and Chan DKS 2014. Meta-analyzing dependent correlations: an SPSS macro and an R script. *Behavior Research Methods* **46**(2), 331–345.

Cochran W 1954. The combination of estimates from different experiments. *Biometrics* **10**(1), 101–129.

Cooper HM 2010. *Research synthesis and meta-analysis: a step-by-step approach*, 4th edn. Sage Publications, Inc., Los Angeles, CA.

Fischer R and Boer D 2011. What is more important for national well-being: money or autonomy? A meta-analysis of well-being, burnout, and anxiety across 63 societies. *Journal of Personality and Social Psychology* **101**(1), 164–184.

Fischer R, Hanke K and Sibley CG 2012. Cultural and institutional determinants of social dominance orientation: a cross-cultural meta-analysis of 27 societies. *Political Psychology* **33**(4), 437–467.

Gleser LJ and Olkin I 1994. Stochastically dependent effect sizes In *The handbook of research synthesis* (ed. Cooper H and Hedges LV). Russell Sage Foundation, New York, pp. 339–355.

Goldstein H 2011. *Multilevel statistical models*, 4th edn. John Wiley & Sons, Inc., Hoboken, NJ.

Hanke K and Fischer R 2013. Socioeconomical and sociopolitical correlates of interpersonal forgiveness: a three-level meta-analysis of the Enright Forgiveness Inventory across 13 societies. *International Journal of Psychology* **48**(4), 514–526.

Hedges LV and Pigott TD 2001. The power of statistical tests in meta-analysis. *Psychological Methods* **6**(3), 203–217.

Hedges LV and Pigott TD 2004. The power of statistical tests for moderators in meta-analysis. *Psychological Methods* **9**(4), 426–445.

Hedges LV, Tipton E and Johnson MC 2010a. Erratum: robust variance estimation in meta-regression with dependent effect size estimates. *Research Synthesis Methods* **1**(2), 164–165.

Hedges LV, Tipton E and Johnson MC 2010b. Robust variance estimation in meta-regression with dependent effect size estimates. *Research Synthesis Methods* **1**(1), 39–65.

Hedges LV and Vevea JL 1998. Fixed- and random-effects models in meta-analysis. *Psychological Methods* **3**(4), 486–504.

Higgins JPT and Thompson SG 2002. Quantifying heterogeneity in a meta-analysis. *Statistics in Medicine* **21**(11), 1539–1558.

Hox JJ 2010. *Multilevel analysis: techniques and applications*, 2nd edn. Routledge, New York.

Konstantopoulos S 2011. Fixed effects and variance components estimation in three-level meta-analysis. *Research Synthesis Methods* **2**(1), 61–76.

Marín-Martínez F and Sánchez-Meca J 1999. Averaging dependent effect sizes in meta-analysis: a cautionary note about procedures. *Spanish Journal of Psychology* **2**(1), 32–38.

Marsh HW, Bornmann L, Mutz R, Daniel HD and O'Mara A 2009. Gender effects in the peer reviews of grant proposals: a comprehensive meta-analysis comparing traditional and multilevel approaches. *Review of Educational Research* **79**(3), 1290–1326.

Mehta PD and Neale MC 2005. People are variables too: multilevel structural equations modeling. *Psychological Methods* **10**(3), 259–284.

Nguyen AMD and Benet-Martínez V 2013. Biculturalism and adjustment: a meta-analysis. *Journal of Cross-Cultural Psychology* **44**(1), 122–159.

Owens C and Ferron J 2012. Synthesizing single-case studies: a Monte Carlo examination of a three-level meta-analytic model. *Behavior Research Methods* **44**(3), 795–805.

Pinheiro JC and Bates D 2000. *Mixed-effects models in S and S-Plus*. Springer-Verlag, New York.

Preacher KJ 2011. Multilevel SEM strategies for evaluating mediation in three-level data. *Multivariate Behavioral Research* **46**(4), 691–731.

Raudenbush SW 2009. Analyzing effect sizes: random effects models In *The handbook of research synthesis and meta-analysis* (ed. Cooper HM, Hedges LV and Valentine JC), 2nd edn. Russell Sage Foundation, New York, pp. 295–315.

Raudenbush SW and Bryk AS 2002. *Hierarchical linear models: applications and data analysis methods*. Sage Publications, Inc., Thousand Oaks, CA.

Rosenthal R and Rubin DB 1986. Meta-analytic procedures for combining studies with multiple effect sizes. *Psychological Bulletin* **99**(3), 400–406.

Scammacca N, Roberts G and Stuebing KK 2013. Meta-analysis with complex research designs dealing with dependence from multiple measures and multiple group comparisons. *Review of Educational Research* **84**(3), 328–364.

Schmidt FL and Hunter JE 2015. *Methods of meta-analysis: correcting error and bias in research findings*, 3rd edn. Sage Publications, Inc., Thousand Oaks, CA.

Shin IS 2009. *Same author and same data dependence in meta-analysis* PhD thesis. Florida State University.

Snijders TAB and Bosker RJ 2012. *Multilevel analysis: an introduction to basic and advanced multilevel modeling*, 2nd edn. Sage Publications, Inc., Los Angeles, CA, London.

Stevens JR and Taylor AM 2009. Hierarchical dependence in meta-analysis. *Journal of Educational and Behavioral Statistics* **34**(1), 46–73.

Tanner-Smith EE and Tipton E 2014. Robust variance estimation with dependent effect sizes: practical considerations including a software tutorial in Stata and SPSS. *Research Synthesis Methods* **5**(1), 13–30.

Tipton E 2013. Robust variance estimation in meta-regression with binary dependent effects. *Research Synthesis Methods* **4**(2), 169–187.

Van den Bussche E, Van den Noortgate W and Reynvoet B 2009. Mechanisms of masked priming: a meta-analysis. *Psychological Bulletin* **135**(3), 452–477.

Van den Noortgate W, López-López JA, Marín-Martínez F and Sánchez-Meca J 2013. Three-level meta-analysis of dependent effect sizes. *Behavior Research Methods* **45**(2), 576–594.

Wilson SJ 2013. An application of multivariate meta-analytic structural equation modeling: longitudinal relations between school readiness skills and later school performance. Paper presented at the 8th Annual Meeting of the Society for Research Synthesis Methodology, Providence, Rhode Island.

Yuan KH and Bentler PM 2007. Robust procedures in structural equation modeling In *Handbook of latent variable and related models* (ed. Lee SY). Elsevier/North-Holland Amsterdam, Boston, MA, pp. 367–397.

7

Meta-analytic structural equation modeling

This chapter covers meta-analytic structural equation modeling (MASEM), a technique that combines meta-analysis and structural equation modeling (SEM) to synthesize correlation or covariance matrices and to fit structural equation models on the pooled correlation (covariance) matrix. We begin this chapter with a discussion on the need to synthesize existing research findings when SEM is applied as the methodology in the primary studies. MASEM is then proposed as a statistical method to synthesize these research findings. Conventional methods based on the univariate approaches and the generalized least squares (GLS) approach are briefly reviewed. The fixed- and the random-effects two-stage structural equation modeling (TSSEM) are introduced in details. Issues related to conducting MASEM are discussed. Several examples are used to illustrate the procedures in the R statistical environment.

7.1 Introduction

As introduced in Chapter 2, SEM is a popular statistical technique to test hypothesized models in the social, educational, and behavioral sciences. What makes SEM so popular in applied research is that theoretical models can be translated into a set of interrelated equations involving latent and observed variables. The proposed models can be path models, confirmatory factor analytic (CFA) models, or general structural equation models. The proposed models can be empirically tested by the use of a likelihood ratio (LR) statistic and various goodness-of-fit indices. If the proposed models are rejected by the test statistics or goodness-of-fit indices, there is evidence that the proposed models are not consistent with the data; otherwise, the proposed models are consistent with the collected data.

Meta-Analysis: A Structural Equation Modeling Approach, First Edition. Mike W. -L. Cheung.

Companion Website: www.wiley.com/go/cheung/meta_analysis

As SEM is so powerful in testing hypothesized models, it may be tempting to believe that the increase in research findings on a research topic using SEM must improve our understanding of that topic. However, this may not generally be true when the research findings are inconsistent even though SEM is used as the methodology. For a set of similar constructs, different researchers may propose different models that are supported by their own data and it is difficult to systematically compare and synthesize these models. It has been recognized that the statistical power of the SEM in rejecting incorrect models may not be high enough when the sample sizes are small. This means that findings supporting different models may not be directly comparable. Moreover, it has also been found that researchers are reluctant to consider alternative models (MacCallum and Austin, 2000). As long as the proposed models are consistent with the researchers' theories and supported by the data, most researchers may not consider the need to test and compare alternative models. This confirmation bias—the prejudice in favor of the model being evaluated—hinders research progress (Greenwald et al., 1986). Hence, conducting more empirical research does not necessarily decrease the uncertainty surrounding a particular topic if the findings from that research are inconsistent (National Research Council, 1992).

7.1.1 Meta-analytic structural equation modeling as a possible solution for conflicting research findings

The limitations of applying SEM to primary research can be partially addressed by MASEM, a technique combining meta-analysis and SEM for the purpose of synthesizing research findings in studies using SEM (e.g., Bergh et al. (2014); Landis, 2013; Viswesvaran and Ones, 1995). The basic idea is to synthesize correlation (or covariance) matrices into a pooled correlation (or covariance) matrix in the first stage of analysis. In the second stage of analysis, the pooled correlation matrix is used to fit and compare different structural models supported by the theories. Several terms for this analysis have been used interchangeably in the literature, for instance, meta-analytic path analysis (Colquitt et al., 2000), meta-analysis of factor analysis (Becker, 1996), meta-analytical structural equations analysis (Hom et al., 1992), path analysis of meta-analytically derived correlation matrices (Eby et al., 1999), SEM of a meta-analytic correlation matrix (Conway, 1999), path analysis based on meta-analytic findings (Tett and Meyer, 1993), and model-based meta-analysis (Becker, 2009). Following Cheung (2002) and Cheung and Chan (2005b), in this book, we use the generic term *meta-analytic structural equation modeling* to describe this class of techniques.

Conventional SEM focuses on primary data, whereas MASEM deals with correlation or covariance matrices from a pool of studies. This allows MASEM to address research questions that may not feasibly be addressed in conventional SEM based on the primary data. First, MASEM enables researchers to test the proposed models across various samples, conditions, and measurements. As the primary studies have been conducted by different researchers, it is likely that different samples and measurements were used. If the proposed models still fit the data well across studies, this provides strong evidence of the validity of the proposed models. If the proposed models do not fit the data, the studies may be grouped according to the study characteristics, such as samples and measurements. The proposed models can then be fitted within different groups. The study characteristics may be used to explain the differences in the findings that different models fit the data. Alternatively, a random-effects model recognizing that studies having their own population correlation matrices may be used to account for the heterogeneity of the studies. Results on a meta-analysis may provide useful information than a single study with large sample size (Schmidt and Hunter, 2015).

Let us illustrate the above ideas with two examples. Norton et al. (2013) studied the factor structure of the Hospital Anxiety and Depression Scale. These authors identified 10 different factor structures that have been suggested in the literature. More importantly, these 10 structures were supported by some empirical data. It is difficult to draw a general conclusion on which model, if any, best fits the available data. Norton et al. (2013) conducted an MASEM on 28 independent samples from 21 studies. They found that the bifactor structure consisting of a general distress factor and anxiety and depression group factors fitted the data best. MASEM resolved the issue of which factor structure is the best model in describing the factor structure of the Hospital Anxiety and Depression Scale.

Another example is from Murayama and Elliot (2012). These authors conducted a meta-analysis on the association between competition and performance. They found that the average correlation between these two constructs was close to zero. Theoretically, it is difficult to explain why the correlation between competition and performance is zero. They proposed two mediators (performance-approach goals and performance-avoidance goals) to explain this apparent zero correlation. They argued that the effects from competition to performance via performance-approach goals and performance-avoidance goals were positive and negative, respectively. Therefore, the total effect from competition to performance is close to zero. They tested this hypothesis empirically with MASEM by synthesizing correlation matrices on studies involving these four variables. A mediation model with performance-approach goals and performance-avoidance goals as specific mediators (e.g., Cheung, 2007) was fitted on the pooled correlation matrix. They found that the specific indirect effects were in opposite directions and significant as predicted by their hypotheses. Therefore, MASEM enabled the researchers to resolve the issue on why the average correlation in the meta-analysis is close to zero.

There are other examples of the application of MASEM to synthesize studies in the literature. For example, Cheung (2014a) and Cheung and Chan (2005a) tested a second-order factor structure on the five-factor model proposed by Digman (1997). Fourteen studies of correlation matrices with a total of 4496 participants were meta-analyzed using MASEM. It was found that the proposed model reasonably fits the data using a random-effects model. Steinmetz et al. (2012) synthesized more than 300 correlation matrices on Schwartz' theory of human values. They found that three clusters of studies adequately fitted the data and theory. Dunst and Trivette (2009) synthesized 15 studies with 2900 participants and tested the influences of family-centered care on the psychological health of parent and children. Yufik and Simms (2010) studied the structure of posttraumatic stress disorder symptoms by synthesizing 40 studies with 14,827 participants. They found that the model comprising intrusions, avoidance, hyperarousal, and dysphoria factors fitted the data best.

Before closing this section, we have to mention a few limitations of MASEM. Similar to conventional meta-analysis, MASEM is usually based on the summary statistics (correlation or covariance matrices). As the raw data are usually not available, techniques involving raw data are generally not feasible in MASEM. These include, for example, analysis of missing data in subject level data with maximum likelihood (ML) estimation, analysis of binary or categorical data, robust test statistics and standard errors (SEs), mixture modeling on subjects, and modeling nonlinear relationship among variables. If there are problematic data such as missing data and nonnormal data in the primary studies, it is hard to correct them in MASEM. In the following sections, we introduce the key concepts in MASEM.

7.1.2 Basic steps for conducting a meta-analytic structural equation modeling

The steps for conducting an MASEM are basically similar to those for conducting a meta-analysis. Viswesvaran and Ones (1995) provided an outline on how to conceptualize an MASEM. Their Table 1 (p. 867) summarizes the key steps as follows:

(i) identify important constructs and relationships;

(ii) identity different measures used to operationalize each constructs;

(iii) obtain all relevant statistics from the studies;

(iv) conduct psychometric meta-analyses and estimate the true score correlations between the measures;

(v) use factor analysis to test the measurement model;

(vi) estimate the correlations between the constructs by forming composite scores of different constructs; and

(vii) use path analysis with the estimated true score correlations to test the proposed theory.

Steps (i)–(iii) involve the conceptualization and operationalization of an MASEM, whereas steps (iv)–(vii) are the statistical steps for conducting an MASEM. Whether a CFA model, path model, or SEM is conducted depends on the availability of the data. If correlation matrices on the item level are available, CFA or SEM with latent variables may be fitted. If only correlation matrices of the composite scores are available, a path model may be fitted.

7.2 Conventional approaches

Both univariate and multivariate methods have been used to conduct MASEM. The univariate approaches treat the correlation matrix as correlation coefficients and synthesize them independently, whereas the multivariate approaches consider the dependence of the correlation coefficients in the correlation matrices. The most popular approaches in the past were univariate methods in which elements of correlation matrices are treated as independent within studies. More recently, there is consensus that researchers should use multivariate methods that take the dependence among the correlation matrices into account in MASEM (e.g., Becker, 2009; Cheung and Chan, 2005b). In this section, we briefly review the univariate approaches and the generalized least squares (GLS) approach, one of the multivariate approaches.

7.2.1 Univariate approaches

There are two popular univariate meta-analytic techniques—that of Hedges and Olkin (1985) and Hunter and Schmidt (1990, 2004). They differ on how to combine the correlation coefficients and how to test the homogeneity of the correlation coefficients. To avoid confusion between the original methods proposed by Hedges and Olkin (1985) and Hunter and Schmidt (1990) in synthesizing correlations and the current approach in conducting MASEM, we use the terms univariate-r and univariate-z methods to, respectively, denote the application of Hunter and Schmidt's and Hedges and Olkin's methods to MASEM. Moreover, the current univariate-r method discussed in this chapter is also a simplified version of the method proposed by Hunter and Schmidt (1990), because it does not involve issues such as correction for unreliability or range restriction.

The univariate approaches synthesize the correlation coefficients in a correlation matrix as if the correlation coefficients were independent. The pooled correlation matrix is treated as if it was an observed correlation matrix in fitting structural equation models. There are a couple of issues related to this practice. As the primary studies have been independently conducted by different researchers, the numbers of variables involved in the studies are likely to be different. One critical issue for meta-analysts is to determine how to combine those correlation matrices that are based on different numbers of variables.

There are usually three methods to handle this problem (e.g., Viswesvaran and Ones, 1995). The first method is to exclude studies that do not contain all of the variables in the model (e.g., Hom et al., 1992). The main drawback to this is that the final number of studies may be sharply reduced. This also limits the generalizability of the meta-analytic results. The second method is to reduce the number of variables used in the model in order to include as many studies as possible. The problem with this approach is that it cannot be used to test complex models as the number of variables is small.

The third method, which is the main method used by applied researchers, is to estimate the elements of the pooled correlation matrix based on different numbers of studies, that is, pairwise aggregation (e.g., Brown and Peterson, 1993; Premack and Hunter, 1988). The advantage of this method is that it includes as many studies as possible. The disadvantage is that the pooled correlation matrix may be nonpositive definite as required in the stage 2 analysis in fitting SEM because each element in the pooled correlation matrix is based on a different number of studies, or different samples (see Section 5.3.2 for the discussion of nonpositive definite matrix). A related issue that will be elaborated upon later is how to decide the sample size in fitting structural equation models in the stage 2 analysis.

After obtaining a pooled correlation matrix, the pooled correlation matrix is used as if it were an observed *covariance* matrix in fitting structural equation models in the stage 2 analysis. Four statistical difficulties may occur in the stage 2 analysis. They involve the following:

(i) choosing the appropriate sample size in SEM;

(ii) a nonpositive definite matrix of the pooled correlation (covariance) matrix;

(iii) the analysis of the correlation matrix; and

(iv) ignoring the sampling variations across studies.

The first difficulty is in deciding on an appropriate sample size to fit the structural equation models. When there is no missing correlation, all of the correlation coefficients in the pooled correlation matrix are based on the same sample sizes. There is no problem about choosing the sample size in fitting structural models in the stage 2 analysis; it is the sum of the individual sample sizes in the analysis. However, the pooled correlation matrix is usually formed by averaging across different studies based on pairwise aggregation. Researchers have to decide on the appropriate sample size for the analysis in SEM. Researchers have used a variety of sample sizes such as the arithmetic mean (e.g., Premack and Hunter, 1988), the harmonic mean (e.g., Colquitt et al., 2000), the median (e.g., Brown and Peterson, 1993), or the total (e.g., Hunter, 1983) of the sample sizes based on the synthesized correlation coefficients.

Using an example, let us illustrate how to calculate these sample sizes. Suppose that there are three variables of interest in the analysis. The pooled correlation matrix is $\begin{matrix} & x_1 & x_2 & x_3 \\ x_1 & 1.0 & & \\ x_2 & .6 & 1.0 & \\ x_3 & .6 & .6 & 1.0 \end{matrix}$. $\bar{r}_{21}$, $\bar{r}_{31}$, and $\bar{r}_{32}$ are based on the sample sizes of 200, 500, and 1000, respectively, across studies. Then, the arithmetic mean, the harmonic mean, and the median are 567, 375, and 500, respectively. The total sample size is the sum of the sample sizes involved in the studies. The total sample size can be larger than 1000 because some studies that report r_{21} and r_{31} may not report r_{32}.

As the Type I error of the chi-square test statistics, the goodness-of-fit indices, the statistical power, and the SEs of parameter estimates are all dependent on the sample size used (Bollen, 1990), using different sample sizes in the analysis can lead to different results and conclusions. If a smaller sample size is used to fit the structural models, the LR statistic will be too small, whereas the SEs will be too large. In contrast, if a larger sample size is used, the SEs will be too small, whereas the LR statistic will be too large. The issue of statistical power is not only a concern of researchers of SEM (e.g., Kaplan, 1990) but recently also of meta-analysts (Hedges and Pigott, 2001, 2004). No matter which sample size is used, it is hard to obtain *correct* test statistics in testing the whole model and SEs in testing the individual parameter estimates. The main problem is that all of the suggestions (arithmetic mean, harmonic mean, median, and total) are simply ad hoc solutions because they are not based on any strong statistical theories.

On the other hand, some may argue that the choice of the sample size is not critical at all. As the MASEM is usually based on a large sample size, the LR statistic and the significance test on the parameter estimates are likely statistically significant regardless of which sample size we used in fitting the structural models. This is only partially correct. The objective of a meta-analysis (and MASEM) is not only to test whether the parameter estimates are statistically significant. Researchers should pay more attention to the *precision* of the parameter estimates. The precision is usually in the form of a confidence interval (CI), which is a function of the sample size. The choice of the sample size in fitting the structural models has a direct impact on the length of the CI of the parameter estimates.

The second difficulty is that the input correlation matrix for SEM can be nonpositive definite. As each study may contain a different set of variables, the pooled correlation matrix from pairwise deletion may be nonpositive definite (Enders, 2010; Wothke, 1993). In such cases, SEM is no longer appropriate. Moreover, even though the pooled correlation matrix based on pairwise deletion is positive definite, its statistical properties are still questionable in SEM because different elements of the pooled correlation matrix are probably based on different samples (Wothke, 2000).

The third difficulty is ignoring the sampling variation across studies. After pooling the correlation matrices, researchers often use the pooled correlation matrix as the observed correlation matrix without considering the sampling variances and covariances of the correlation coefficients. There are sampling variations in individual correlation matrices even when they share the same population correlation matrix. Some estimated pooled correlation coefficients may contain more sampling variation than others. However, the sampling variation associated with the pooled correlation matrix is not accurately reflected when fitting SEM under the univariate approaches in which their SEs are ignored. By using a single sample size, the sampling error of the pooled correlation matrix is primarily determined by the sample size used in the analysis. Moreover, the covariation among the correlations is totally ignored in the univariate approaches in spite of the fact that the correlations are indeed correlated to a certain extent.

The fourth difficulty is analyzing a correlation matrix instead of a covariance matrix. It is generally incorrect to analyze the correlation matrix in SEM, although most published articles using MASEM have treated the pooled correlation matrix as a covariance matrix. Many SEM experts (e.g., Cudeck, 1989) have warned about the problems of analyzing the correlation matrix instead of the covariance matrix in primary research applications of SEM. Specifically, the chi-square statistics and (or) the SEs of parameter estimates may be incorrect. It should be noted that there is nothing wrong with the analysis of correlation matrix if the research questions are related to the standardized variables of the constructs (e.g., Bentler, 2007). The main issue here is that many researchers incorrectly treat the correlation matrix as a covariance matrix without using the correct methods to analyze the data.

The first three difficulties may be encountered only when pairwise deletion is used in handling missing correlations, while the fourth difficulty may occur no matter whether pairwise or listwise deletion is used. Although fewer technical problems are encountered with listwise deletion than with pairwise deletion, listwise deletion is less popular in MASEM because many studies would be deleted because of missing correlations. Although the univariate methods suffer from these technical difficulties and are generally not recommended (e.g., Becker, 2000, 2009; Cheung and Chan, 2005b), they are still popular in some disciplines because of their ease of application.

7.2.2 Generalized least squares approach

Besides the univariate methods, another popular multivariate approach is the GLS approach (Becker, 1992, 1995). We focus on the fixed-effects model here. Readers may refer to Section 5.3 for the generalization to the random-effects model. Suppose that there are $p \times p$ correlation matrices involved in the analysis where p is the number of variables; the ith sample correlation matrix is $\boldsymbol{R}_i$, whereas the common population correlation matrix is $\boldsymbol{P}_\mathrm{F}$ (assuming a fixed-effects model). It is more convenient to represent the correlation matrices using the vector notation. We may stack the nonduplicate columns of the correlation matrix into vectors using $\boldsymbol{r}_i = \mathrm{vechs}(\boldsymbol{R}_i)$ and $\boldsymbol{\rho}_\mathrm{F} = \mathrm{vechs}(\boldsymbol{P}_\mathrm{F})$ (see Section 2.4.2 for explanations of the vechs() and vech() operators). The model for the ith study is

$$\boldsymbol{r}_i = \boldsymbol{X}_i \boldsymbol{\rho}_\mathrm{F} + \boldsymbol{e}_i, \tag{7.1}$$

where $\boldsymbol{X}_i$ is a design matrix of 0 and 1 indicating whether the correlation coefficients are present or missing and $\boldsymbol{e}_i$ is the sampling error. The conditional sampling covariance matrix $\boldsymbol{e}_i \sim \mathcal{N}(0, \boldsymbol{V}_i)$ can be estimated by the methods discussed in Section 3.3.2. We treat $\boldsymbol{V}_i$ as known values in the GLS approach.

Suppose that there are k studies involved in the MASEM; we combine the matrices together into a single model:

$$\boldsymbol{r} = \boldsymbol{X} \boldsymbol{\rho}_\mathrm{F} + \boldsymbol{e}, \tag{7.2}$$

where $\boldsymbol{r} = \begin{bmatrix} \boldsymbol{r}_1 \\ \boldsymbol{r}_2 \\ \vdots \\ \boldsymbol{r}_k \end{bmatrix}$, $\boldsymbol{X} = \begin{bmatrix} \boldsymbol{X}_1 \\ \boldsymbol{X}_2 \\ \vdots \\ \boldsymbol{X}_k \end{bmatrix}$, and $\boldsymbol{e} = \begin{bmatrix} \boldsymbol{e}_1 \\ \boldsymbol{e}_2 \\ \vdots \\ \boldsymbol{e}_k \end{bmatrix}$. As the effect sizes are independent across studies, the known sampling covariance matrix $\boldsymbol{V}$ for all studies is a block diagonal matrix:

$$\boldsymbol{V} = \mathsf{Cov}(\boldsymbol{e}) = \begin{bmatrix} \boldsymbol{V}_1 & & & \\ 0 & \boldsymbol{V}_2 & & \\ \vdots & \ddots & \ddots & \\ 0 & \cdots & 0 & \boldsymbol{V}_k \end{bmatrix}. \tag{7.3}$$

Let us illustrate the above model with an example. Suppose that there are a total of three variables involved in the analysis. The first study is complete, whereas variable 2 and variable 3 are missing in Studies 2 and 3, respectively. The sample correlation vectors are $\boldsymbol{r}_1 = \begin{bmatrix} r_{21} & r_{31} & r_{32} \end{bmatrix}_1^{\mathrm{T}}$, $\boldsymbol{r}_2 = \begin{bmatrix} r_{31} \end{bmatrix}_2$, and $\boldsymbol{r}_3 = \begin{bmatrix} r_{21} \end{bmatrix}_3$ where the subscripts represent the studies. The design matrices are $\boldsymbol{X}_1 = \begin{bmatrix} 1 & 0 & 0 \\ 0 & 1 & 0 \\ 0 & 0 & 1 \end{bmatrix}$, $\boldsymbol{X}_2 = \begin{bmatrix} 0 & 1 & 0 \end{bmatrix}$, and $\boldsymbol{X}_3 = \begin{bmatrix} 1 & 0 & 0 \end{bmatrix}$, whereas the sampling errors are $\boldsymbol{e}_1 = \begin{bmatrix} e_{21} & e_{31} & e_{32} \end{bmatrix}_1^{\mathrm{T}}$, $\boldsymbol{e}_2 = \begin{bmatrix} e_{31} \end{bmatrix}_2$, and $\boldsymbol{e}_3 = \begin{bmatrix} e_{21} \end{bmatrix}_3$.

The vector of the parameter estimates and its asymptotic covariance matrix can be obtained by

$$\begin{aligned} \hat{\boldsymbol{\rho}}_{\mathrm{F}} &= (\boldsymbol{X}^{\mathrm{T}}\boldsymbol{V}^{-1}\boldsymbol{X})^{-1}\boldsymbol{X}^{\mathrm{T}}\boldsymbol{V}^{-1}\boldsymbol{r}, \\ \hat{\boldsymbol{V}}_{\mathrm{F}} &= (\boldsymbol{X}^{\mathrm{T}}\boldsymbol{V}^{-1}\boldsymbol{X})^{-1}. \end{aligned} \tag{7.4}$$

To test the homogeneity of all correlation matrices across k studies, the test statistic

$$Q_{\mathrm{GLS}} = (\boldsymbol{r} - \boldsymbol{X}\hat{\boldsymbol{\rho}}_{\mathrm{F}})^{\mathrm{T}}\boldsymbol{V}^{-1}(\boldsymbol{r} - \boldsymbol{X}\hat{\boldsymbol{\rho}}_{\mathrm{F}}) \tag{7.5}$$

is approximately distributed as a chi-square variate with $(\sum_{i=1}^{k}(p_i(p_i - 1)/2) - p(p - 1)/2)$ degrees of freedom (dfs) in large samples, where p_i is the number of observed variables in the ith study (see Becker, 1992; Cheung and Chan, 2005b; Hedges and Olkin, 1985). Readers may refer to Section 5.2 for more details on the GLS approach.

Becker (1992) proposed a method to fit regression models on the estimated common correlation matrix $\boldsymbol{R}$, where $\hat{\boldsymbol{\rho}}_{\mathrm{F}} = \mathrm{vechs}(\boldsymbol{R})$. As $\boldsymbol{R}$ is used as the input in fitting regression models, we remove the hat notation here. Assuming that the first variable is the dependent variable and that the remaining variables are the predictors, $\boldsymbol{P}_{\mathrm{F}}$ can be partitioned into

$$\boldsymbol{P}_{\mathrm{F}} = \begin{bmatrix} 1 & \\ \boldsymbol{P}_{01} & \boldsymbol{P}_{11} \end{bmatrix}, \tag{7.6}$$

where $\boldsymbol{P}_{01}$ are the correlations between the dependent variable and the predictors and $\boldsymbol{P}_{11}$ is the correlation matrix among the predictors.

Becker (1992) showed that the (population) standardized regression coefficients $\boldsymbol{\beta} = [\beta_1, \beta_2, \ldots, \beta_{p-1}]^{\mathrm{T}}$ of regressing y (the first variable in the correlation matrix) on $x_1, x_2, \ldots, x_{p-1}$ (the remaining variables in the correlation matrix) are given by

$$\boldsymbol{\beta} = \boldsymbol{P}_{11}^{-1}\boldsymbol{P}_{01}. \tag{7.7}$$

By using $\boldsymbol{R}$ as a sample estimate of $\boldsymbol{P}$, we estimate the standardized regression coefficients by

$$\hat{\boldsymbol{\beta}} = \boldsymbol{R}_{11}^{-1}\boldsymbol{R}_{01}. \tag{7.8}$$

As $\hat{\boldsymbol{\beta}}$ is a function of $\boldsymbol{R}_{11}$ and $\boldsymbol{R}_{01}$, multivariate delta method can be used to obtain the sampling covariance matrix of $\hat{\boldsymbol{\beta}}$ (see Section 3.4.1). Once we have estimated $\mathsf{Cov}(\hat{\boldsymbol{\beta}})$, statistical inferences and approximate confidence intervals on the on $\hat{\boldsymbol{\beta}}$ can be performed (also see Card (2012) for an illustration).

The GLS approach seems appealing. There is one major limitation, however. The models that can be fitted are limited to regression models. When the proposed models are CFA models or structural equation models, there is no closed form solution for these models similar to the above regression model. SEM packages are then required to fit these models.

Some researchers use the GLS approach as the first stage of MASEM in pooling correlation matrices (e.g., Geyskens et al., 1998; Smith et al., 1999). The pooled correlation matrix is then treated as the observed covariance matrix in fitting structural equation models by using conventional SEM packages. Although this approach addresses the limitation of the GLS approach in fitting regression models only, it does not address some of the issues discussed in Section 7.2.1. For example, researchers may use the harmonic mean in fitting the structural models by treating the pooled correlation matrix as the observed covariance matrix and ignoring the sampling variations across studies.

As $\hat{\rho}_{\mathrm{F}}$ and its asymptotic sampling covariance matrix $\hat{V}_{\mathrm{F}}$ are available after pooling the correlation matrices, Cheung and Chan (2005b) showed how conventional SEM packages can be used to fit structural models on $\hat{\rho}_{\mathrm{F}}$ by the use of the weighted least squares (WLS) estimation method with $\hat{V}_{\mathrm{F}}$ as the weight matrix (see Section 7.3.2 for details). This approach corrects three main limitations of conventional applications of the GLS approach. First, a single *correct* sample size is used. Second, the correlation matrix can be correctly analyzed. Third, it includes the sampling variations of the pooled correlation matrix in the stage 2 analysis.

7.3 Two-stage structural equation modeling: fixed-effects models

There are two classes of models in meta-analysis—fixed-effects models and random-effects models (Becker, 1992, 1995; Hedges and Vevea, 1998; Schmidt et al., 2009; and see Section 4.4 for a general introduction). Fixed-effects models are used for conditional inferences based on the selected studies. They are intended to draw conclusions on the studies included in the meta-analysis. Researchers are mainly interested in the studies used in the analysis. The assumption in fixed-effects models is usually, but not always, that all studies share common effect sizes. In the context of MASEM, the assumption behind the fixed-effects model is that the population correlation matrices are assumed equal for all studies. In this section, we first introduce fixed-effects TSSEM. The extension to random-effects TSSEM is presented in the next section.

Although other MASEM procedures discussed here also entail two stages (e.g., Becker, 1992; Viswesvaran and Ones, 1995), they use different methods in the two stages of analysis. In the first stage of analysis, conventional meta-analytic techniques are used to synthesize correlation matrices, whereas SEM is only used for fitting proposed models in the stage 2 analysis. In this book, we use the label of TSSEM to highlight the fact that SEM is used as the sole statistical framework for all stages (Cheung, 2002; Cheung and Chan, 2005b).

There are two distinct features in the fixed-effects TSSEM when compared to the univariate or the GLS approaches. First, multiple-group SEM is used to pool correlation or covariance matrices in the first stage of the analysis. LR statistic and goodness-of-fit indices in SEM are used to test the homogeneity of correlation or covariance matrices. Second, the WLS estimation method is used to weigh the precision of the pooled correlation or covariance matrix in fitting structural models in the second stage of analysis. Thus, it allows different elements of the pooled correlation matrix to be weighted differently in fitting the structural equation models.

7.3.1 Stage 1 of the analysis: pooling correlation matrices

As studies are likely to be different in terms of the measures used, correlation matrices are usually used in MASEM and meta-analysis in general. When the same measurement is used in all studies and the covariance matrices are available, meta-analysts may choose between correlation or covariance matrices in conducting MASEM. Generally speaking, researchers may synthesize correlation matrices even though the covariance matrices are available. If covariance matrices are synthesized, researchers may address research questions related to the scaling of the variables. In this chapter, we mainly focus on synthesizing correlation matrices, while the extensions to analysis of covariance structure are trivial under the TSSEM approach.

7.3.1.1 Analysis of correlation matrices

The distribution theory of SEM is based on covariance matrices, whereas correlation matrices are usually used in MASEM. It is usually not appropriate to treat a correlation matrix as a covariance matrix in the analysis (Cudeck, 1989). Let us illustrate the problems with an example. Suppose that we are fitting the two-factor CFA model with two indicators per factor displayed in Figure 2.3. When a covariance matrix is analyzed, the total number of pieces of information (nonduplicate elements in the covariance matrix) is $10 = 4(4 + 1)/2$. The number of parameters is 9. Thus, the df of the model is 1, which is overidentified. The model-implied covariance matrix is $\hat{\boldsymbol{\Sigma}} = \hat{\boldsymbol{\Lambda}}\hat{\boldsymbol{\Phi}}\hat{\boldsymbol{\Lambda}}^{\mathrm{T}} + \hat{\boldsymbol{\Psi}}$, where $\hat{\boldsymbol{\Lambda}}$, $\hat{\boldsymbol{\Phi}}$, and $\hat{\boldsymbol{\Psi}}$ are the estimated factor loadings, the factor covariance matrix, and the covariance matrix of the measurement errors, respectively. The diagonals of $\hat{\boldsymbol{\Sigma}}$ are close but not necessarily equal to the variances of the variables.

When a correlation matrix is analyzed as a covariance matrix, at least two issues arise. As the correlation matrix is used as a covariance matrix, the sample variances are always ones by definition. Although the diagonals of $\hat{\boldsymbol{\Sigma}}$ may be very close to the sample variances (1 here), they may not be *exactly* 1. If they are not exactly 1, the model-implied matrix is no longer qualified as a correlation matrix. It is hard to tell whether the results are still correct. The second issue is about the uncertainty in the estimation. When a correlation matrix is used as a covariance matrix in the input, the SEM package counts that there are 10 pieces of information in the data. In reality, there are only $6 = 4(4-1)/2$ pieces of information, because the diagonals do not carry any useful information. This may lead to incorrect statistical inferences and SEs.

There are two methods to correctly analyze correlation matrices (or correlation structures). One method is to analyze the correlation matrix directly by the use of the WLS estimation method that will be discussed later. The second method, which is introduced here, is to pretend that the correlation matrix is a covariance matrix and to take this into account in the estimation. The model of a correlation structure in a single group analysis is

$$\boldsymbol{\Sigma}(\boldsymbol{\theta}) = \boldsymbol{D}\boldsymbol{P}(\boldsymbol{\theta})\boldsymbol{D}, \tag{7.9}$$

where $\boldsymbol{\Sigma}(\boldsymbol{\theta})$ is the structural model on the covariance matrix, $\boldsymbol{D}$ is a diagonal matrix, and $\boldsymbol{P}(\boldsymbol{\theta})$ is the structural model on the correlation matrix with the constraints that $\mathrm{Diag}(\boldsymbol{P}(\boldsymbol{\theta})) = \mathbf{1}$, where $\mathbf{1}$ is a vector of ones (Jöreskog, 1978; Jöreskog and Sörbom, 1996).

This model addresses the two aforementioned issues. Because of the constraints on the diagonals, $\boldsymbol{P}(\boldsymbol{\theta})$ is always a correlation matrix. Second, the model always takes into account the uncertainty in analyzing a correlation matrix as a covariance matrix. Let us check the df of the model for the CFA model in Figure 2.3. Because of the constraints on the diagonals, the error variances are *not* parameters—they are computed from the constraints. There are only four factor loadings, one factor correlation, and four scaling variances in $\boldsymbol{D}$. Thus, there are a total of nine parameters. As we are treating the correlation matrix as a covariance matrix, the SEM package counts that there are 10 pieces of information in the input. The df for this model is 1.

The first stage of the fixed-effects TSSEM is based on the above approach. The correlation matrix in the ith study can be decomposed as

$$\boldsymbol{\Sigma}_i = \boldsymbol{D}_i\boldsymbol{P}_i\boldsymbol{D}_i, \tag{7.10}$$

where $\boldsymbol{\Sigma}_i$ is the population covariance matrix, $\boldsymbol{D}_i$ is a diagonal matrix, and $\boldsymbol{P}_i$ is the correlation matrix (Cheung and Chan, 2004, 2005b). For example, we may decompose the covariance matrix into the correlation matrix and matrices of standard deviations: $\begin{bmatrix} 4.0 & & \\ 1.8 & 9.0 & \\ 3.2 & 6.0 & 16.0 \end{bmatrix} = \begin{bmatrix} 2 & & \\ 0 & 3 & \\ 0 & 0 & 4 \end{bmatrix} \begin{bmatrix} 1 & & \\ 0.3 & 1 & \\ 0.4 & 0.5 & 1 \end{bmatrix} \begin{bmatrix} 2 & & \\ 0 & 3 & \\ 0 & 0 & 4 \end{bmatrix}$. If the sample covariance matrix is used as the input, $\boldsymbol{D}_i$ represents the standard deviations of the variables. If the sample correlation matrix is used as a covariance matrix, $\boldsymbol{D}_i$ represents estimates close to an identity matrix, that is, ones. When a covariance matrix is used in the analysis, the diagonals also carry important information (variances). These extra parameters $\boldsymbol{D}_i$ allow the theory of covariance structure analysis to apply to the analysis of correlation matrix (Jöreskog, 1978; Jöreskog and Sörbom, 1996). The model can be viewed as a special case of a CFA model with $\boldsymbol{P}_i$ as the correlation matrix of the latent factors, $\boldsymbol{D}_i$ as the factor loadings with only diagonal elements, and a zero matrix of the variance–covariance matrix of the measurement errors. Under the fixed-effects model with the assumption of the homogeneity of correlation matrices, we estimate a common correlation matrix $\hat{\boldsymbol{P}}_{\mathrm{F}}$ by imposing the constraints $\boldsymbol{P}_{\mathrm{F}} = \boldsymbol{P}_1 = \boldsymbol{P}_2 = \cdots = \boldsymbol{P}_k$, where $\boldsymbol{D}_i$ may vary across studies. When there are missing correlations, the missing data are filtered out before the analysis. After fitting this model, an LR statistic may be used to test the homogeneity of correlation matrices $H_0 : \boldsymbol{P}_1 = \boldsymbol{P}_2 = \cdots = \boldsymbol{P}_k$ (Cheung and Chan, 2005b).

Figure 7.1 shows an example of two studies with three variables. All variables are complete in Study 1, while x_3 is missing in Study 2. To pool the correlation matrices, we may impose the equality constraint $\rho_{2,1}^{(1)} = \rho_{2,1}^{(2)}$, while $\sigma_1^{(1)}$, $\sigma_2^{(1)}$, $\sigma_3^{(1)}$, $\sigma_1^{(2)}$, and $\sigma_2^{(2)}$ are freely estimated. If the input matrices are covariances, the estimated σ's are close to their correspondent standard deviations. If the input matrices are correlations, they are close to 1.

This approach has several advantages. First, missing or incomplete correlation elements can be easily handled by the ML method (e.g., Allison, 1987; Muthén

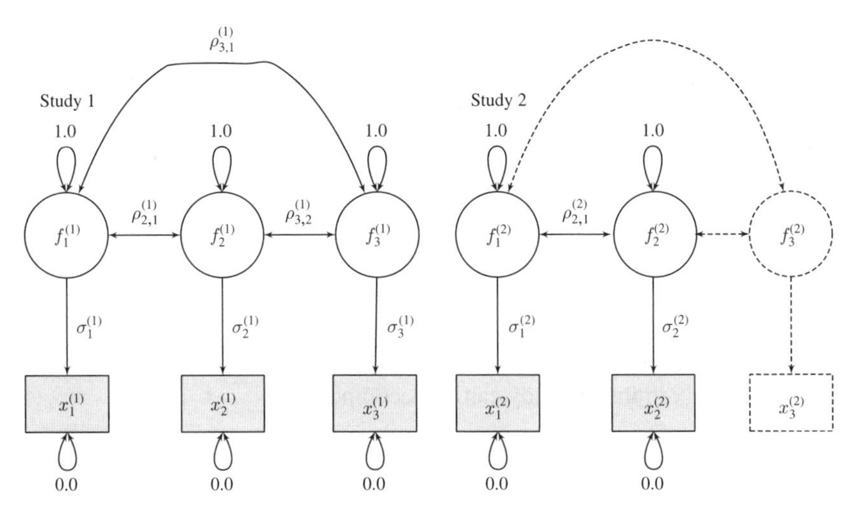

Figure 7.1 A fixed-effects model of the first stage of the TSSEM approach.

et al., 1987). When the missing mechanism is missing completely at random (MCAR) or missing at random (MAR), the parameter estimates using the ML estimation are unbiased and efficient (see Section 8.2). Second, the asymptotic covariance matrix of $\hat{\boldsymbol{P}}_{\text{F}}$, which indicates the precision of the estimates, is routinely available after the analysis. SEs may be used to test the significance of the estimated common correlation coefficients or to construct the approximate CIs of the estimated common correlation or covariance matrices. Third, besides testing the assumption of the homogeneity of correlation matrices using an LR statistic, many goodness-of-fit indices, such as root mean square error of approximation (RMSEA) and standardized root mean square residual (SRMR), may also be used to test the close or approximate fit of the homogeneity of correlation matrices.

7.3.1.2 Analysis of covariance matrices

When the covariance matrices are available and the scales are comparable across studies, researchers have the option of synthesizing the covariance matrices. The model in Equation 7.10 can directly be used to obtain a common covariance matrix by imposing $\boldsymbol{P}_{\text{F}} = \boldsymbol{P}_1 = \boldsymbol{P}_2 = \cdots = \boldsymbol{P}_k$ and $\boldsymbol{D}_{\text{F}} = \boldsymbol{D}_1 = \boldsymbol{D}_2 = \cdots = \boldsymbol{D}_k$. As both the correlation matrices and the standard deviations are imposed equally, the resultant matrix is a common covariance matrix $\boldsymbol{\Sigma}_{\text{F}}$ under a fixed-effects model. One issue remains, however. The asymptotic covariance matrix of the elements of the correlation matrix and the standard deviations are estimated separately. In the stage 2 analysis, we need the asymptotic covariance matrix of the pooled covariance matrix, not the asymptotic covariance matrix of the pooled correlation matrix and the pooled standard deviations. It is not easy to transform the latter into the asymptotic covariance matrix for the pooled covariance matrix.

A better approach is to model the covariance matrix directly. We may fit the model with the constraint $\boldsymbol{\Sigma}_{\text{F}} = \boldsymbol{\Sigma}_1 = \boldsymbol{\Sigma}_2 = \cdots = \boldsymbol{\Sigma}_k$. When there are missing covariances or variances, they are filtered out before the analysis. After fitting this model, an LR statistic may be used to test the homogeneity of covariance matrices $H_0 : \boldsymbol{\Sigma}_1 = \boldsymbol{\Sigma}_2 = \cdots = \boldsymbol{\Sigma}_k$. Moreover, various goodness-of-fit indices may also be used to evaluate the *approximate* fit of the homogeneity of covariance matrices (Cheung and Chan, 2009). The estimated common covariance matrix $\hat{\boldsymbol{\Sigma}}_{\text{F}}$ and its asymptotic sampling covariance matrix $\hat{\boldsymbol{V}}_{\text{F}}$ are estimated after the stage 1 analysis.

7.3.2 Stage 2 of the analysis: fitting structural models

Although we focus on the analysis of correlation matrices and correlation structures, the same theory applies to the analysis of covariance matrices and covariance structures. In the stage 1 analysis based on the fixed-effects model, a vector of the pooled correlation matrix $\hat{\rho}_{\text{F}} = \text{vechs}(\hat{\boldsymbol{P}}_{\text{F}})$ and its asymptotic sampling covariance matrix $\hat{\boldsymbol{V}}_{\text{F}} = \text{Cov}(\hat{\rho}_{\text{F}})$ are estimated. $\hat{\rho}_{\text{F}}$ from the stage 1 analysis is treated as the sample correlation vector in the stage 2 analysis, whereas $\hat{\boldsymbol{V}}_{\text{F}}$ is treated as a known matrix. Therefore, we take out the hat from them, that is, $\boldsymbol{r}_{\text{F}} = \hat{\rho}_{\text{F}}$ and $\boldsymbol{V}_{\text{F}} = \hat{\boldsymbol{V}}_{\text{F}}$. We will use the notations $\boldsymbol{r}_{\text{F}}$ and $\boldsymbol{V}_{\text{F}}$ here.

If all studies are complete in the stage 1 analysis, $\boldsymbol{r}_\mathrm{F}$ can be treated as observed correlations in fitting structural models. As there are no missing data, all sample sizes (arithmetic mean, harmonic mean, median, and total) are the same. There is no ambiguity in choosing the sample size in fitting the structural models. When there are missing correlations, however, any choices of the above sample sizes are not optimal.

Let us illustrate this case with the example in Section 7.2.1. Recall that the pooled correlation matrix is $\begin{matrix} & x_1 & x_2 & x_3 \\ x_1 & 1.0 & & \\ x_2 & .6 & 1.0 & \\ x_3 & .6 & .6 & 1.0 \end{matrix}$, and the sample sizes for $\bar{r}_{21}$, $\bar{r}_{31}$, and $\bar{r}_{32}$ are 200, 500, and 1000, respectively. The arithmetic mean, the harmonic mean, and the median are 567, 375, and 500, respectively. If the harmonic mean is used, this sample size is larger than the actual sample size for $\bar{r}_{21}$ but smaller than those for $\bar{r}_{31}$ and $\bar{r}_{32}$. As only *one* sample size can be used in fitting the structural model, it is unlikely to find the *correct* sample size. The precisions of some of the pooled correlation elements are overestimated, whereas the those of the other elements are underestimated.

To resolve the above problems in the stage 2 analysis, Cheung (2002) and Cheung and Chan (2005b, 2009) proposed to use the WLS estimation method to fit the structural equation models. Suppose that we are fitting a structural model on the population correlation matrix $\boldsymbol{P}(\boldsymbol{\theta})$, the discrepancy function (e.g., Bentler and Savalei, 2010; Fouladi, 2000) for the proposed structural model $\boldsymbol{\rho}(\boldsymbol{\theta}) = \mathrm{vechs}(\boldsymbol{P}(\boldsymbol{\theta}))$ is

$$F_{\mathrm{WLS}}(\boldsymbol{\theta}) = (\boldsymbol{r}_\mathrm{F} - \boldsymbol{\rho}_\mathrm{F}(\boldsymbol{\theta}))^\mathrm{T} \boldsymbol{V}_\mathrm{F}^{-1} (\boldsymbol{r}_\mathrm{F} - \boldsymbol{\rho}_\mathrm{F}(\boldsymbol{\theta})). \tag{7.11}$$

The logic of the WLS estimation method is to weigh the correlation elements by the inverse of its sampling covariance matrix. Different weights are assigned into the elements in the estimated common correlation matrix depending on their precisions. For example, the asymptotic variance on $\bar{r}_{21}$ is larger than those on $\bar{r}_{31}$ and $\bar{r}_{32}$ because $\bar{r}_{21}$ is based on a smaller sample size (and thus a larger SE). Because $\boldsymbol{V}_\mathrm{F}$ is inverted, less weight is given to $\bar{r}_{21}$ than to $\bar{r}_{32}$. Therefore, the TSSEM approach takes the precision of the estimates from the stage 1 analysis into account by using different weights. This principle is the same as those used in meta-analysis. In fact, Cheung (2010) demonstrated the equivalence of the WLS estimation function in SEM to the fixed-effects meta-analysis with the GLS approach. By using the WLS estimation function, parameter estimates with appropriate SEs, test statistics, and goodness-of-fit indices can be obtained in the stage 2 analysis.

It is of relevance to explain how Equation 7.11 was implemented in the `metaSEM` package (Cheung, 2014b) and in `LISREL` (Cheung, 2009; Jöreskog and Sörbom, 1996; Jöreskog et al., 1999). Equation 7.11 was directly implemented in the `metaSEM` package without any modifications. That is, the sample size is not involved in the discrepancy function. The sample size only indirectly affects the estimation by affecting the value of $\boldsymbol{V}_\mathrm{F}$ obtained from the stage 1 analysis. After the optimization, the sample size is used to calculate various goodness-of-fit indices.

When LISREL is used to conduct the analysis (Cheung, 2009), $\boldsymbol{V}_{\mathrm{F}}$ obtained from the stage 1 analysis has to be multiplied by the total sample size N. This is because LISREL follows the SEM tradition by using the discrepancy functions in Equations 2.15 and 2.16. The sample size is multiplied into the minimum of the discrepancy function in order to calculate the LR statistic in Equation 2.18. That is,

$$\begin{aligned} F_{\mathrm{WLS}}(\boldsymbol{\theta}) &= (\boldsymbol{r}_{\mathrm{F}} - \boldsymbol{\rho}_{\mathrm{F}}(\boldsymbol{\theta}))^{\mathrm{T}}(N\boldsymbol{V}_{\mathrm{F}})^{-1}(\boldsymbol{r}_{\mathrm{F}} - \boldsymbol{\rho}_{\mathrm{F}}(\boldsymbol{\theta})), \quad \text{and} \\ T &= (N-1)F_{\min}(\hat{\boldsymbol{\theta}}). \end{aligned} \tag{7.12}$$

As N is inverted in the discrepancy function in the estimation and then $N - 1$ is multiplied in calculating the test statistic T, the effect of the sample size is cancelled out (see Cheung and Chan (2009) for a discussion).

Another issue is how the parameters in the diagonals of the correlation structure are handled. Special measures have been taken to ensure that the diagonals of the model-implied matrix are always ones to which a correlation structure is fitted. Let us illustrate the issues with the model shown in Figure 7.2 (see Section 7.6.1 for a description of the example). The proposed model is a CFA model of five observed variables with two latent factors. The reticular action model (RAM) specification (see Section 2.2.3) of the model is

$$\begin{aligned}
\boldsymbol{A} &= \begin{array}{c} \\ A \\ C \\ ES \\ E \\ I \\ f_{\mathrm{Alpha}} \\ f_{\mathrm{Beta}} \end{array} \begin{array}{c} \begin{array}{ccccccc} A & C & ES & E & I & f_{\mathrm{Alpha}} & f_{\mathrm{Beta}} \end{array} \\ \begin{bmatrix} 0 & 0 & 0 & 0 & 0 & \lambda_{1,1} & 0 \\ 0 & 0 & 0 & 0 & 0 & \lambda_{2,1} & 0 \\ 0 & 0 & 0 & 0 & 0 & \lambda_{3,1} & 0 \\ 0 & 0 & 0 & 0 & 0 & 0 & \lambda_{4,2} \\ 0 & 0 & 0 & 0 & 0 & 0 & \lambda_{5,2} \\ 0 & 0 & 0 & 0 & 0 & 0 & 0 \\ 0 & 0 & 0 & 0 & 0 & 0 & 0 \end{bmatrix} \end{array}, \\
\boldsymbol{S} &= \begin{array}{c} \\ A \\ C \\ ES \\ E \\ I \\ f_{\mathrm{Alpha}} \\ f_{\mathrm{Beta}} \end{array} \begin{array}{c} \begin{array}{ccccccc} A & C & ES & E & I & f_{\mathrm{Alpha}} & f_{\mathrm{Beta}} \end{array} \\ \begin{bmatrix} \psi_{1,1} & 0 & 0 & 0 & 0 & 0 & 0 \\ 0 & \psi_{2,2} & 0 & 0 & 0 & 0 & 0 \\ 0 & 0 & \psi_{3,3} & 0 & 0 & 0 & 0 \\ 0 & 0 & 0 & \psi_{4,4} & 0 & 0 & 0 \\ 0 & 0 & 0 & 0 & \psi_{5,5} & 0 & 0 \\ 0 & 0 & 0 & 0 & 0 & 1 & \phi_{2,1} \\ 0 & 0 & 0 & 0 & 0 & \phi_{2,1} & 1 \end{bmatrix} \end{array} \quad \text{and} \\
\boldsymbol{F} &= \begin{array}{c} \\ A \\ C \\ ES \\ E \\ I \end{array} \begin{array}{c} \begin{array}{ccccccc} A & C & ES & E & I & f_{\mathrm{Alpha}} & f_{\mathrm{Beta}} \end{array} \\ \begin{bmatrix} 1 & 0 & 0 & 0 & 0 & 0 & 0 \\ 0 & 1 & 0 & 0 & 0 & 0 & 0 \\ 0 & 0 & 1 & 0 & 0 & 0 & 0 \\ 0 & 0 & 0 & 1 & 0 & 0 & 0 \\ 0 & 0 & 0 & 0 & 1 & 0 & 0 \end{bmatrix} \end{array}.
\end{aligned} \tag{7.13}$$

When a covariance structure is fitted, all of the parameters in the above model are free. The number of nonduplicate elements in the covariance matrix is $(15 = 5(5+1)/2)$, whereas the number of parameters is 11. The model is overidentified with 4 dfs. No special measures are required. When a correlation structure is fitted, the diagonals do not carry any information (they are always one). The number of nonduplicate elements is only $(10 = 5(5-1)/2)$. If we estimate all parameters, the model is underidentified with -1 df. Therefore, special care has to be taken so that the correlation structure is identified. The next section discusses two approaches to fitting correlation structure analysis.

7.3.2.1 Two approaches to fitting correlation structures

As the diagonals of the model-implied correlation matrix have to be one, we need to discuss how this can be achieved. There are two approaches to ensure that the diagonals of the model-implied matrix are always ones. The first approach is to exclude the error variances from the estimation, while the second approach applies nonlinear constraints on the error variances to ensure that the diagonals of the model-implied correlation matrix are always ones. We discuss them one by one here. As indicated in the discrepancy function in Equation 7.11, the diagonals of the model-implied correlation matrix (variances of the observed variables) are not involved in the discrepancy function when a correlation structure is analyzed. This means that the parameters $\psi_{1,1}$ to $\psi_{5,5}$ in the $\boldsymbol{S}$ matrix are not estimable in the analysis.

The first approach estimates all parameters except the error variances. This approach works for regression and CFA models. For both the regression and CFA models, the error variances are not involved in the off-diagonals of the model-implied correlation matrix. Therefore, we can estimate the other parameters even without estimating the error variances. After the estimation, we *compute* the error variances, for example, $\psi_{1,1}$ to $\psi_{5,5}$ in our previous example, based on the fact that the diagonals of the model-implied correlation matrix have to be one. In the `metaSEM` package, the parameters in the diagonals of the $\boldsymbol{S}$ matrix are replaced with zero before the analysis. After the analysis, we compute the error variances again. For example, the error variances of the above model can be computed as

$$
\begin{aligned}
&\text{Diag}(\hat{\boldsymbol{\Psi}}) = \mathbf{1} - \text{Diag}(\hat{\boldsymbol{\Lambda}}\hat{\boldsymbol{\Phi}}\hat{\boldsymbol{\Lambda}}^{\mathrm{T}}), \quad \text{or} \\
&\begin{bmatrix} \hat{\psi}_{1,1} \\ \hat{\psi}_{2,2} \\ \hat{\psi}_{3,3} \\ \hat{\psi}_{4,4} \\ \hat{\psi}_{5,5} \end{bmatrix} = \begin{bmatrix} 1 \\ 1 \\ 1 \\ 1 \\ 1 \end{bmatrix} - \text{Diag}\left(\begin{bmatrix} \hat{\lambda}_{1,1} & 0 \\ \hat{\lambda}_{2,1} & 1 \\ \hat{\lambda}_{3,1} & 0 \\ 0 & \hat{\lambda}_{4,2} \\ 0 & \hat{\lambda}_{5,2} \end{bmatrix} \begin{bmatrix} 1 & \hat{\phi}_{2,1} \\ \hat{\phi}_{2,1} & 1 \end{bmatrix} \begin{bmatrix} \hat{\lambda}_{1,1} & 0 \\ \hat{\lambda}_{2,1} & 1 \\ \hat{\lambda}_{3,1} & 0 \\ 0 & \hat{\lambda}_{4,2} \\ 0 & \hat{\lambda}_{5,2} \end{bmatrix}^{\mathrm{T}} \right).
\end{aligned}
\tag{7.14}
$$

The advantage of this approach is its ease of implementation. Moreover, SEs on the parameter estimates, except for the error variances, are reported. The disadvantage of this approach is that it does not work for models involving latent or observed *mediators*—variables are serving as both independent and dependent variables at the same time. This approach also does not work for structural equation models with directions among the latent variables. It is because the dependent latent variables are in fact mediators—they are predicted by other latent variables, while they predict their indicators. The first approach does not work for these models because the error variances are also involved in the off-diagonals of the model-implied matrix. Thus, we need to simultaneously estimate both error variances and other parameters.

Let us consider the path model in Figure 7.4 (see Section 7.6.3 for a description of the example). *Job knowledge* and *Work sample* are intermediate mediators; they are both dependent variables and predictors. Thus, we cannot use the first approach to exclude the error variances in the estimation. We need to include their error variances as parameters in the model.

The second approach to fitting correlation structures is to include all parameters in the model. We apply nonlinear constraints to ensure that the diagonals of the model-implied matrix are always one. Suppose that the model-implied correlation matrix is $\underset{p\times p}{\hat{\boldsymbol{P}}(\hat{\boldsymbol{\theta}})}$, we may apply the following nonlinear constraints:

$$\underset{p\times 1}{\text{Diag}(\hat{\boldsymbol{P}}(\hat{\boldsymbol{\theta}}))} = \underset{p\times 1}{\boldsymbol{1}}, \tag{7.15}$$

where $\underset{p\times 1}{\mathbf{1}}$ is a $p \times 1$ vector of ones. In this example, the variance of *Ability* is fixed at 1 as it is an independent variable, whereas the variances of the model-implied matrix of the other three variables are constrained at 1. Therefore, three nonlinear constraints are imposed in this example.

When there are latent variables in the model, we also need to ensure that the variances (not the error variances) of the latent variables are also constrained at 1 (e.g., Steiger, 2002). Therefore, both the observed and the latent variables have to be standardized. It can be shown that the model-implied covariance matrix for *all* observed and latent variables is $(\boldsymbol{I} - \boldsymbol{A})^{-1}\boldsymbol{S}((\boldsymbol{I} - \boldsymbol{A})^{-1})^{\text{T}}$ under the RAM formulation. We may impose the following constraints to ensure that *all* variables are standardized and have a variance of 1:

$$\text{Diag}((\boldsymbol{I} - \boldsymbol{A})^{-1}\boldsymbol{S}((\boldsymbol{I} - \boldsymbol{A})^{-1})^{\text{T}}) = \boldsymbol{1}. \tag{7.16}$$

This approach is more general than the previous one. The advantage of this approach is that it works for all models regardless of whether there are mediators. There is one main disadvantage—SEs are not available for the parameter estimates. This is because SEs may not be accurate when there are nonlinear constraints. Thus, `OpenMx` does not report them. This limitation can be addressed by using either the parametric bootstrap or the likelihood-based confidence interval (LBCI).

The parametric bootstrap begins by generating B, for example, $B = 1000$, replications of data from $\boldsymbol{x}_i \sim \mathcal{N}(\boldsymbol{r}_{\text{F}}, \boldsymbol{V}_{\text{F}})$ (the estimated common correlation vector and its asymptotic covariance matrix from the stage 1 analysis), where $\boldsymbol{x}_i$ is the ith replication. The correlation structure is then fitted on the correlation vector $\boldsymbol{x}_i$ with the WLS estimation method and with $\boldsymbol{V}_{\text{F}}$ as the known sampling covariance matrix. Therefore, $\boldsymbol{x}_i$ may vary across the replications, whereas $\boldsymbol{V}_{\text{F}}$ is constant. After each replication, $\hat{\theta}_i$ is estimated. By repeating this process B times, we have B samples of $\hat{\theta}$. The empirical sampling covariance matrix on θ can be calculated. If we take the square root of the diagonal elements of this matrix, they represent the SEs of $\hat{\theta}$. Although this procedure is intuitive, it is computationally intensive. A better alternative is to use the LBCI unless the empirical sampling covariance matrix on $\hat{\theta}$ is required in further calculations.

All of the aforementioned approaches were implemented in the `wls()` and the `tssem2()` functions in the `metaSEM` package. The first approach is used when `diag.constraints=FALSE` is specified in `wls()` or `tssem2()`. If `diag.constraints=TRUE` is specified, the second approach is applied. When `intervals.type="LB"` is specified, the LBCI is reported; otherwise, the parametric bootstrap is used to approximate the sampling covariance matrix of the parameter estimates.

Before closing this section, it is also of importance to discuss some issues of applying equality constraints in a correlation structure analysis. When we are fitting covariance structures, both the factor loadings and the error variances are parameters. We sometimes may want to compare whether the factor loadings (or error variances) are the same across items. For example, we can impose equality constraints on $\lambda_{1,1} = \lambda_{2,1} = \lambda_{3,1}$ to test the equality of factor loadings or on $\psi_{1,1} = \psi_{2,2} = \psi_{3,3}$ to test the equality of error variances for the model in Figure 7.2. The hypothesis of equality of factor loadings tests whether the true score variances are the same for these items, while the hypothesis of equality of error variances tests whether the variances of the measurement errors are the same for these items. These two hypotheses can be independently tested in covariance structure analysis. However, this is not the case in a correlation structure analysis. As the variances of the variables are constrained at one, the factor loading and the error variance are related by $1 = \lambda^2 + \psi$. If we impose equality constraints on the factor loadings, we are also imposing equality constraints on the error variances at the same time and vice versa. Researchers should be careful when they are specifying equality constraints on the parameters in the correlation structure analysis.

7.3.2.2 Goodness-of-fit indices

As WLS estimation is used as the estimation method in fitting the structural models, it is of relevance to discuss how the WLS estimation method might affect the goodness-of-fit indices in the stage 2 analysis. Both Cheung et al. (2006) and Cheung and Chan (2009) have noted that the RMSEA and SRMR may indicate the proposed models are reasonable in fitting the data in the stage 2 analysis, whereas the comparative fit index (CFI) and Tucker–Lewis index (TLI) may indicate the other way. Readers may wonder why there is such a discrepancy.

When the model is correctly specified, the test statistics based on the ML and the WLS estimation methods are asymptotically equal. An baseline model, usually an independence model, is required to calculate the incremental fit indices, such as the CFI and TLI (see Equations 2.29 and 2.30). Both CFI and TLI indicate the model fit improvement *comparing* to the baseline model. In applied research, the independence model is likely wrong, that is, it is misspecified. Yuan and Chan (2005) showed that there are substantial systematic differences among the chi-square test statistics derived from different estimation methods when a model is misspecified. When the test statistic on the independence model based on the WLS estimation method is larger than that based on the ML estimation method, the CFI and TLI calculated based on the WLS estimation method will be smaller than those based on the ML estimation method.

Both RMSEA and SRMR are not affected by the test statistic of the baseline model. Tentatively, it appears that RMSEA and SRMR are preferable in assessing the model fit in the stage 2 analysis with the TSSEM approach. As there is only limited studies on this issue, more studies are required to compare the empirical performance in CFI and TLI versus RMSEA and SRMR in assessing the model fit when the WLS estimation is used as the estimation method.

7.3.3 Subgroup analysis

The fixed-effects TSSEM approach assumes that the population correlation matrices are homogeneous. If the population correlation matrices are heterogeneous, the above approach may not be appropriate. The fixed-effects TSSEM may be modified to handle the heterogeneity by grouping the studies into relatively homogeneous subgroups. Then, the fixed-effects TSSEM may be applied to each subgroup. If the correlation matrices are homogeneous within the subgroups, the grouping variable may be used to explain the heterogeneity. The categorical variable of grouping variable can be considered as a moderator. It should be noted, however, that if the number of studies is too small, the test statistic may not be powerful enough to reject the null hypothesis of the homogeneity of correlation matrices.

Two types of grouping variables should be distinguished—a priori versus ad hoc (Cheung and Chan, 2005a). A priori moderators are grouping variables according to theory and (or) study characteristics, whereas ad hoc moderators may be found by grouping the studies with the use of cluster analysis or mixture models. For example, Cheung and Chan (2005a) tested the second-order factor model on a five-factor model of personality with 14 studies reported by Digman (1997). The fixed-effects model did not fit the data well. Digman (1997) grouped the studies into younger (five studies) and adult (nine studies) samples (see Section 7.6.1 for the illustrations in R). On the basis of the cluster analysis, Cheung and Chan (2005a) found that a three-cluster solution fitted the data best. As there is no theoretical support for the ad hoc moderators, the ad hoc moderators should be interpreted with cautions. A better approach to deal with heterogeneity is to apply a random-effects approach that will be addressed in the next section.

7.4 Two-stage structural equation modeling: random-effects models

A fixed-effects model assumes that the population correlation matrices are homogeneous. This assumption may not be realistic in applied research. If a fixed-effects model applies to heterogeneous data, the estimated SEs are underestimated. This section extends the above fixed-effects TSSEM to a random-effects TSSEM. The main difference between a fixed-effects and a random-effects models is on the stage 1 analysis. The procedures are exactly the same in the stage 2 analysis. Random-effects models allow studies to have their study-specific correlation matrices even though the proposed structural models remain the same across studies (see Becker, 1992; Cheung, 2014a).

7.4.1 Stage 1 of the analysis: pooling correlation matrices

Suppose that $\boldsymbol{P}(\boldsymbol{\theta})$ is the proposed correlation structure on the average population correlation matrix $\boldsymbol{P}$ under a random-effects model, that is, $\boldsymbol{P} = \boldsymbol{P}(\boldsymbol{\theta})$, in the analysis. $\boldsymbol{P}(\boldsymbol{\theta})$ can be a regression model, a path model, a CFA model, or an SEM. The random-effects model assumes that each study has its own population correlation matrix $\boldsymbol{P}_i$ in the ith study. The ith study sample correlation matrix is $\boldsymbol{R}_i$.

It is more convenient to work with vectors. We may vectorize these matrices by stacking the off-diagonal elements, that is, $\boldsymbol{\rho}_{\mathrm{R}} = \mathrm{vechs}(\boldsymbol{P})$ (with the subscript R indicating that it is based on a random-effects model), $\boldsymbol{\rho}_i = \mathrm{vechs}(\boldsymbol{P}_i)$, and $\boldsymbol{r}_i = \mathrm{vechs}(\boldsymbol{R}_i)$. The random-effects model for the ith study is

$$\begin{aligned} \text{Level 1:} \quad & \boldsymbol{r}_i = \boldsymbol{\rho}_i + \boldsymbol{e}_i, \\ \text{Level 2:} \quad & \boldsymbol{\rho}_i = \boldsymbol{\rho}_{\mathrm{R}} + \boldsymbol{u}_i, \end{aligned} \tag{7.17}$$

where $\boldsymbol{e}_i \sim \mathcal{N}(\boldsymbol{0}, \boldsymbol{V}_i)$ is the known sampling covariance matrix and $\boldsymbol{u}_i \sim \mathcal{N}(\boldsymbol{0}, \boldsymbol{T}^2)$ is the heterogeneity covariance matrix that has to be estimated.

When we compare the above model with the model in Equation 5.12, we will notice that the above model is as same as the multivariate meta-analysis model with $\boldsymbol{\beta}_{\mathrm{R}} = \boldsymbol{\rho}_{\mathrm{R}}$. Under the above model, it is of importance to note that the structural model $\boldsymbol{P}(\boldsymbol{\theta})$ holds across all studies. In other words, the parameters in the structural model are the same across all studies. The heterogeneity of the effect sizes is captured by the random effects $\boldsymbol{u}_i$. The random effects are treated as *noise* rather than as meaningful *signals*. Alternative random-effects models on MASEM are discussed in Section 7.5.3.

Following are the steps required to conduct the first stage of the random-effects TSSEM. The first step is to estimate the known sampling covariance matrix $\boldsymbol{V}_i$ in each study via a CFA model outlined in Section 3.3.2. The estimated $\boldsymbol{V}_i$ is treated as known values in the subsequent analyses. The second step is to conduct the multivariate meta-analysis on the correlation vectors. Two issues are worth noting here. First, the number of effect sizes in TSSEM is usually much larger than the typical applications of a multivariate meta-analysis. For example, it is not uncommon to synthesize correlation matrices with the dimensions of 5×5. In this example, there are a total of 10 averaged correlation coefficients with 55 elements in $\boldsymbol{T}^2$. There may not be sufficient data to estimate all of the elements in $\boldsymbol{T}^2$. A workaround solution is to restrict the structure of $\boldsymbol{T}^2$ to a diagonal matrix. Instead of estimating all of the elements in $\boldsymbol{T}^2$, we may only estimate 10 variances in $\boldsymbol{T}^2$ in this example (see Section 5.3.1 for the discussion). Second, if $\boldsymbol{T}^2 = 0$ is used, the model is equivalent to the GLS approach discussed in Section 7.2.2. It then becomes a fixed-effects model.

In the analysis, the average correlation matrix $\hat{\rho}_R$ based on a random-effects model and its asymptotic sampling covariance matrix $\hat{V}_R$ are estimated, and $\hat{T}^2$. $\hat{\rho}_R$ is the vector of the average correlation coefficients, whereas $\hat{V}_R$ indicates the precision of $\hat{\rho}_R$. $\hat{T}^2$ shows the heterogeneity of the correlation vectors. As it is difficult to judge how large or small the degree of heterogeneity is by inspecting $\hat{T}^2$, it is of practical relevance to check the I^2 of the correlation coefficients (see Section 4.3.3). If the I^2s on the correlation coefficients are all very small, it is expected that the results based on the fixed-effects model and the random-effects model will be similar.

7.4.2 Stage 2 of the analysis: fitting structural models

Similar to the second stage of the fixed-effects TSSEM, the estimated values from stage 1 are treated as the sample statistics in the stage 2 analysis. Therefore, there is no hat on them, that is, $r_R = \hat{\rho}_R$ and $V_R = \hat{V}_R$. We will use them as the inputs in fitting the structural model $\rho_R(\theta) = \text{vechs}(P(\theta))$ in the stage 2 analysis. The discrepancy function in the stage 2 analysis is

$$F_{\text{WLS}}(\theta) = (r_R - \rho_R(\theta))^{\text{T}} V_R^{-1} (r_R - \rho_R(\theta)). \tag{7.18}$$

The analysis is the same as those listed in the second stage of the fixed-effects TSSEM, except that r_F and V_F are replaced by r_R and V_R in the fit function.

It should also be noted that the estimated variance component $\hat{T}^2$ is not directly involved in the above fit function in the stage 2 analysis. As V_R is estimated after controlling for the random effects T^2, V_R has already taken the random effects into account. Therefore, V_R is usually larger than V_F based on a fixed-effects model. The SEs of the parameter estimates based on a random-effects model are also usually larger than those based on a counterpart fixed-effects model. Therefore, it is important to use V_R as the weight matrix in a random-effects model. Analyses and interpretations are similar to those for the fixed-effects model except that the analyses are based on the average population correlation matrix.

7.5 Related issues

Before moving to the demonstrations with the `metaSEM` package in R, this section discusses some issues related to MASEM. We first compare the multiple-group SEM versus MASEM in analyzing a pool of correlation (or covariance) matrices and then compare and contrast the two fixed-effects models—the TSSEM and the GLS approaches. After this, we address alternative random-effects models for MASEM. Finally, we discuss topics such as the use of ML or restricted (or residual) maximum likelihood (REML) in MASEM, the use of correlation coefficient versus Fisher's z score, and the correction for unreliability in MASEM.

7.5.1 Multiple-group structural equation modeling versus meta-analytic structural equation modeling

Multiple-group SEM may be used to analyze data with missing data for independent groups (e.g., Allison, 1987; Muthén et al., 1987). The general idea is to partition the data into data sets that contains both complete data and several data sets with different missing-data patterns. By hypothesizing that the same model holds across the complete and incomplete data sets, the whole model can be estimated by applying appropriate equality constraints among different samples.

Let us illustrate the idea with an example. Suppose that we are fitting a two-factor CFA model with two indicators per factor. x_2 and x_4 are missing in some cases. We may partition the data into two samples—Sample 1 is complete and Sample 2 has missing data. The CFA models in them are $\Phi^{(1)} = \begin{bmatrix} \phi_{11}^{(1)} & \\ \phi_{21}^{(1)} & \phi_{22}^{(1)} \end{bmatrix}$, $\Lambda^{(1)} = \begin{bmatrix} 1 & \lambda_{21}^{(1)} & 0 & 0 \\ 0 & 0 & 1 & \lambda_{42}^{(1)} \end{bmatrix}^{\mathrm{T}}$, and $\Psi^{(1)} = \mathrm{Diag}\begin{bmatrix} \psi_{11}^{(1)} & \psi_{22}^{(1)} & \psi_{33}^{(1)} & \psi_{44}^{(1)} \end{bmatrix}$ and $\Phi^{(2)} = \begin{bmatrix} \phi_{11}^{(2)} & \\ \phi_{21}^{(2)} & \phi_{22}^{(2)} \end{bmatrix}$, $\Lambda^{(2)} = \begin{bmatrix} 1 & \mathrm{NA} & 0 & 0 \\ 0 & 0 & 1 & \mathrm{NA} \end{bmatrix}^{\mathrm{T}}$, and $\Psi^{(2)} = \mathrm{Diag}\begin{bmatrix} \psi_{11}^{(2)} & \mathrm{NA} & \psi_{33}^{(2)} & \mathrm{NA} \end{bmatrix}$, where Λ, Φ, and Ψ are the factor loadings, factor covariance, and error variance matrices, respectively, and NA represents the missing parameters. The model in Sample 1 is identified with 1 df, whereas the model in Sample 2 is not identified by itself because of the missing values. If we assume that the parameter estimates are the same in these samples under the assumption of MAR, we may combine the samples by applying the constraints, namely, $\phi_{11}^{(1)} = \phi_{11}^{(2)}$, $\phi_{21}^{(1)} = \phi_{21}^{(2)}$, $\phi_{22}^{(1)} = \phi_{22}^{(2)}$, $\psi_{11}^{(1)} = \psi_{11}^{(2)}$, and $\psi_{33}^{(1)} = \psi_{33}^{(2)}$. The model is then identified with four dfs.

The multiple-group SEM may be applied to conduct MASEM. However, several issues limit the usefulness of this approach in MASEM. The first issue is that this approach assumes that a fixed-effects model is used. The use of fixed-effects models in MASEM may be questionable on theoretical and empirical grounds. The second issue is that it is challenging to formulate submodels for all patterns of missing data. The third issue is that the homogeneity of correlation and covariance matrices is assumed in the proposed model. If the proposed model does not fit the data, it is not clear whether the misfit should be attributed to the heterogeneity of the correlation matrices or to the misspecification of the structural equation model or even both.

Multiple-group SEM may also be used to test the measurement invariance (e.g., Vandenberg and Lance, 2000) of the instrument across studies. Hafdahl (2001) also discussed the benefits and limitations of this approach to conducting MASEM for exploratory factor analysis. For example, we may test the configural invariance (same pattern of fixed and free factor loadings as the proposed model), the metric invariance (factor loadings are the same across groups), and invariance of the factor variance–covariance matrices of the data. Some researchers may apply techniques on measurement invariance in MASEM. That is, we may treat the studies as groups and test whether the measurement is invariant across these studies. There are a couple of limitations to this approach. First, covariance rather than correlation matrices are used in testing measurement invariance. Unless the same measurement is used in all studies, it may not make sense to use covariance matrices in MASEM. Second, equality constraints have to be applied in some or all parameters in order to identify the model when there are missing variables. Some of these models, for example, the metric invariance, may not be testable.

Another difficulty is that measurement invariance is usually applied to studies with only a small number of groups. When there are many (heterogeneous) studies as in MASEM, it may not be reasonable to expect that some models, for example, the metric invariance, work for all studies. Researchers may need to look for partially measurement invariance (Byrne et al., 1989) by freeing some equality constraints on the factor loadings in an ad hoc manner. It is not clear whether the findings are still interpretable and replicable. Generally speaking, MASEM would be a better choice than multiple-group SEM to address these issues.

7.5.2 Fixed-effects model: two-stage structural equation modeling versus generalized least squares

Both the stage 1 analysis of the fixed-effects TSSEM approach (`tssem1(..., method='FEM')`) and the GLS approach discussed in Section 7.2.2 (`tssem1(..., method='REM', RE.type='Zero')` implemented in the `metaSEM` package) are assumed to be a fixed-effects model. This section compares and contrasts these two approaches to conducting a fixed-effects MASEM.

One main difference is whether the conditional sampling covariance matrix $\boldsymbol{V}_i$ has to be estimated. The stage 1 analysis of the TSSEM approach does not involve estimating the sampling covariance $\boldsymbol{V}_i$ of the correlation vector $\boldsymbol{r}_i = \text{vechs}(\boldsymbol{R}_i)$. In the TSSEM approach, the sample correlation matrices are treated as sample statistics with sampling errors. Thus, the TSSEM approach is more stable and accurate. In contrast, in the GLS approach, it is necessary to estimate the sampling covariance matrix of the correlation coefficients. If $\boldsymbol{R}_i$ is a 5×5 correlation matrix, for example, $\boldsymbol{V}_i$ is a 10×10 matrix with 45 elements. Because of the fact that $\boldsymbol{V}_i$ is estimated under the GLS approach and a multiple-group SEM is used in the fixed-effects TSSEM, the test statistics on testing the null hypothesis of the homogeneity of correlation matrices are slightly different. Treating $\boldsymbol{V}_i$ as known values affects the accuracy of the estimation, especially when the sample sizes are small. Thus, the GLS approach does not perform well empirically (e.g., Cheung and Chan, 2005b, 2009).

Several modifications have been made in the attempt to improve the empirical performance of the GLS approach (e.g., Becker and Fahrbach, 1994; Cheung, 2000; Furlow and Beretvas, 2005; Hafdahl, 2007). The basic idea of these modifications is to use some estimates that are more stable than $\boldsymbol{V}_i$. If we assume that the studies are homogeneous (a fixed-effects model), we may calculate a simple (or weighted average) correlation matrices (or Fisher's z scores). These mean correlation matrices (or Fisher's z scores) are used to derive $\boldsymbol{V}_i$. As the mean correlation matrices (or Fisher's z scores) are based on a much larger sample size, they are much more stable than a $\boldsymbol{V}_i$ that is based on individual studies.

Another difference is that an LR statistic and various goodness-of-fit indices are available in the fixed-effects TSSEM approach. Besides testing the hypothesis on the exact fit, we may also use the goodness-of-fit indices to evaluate the close or approximate fit on the correlation matrices. In contrast, only the Q statistic can be used to evaluate the appropriateness of the fixed-effects model in the GLS approach.

Another key difference between the GLS approach and the fixed-effects TSSEM approach is how the missing values are handled. The GLS approach does not assume that the input effect sizes are correlation matrices—they are simply generic-dependent effect sizes. Suppose that there are 10 correlation coefficients involved in synthesizing the correlation matrices of five variables. These 10 correlation coefficients are treated as 10 effect sizes in a multivariate meta-analysis. Thus, some of these correlation coefficients can be arbitrarily missing. Moreover, the input correlation matrices do not need to be positive definite.

In contrast, the fixed-effects TSSEM approach uses multiple-group SEM to pool correlation matrices. The unit of the analysis is *variables*. When there are missing correlation coefficients, the variables associated with the missing values are also assumed to be missing. For example, if r_{21} is missing, we may have to treat either Variable 1 or Variable 2 as missing. The main advantage of this approach is that the input matrices are always correlation (or covariance) matrices. This may reduce the chances of producing nonpositive definite matrix on the pooled correlation matrix. The disadvantage is that this approach is less flexible in handling missing correlation coefficients. When there are missing correlation coefficients, some of the correlation coefficients have to be excluded in the analysis. Moreover, the input correlation matrices have to be positive definite.

In principle, it is possible to modify the fixed-effects TSSEM approach to handle missing correlations in a more flexible manner. There are two such approaches. For both approaches, the missing correlations are replaced by some valid correlation values, for example, 0. Therefore, the dimensions of the input matrices are always the same. For the first approach, the equality constraints do not apply to the elements with missing correlations in the studies. As there is no equality constraint in those studies, the estimated correlations are the same as the input correlations, for example, 0 for the missing values. Effectively, the imputed values for the missing correlations neither affect the estimated common correlation matrix nor contribute to the test statistic on the homogeneity of correlation matrices. The stand-alone program to conduct fixed-effects TSSEM with `LISREL` (Cheung, 2009) uses this approach to handle missing correlations.

Jak et al. (2013) proposed another approach to address this issue. Similar to the above approach, the missing correlations are replaced with some valid correlation values, for example, 0. The missing correlations are filtered out from the analysis of equality constraints by using a selection matrix. In order to off-set the fact that "0" is used to represent the missing values, an additional matrix is created in each group. The elements of this matrix are free when correlation coefficients are missing; otherwise, they are fixed parameters with 0. The elements of this additional matrix will off-set the value of the constraints. In other words, the estimated values for the pooled correlation matrix are always the same regardless of what values are used to replace the missing correlations.

These two approaches are more flexible than the approach implemented in the `metaSEM` package for handling missing correlations. One issue arises, however. As the missing values are replaced by some valid correlation values, for example, 0, this imputed value will also be used in calculating the independence model. The calculated chi-square statistic and the dfs of the independence model may be incorrect. This may affect the calculated goodness-of-fit indices, for example, TLI and CFI, because they use the independence model as the reference model in testing the homogeneity of the correlation matrices. Special care has to be taken so that the calculated goodness-of-fit indices are correct.

7.5.3 Alternative random-effects models

As studies are rarely direct replicates of each other, they are likely different in a number of ways, for example, research design, samples, measures, and time of the data collected. It is reasonable to expect that studies may have their own population correlation matrices. Therefore, the random-effects model is usually the preferred choice in meta-analysis. The random-effects models are well defined in conventional meta-analysis. Suppose that we are analyzing a random-effects meta-analysis for $p \times p$ correlation matrices; the model for the ith study (without any missing effect size) is

$$\boldsymbol{r}_i = \boldsymbol{\rho}_\mathrm{R} + \boldsymbol{u}_{ri} + \boldsymbol{e}_{ri}, \tag{7.19}$$

where $\boldsymbol{r}_i$ is the sample correlation vector, $\boldsymbol{\rho}_\mathrm{R}$ is the average population correlation vector under a random-effects model, $\boldsymbol{T}_r^2 = \mathsf{Cov}(\boldsymbol{u}_{ri})$ is the variance component of the random effects, and $\boldsymbol{V}_{ri} = \mathsf{Cov}(\boldsymbol{e}_{ri})$ is the known sampling covariance matrix of the sampling error. The random effects $\boldsymbol{u}_{ri}$ are defined as the study-specific effects *deviated* from the *average* population correlation vector or the fixed effects. Each study has its own *true* effect sizes $\boldsymbol{\rho}_i = \boldsymbol{\rho}_\mathrm{R} + \boldsymbol{u}_{ri}$.

The situation is more complicated in MASEM. It is because we may define the random effects in terms of either the *average* population correlation vector or the *average* population parameter vector in the structural model (Cheung and Cheung, 2010). We introduce them here one by one. Following the literature on the conventional meta-analysis, we may define the random effects as the study-specific effects deviated from the average population correlation vector. For example, Becker (1992); Cheung (2014a), and Section 7.4 in this chapter follow this practice. This average population correlation vector is used to fit the structural model $\boldsymbol{\rho}_\mathrm{R} = \boldsymbol{\rho}(\boldsymbol{\theta})$ in the stage 2 analysis. Therefore, the model for the ith study is

$$\boldsymbol{r}_i = \boldsymbol{\rho}(\boldsymbol{\theta}) + \boldsymbol{u}_{ri} + \boldsymbol{e}_{ri}. \tag{7.20}$$

This approach is termed the *correlation-based MASEM* here because meta-analysis is applied to the correlation coefficients.

By comparing the two above equations, it is apparent that the structural model is fitted on the average population correlation vector $\boldsymbol{\rho}_\mathrm{R}$. It should be noted that there is no guarantee that the proposed model will also fit well in the *true* correlation vectors of the individual studies $\boldsymbol{\rho}_i$ even the proposed model fits well in the *average* population correlation matrix $\boldsymbol{\rho}_{\mathrm{R}}$. Because of this fact, it is reasonable to question the usefulness of the structural model fitted on $\boldsymbol{\rho}_{\mathrm{R}}$. We believe that the correlation-based MASEM is still useful in MASEM. The main argument to defend the usefulness of the structural model fitted on $\boldsymbol{\rho}_{\mathrm{R}}$ is the same as the justifications of the fixed effects in mixed-effects model and random-effects meta-analysis. There are lots of examples involving the *average* effect in the literature, for example, the average population effect sizes $\boldsymbol{\rho}_{\mathrm{R}}$ in the random-effects meta-analysis, the average intercept and slopes in multilevel modeling, and latent growth modeling.

All the fixed effects are in fact the *average* effects. The average effect may not apply to any individual study in a meta-analysis, any specific level 2 unit in a multilevel model, or any particular individual in a latent growth model. Researchers, however, still find the average effect useful because it gives us an idea on the *typical* effect. By considering both the average effect (the fixed effects) and the variability around the average effect (the variance component of the random effects), researchers can get an idea on the phenomena being studied. As the correlation-based MASEM is just yet another example of the fixed effects in mixed-effects model, arguments for the fixed effects in the mixed-effects models are also applicable to the correlation-based MASEM. That said, researchers should also consider the variance component of the random effects when interpreting the fixed effects.

In terms of statistical analyses, a multivariate meta-analysis is conducted on the correlation matrices in the first stage of analysis. After the analysis, an average correlation matrix and its asymptotic covariance matrix are estimated. The average correlation matrix and its asymptotic covariance matrix are used as input in fitting the structural models in the second stage of analysis (see Section 7.4 for details).

An alternative model is to consider the parameter estimates or functions of parameter estimates as the effect sizes. Several authors have proposed using the intercepts and regression slopes as the effect sizes in a meta-analysis. Becker and Wu (2007) compared the univariate and the multivariate approaches to synthesizing regression coefficients. Paul et al. (2006) described a meta-analysis on using the estimated intercepts and regression slopes as the effect sizes. The estimated intercepts and regression slopes were then subjected to a mixed-effects multivariate meta-analysis by using the study characteristics as moderators. Gasparrini et al. (2012) also proposed a similar approach to synthesize parameter estimates. A generalized linear model was first fitted in each study. The parameter estimates were then treated as effect sizes for a multivariate meta-analysis in the second stage of analysis. Hafdahl (2009) proposed an approach that allows researchers to meta-analyze functions of correlation elements.

The key assumption of this approach is that the same *structural model* holds across all studies. However, the parameters may vary across studies, that is, $\boldsymbol{\rho}_i = \boldsymbol{\rho}(\boldsymbol{\theta}_i)$, where $\boldsymbol{\theta}_i$ is the *true* parameter vector in the ith study. Thus, the model-implied correlation matrices may be different in each study. The estimated parameter vector based on the samples are $\boldsymbol{t}_i = \hat{\boldsymbol{\theta}}_i$. As there is a subscript i in $\boldsymbol{\theta}_i$

and $\boldsymbol{t}_i$, the parameters and their sample estimates may vary across studies. The model for the ith study is

$$\boldsymbol{t}_i = \boldsymbol{\theta}_{\mathrm{R}} + \boldsymbol{u}_{\theta i} + \boldsymbol{e}_{\theta i}, \tag{7.21}$$

where $\boldsymbol{\theta}_{\mathrm{R}}$ is the average population vector of parameters under a random-effects model, $\boldsymbol{T}^2_\theta = \mathsf{Cov}(\boldsymbol{u}_{\theta i})$ is the variance component of the random effects, and $\boldsymbol{V}_{\theta i} = \mathsf{Cov}(\boldsymbol{e}_{\theta i})$ is the known sampling covariance matrix of the sampling error estimated from the first stage of the analysis. This approach can be termed the *parameter-based MASEM* because meta-analysis is applied to the parameter estimates or functions of the parameter estimates.

To conduct the parameter-based MASEM, we first fit the correlation structure $\rho(\boldsymbol{\theta}_i)$ in each study and obtain the parameter estimates $\boldsymbol{t}_i$ with its sampling covariance matrix $\boldsymbol{V}_{ti}$ in the first stage of the analysis. $\boldsymbol{t}_{ti}$ are treated as the effect sizes subjected to a multivariate meta-analysis with the inverse of the asymptotic covariance matrix $\boldsymbol{V}_{ti}$ as the weight. If study characteristics are available, a mixed-effects multivariate meta-analysis may also be conducted (see Section 5.4 for details). After the second stage of analysis, we may obtain the average effect on $\hat{\boldsymbol{\theta}}_{\mathrm{R}}$ and the heterogeneity variance–covariance matrix $\hat{\boldsymbol{T}}^2_\theta$ on $\boldsymbol{\theta}_{\theta i}$.

Comparing to the correlation-based MASEM, the parameter-based MASEM has several advantages. Research questions may involve functions of the correlation coefficients. For example, researchers conducting a meta-analysis on a mediation model may be interested in how the direct effect and the indirect effect as effect sizes vary across studies. The parameter-based MASEM enables us to empirically address these research questions. Second, the parameter-based MASEM quantifies the heterogeneity of the parameter estimates across studies. Third, continuous predictors may be used to model the estimated parameter estimates. This helps to address why some studies have larger (or smaller) parameter estimates.

It appears that the parameter-based MASEM is always preferable to the correlation-based MASEM. However, there are a couple of issues limiting the usefulness of the parameter-based MASEM. First, the parameter-based MASEM may fail when there are missing effect sizes. As the structural model is fitted in each study, the presence of missing effect sizes makes this approach difficult to apply. For example, we need the predictor, mediator, and dependent variable in order to estimate the indirect and direct effects. If any of these variables are not reported in the studies, we cannot estimate the required effect sizes. Second, the parameter-based MASEM may not work for the overidentified models. When the models are just identified such as the regression models, the models perfectly reproduce the data. The model-implied correlation matrix is exactly the same as the sample correlation matrix. When the models are overidentified such as the CFA or SEM, the proposed models rarely fit the data. The model-implied correlation matrix may be quite different from the sample correlation matrix. If the proposed model does not fit the sample correlation matrix well, the validity of the parameter estimates is questionable. It is not clear whether the meta-analysis is still appropriate.

Cheung and Cheung (2010) used an empirical data set to compare these two approaches under a justidentified model (mediation model). Data from 42 nations were collected between 1990 and 1993 (World Values Study Group, 1994). The data set included correlation matrices on three variables. A mediation model from job autonomy to life satisfaction with job satisfaction as the mediator was proposed and tested. Under the correlation-based MASEM, the random-effects TSSEM was fitted. Under the parameter-based MASEM, the direct and the indirect effects were first estimated in each study. The estimated direct and indirect effects were then subjected to a multivariate meta-analysis. The results showed that the estimated fixed effects on the direct and indirect effects were nearly identical, whereas the variance components were not directly comparable—one was based on correlation coefficients and the other was based on the direct and indirect effects. As this topic is new to MASEM, more studies are required to address the pros and cons of these approaches.

7.5.4 Maximum likelihood estimation versus restricted (or residual) maximum likelihood estimation

In the literature of meta-analysis, the REML is sometimes preferred to the ML estimation method. It is because the estimated variance component of the random effects is negatively biased in the ML estimation (see Section 8.1). The estimated fixed effects are unbiased for the ML estimation even though the sample sizes are small (Demidenko, 2013). As we only use the estimated fixed effects (average correlation matrix) in the stage 2 analysis, whether or not the estimated variance component is (slightly) biased is not a concern in MASEM. Therefore, the ML estimation should be used in the TSSEM approach.

7.5.5 Correlation coefficient versus Fisher's z score

The discussion so far has been on pooling correlation matrices and on fitting correlation structure on the pooled correlation matrix. In the literature of meta-analysis, there are debates on whether the correlation coefficient or its Fisher's z score should be used. Fisher's z score may also be used in MASEM. For example, the correlation vector can be transformed into a vector of Fisher's z scores by using the Fisher's z transformation with Equation 3.25. The sampling variances of the Fisher's z scores can be estimated as $1/(n-3)$, whereas the sampling covariance between z_{ij} (the Fisher's z score of r_{ij}) and z_{kl} (the Fisher's z score of r_{kl}) can be estimated as

$$\mathsf{Cov}(z_{ij}, z_{kl}) = \frac{n\mathsf{Cov}(r_{ij}, r_{kl})}{(n-3)(1-r_{ij}^2)(1-r_{kl}^2)}, \tag{7.22}$$

where $\mathsf{Cov}(r_{ij}, r_{kl})$ is the sampling covariance between r_{ij} and r_{kl} and n is the sample size in the study (Steiger, 1980; see also Section 3.3.2).

Once the vector of Fisher's z scores and its asymptotic sampling covariance matrix are available, the GLS approach may be used to synthesize the Fisher's

z scores. Hafdahl (2007, 2008) compared several modifications on the use of the Fisher's z score in pooling correlation matrices under the assumption of the homogeneity and heterogeneity of correlation matrices. He found that the modified approaches generally work well on combining correlation matrices using the Fisher's z scores.

Structural equation models may also be fitted on the pooled Fisher's z scores with a discrepancy function similar to the one in Equation 7.18 (see Fouladi (2000) for details). One issue is that it is less intuitive in interpreting the parameter estimates because the models are based on the Fisher's z scores rather than on the correlation coefficients. Therefore, applications of the Fisher's z score on fitting correlation matrices are still rare. An alternative approach is to transform the pooled Fisher's z scores and its asymptotic covariance matrix back into the correlation vector and its asymptotic covariance matrix (see Hafdahl (2007) for the proposed conversions). There are only a limited number of studies addressing the empirical performance of the Fisher's z score in MASEM. Further studies may compare whether the Fisher's z scores perform better than the correlation matrices in MASEM.

7.5.6 Correction for unreliability

When item-level data are available, there is no need to correct for the unreliability in MASEM. It is because fitting CFA and SEM on the item-level data will account for the measurement errors. When MASEM is conducted on the composite scores, some researchers prefer to apply the correction for unreliability.

There is some controversy over whether it is necessary to correct for attenuation and the statistical artifacts in a meta-analysis. As the measurements are liable to contain measurement error, the observed correlation coefficients are usually smaller than the actual correlations without measurement error. Rosenthal (1991) criticized the use of correction for attenuations because the corrected values are different from the *typical* research findings and the corrected values are not as useful as the uncorrected values in realistic settings. Other researchers (e.g., Schmidt and Hunter, 2015) argued for correcting for attenuations before combining them. Schmidt and Hunter (2015) identified 11 artifacts that could be corrected before combining the correlation coefficients. These include sampling error, error of measurement in the dependent and independent variables, restriction of range, and so on. However, it is unlikely that the published articles would include all the information for correction.

One type of measurement error is unreliability, which can be corrected by

$$r_{\text{Cor}} = \frac{r_{xy}}{\sqrt{r_{xx'} r_{yy'}}}, \tag{7.23}$$

where r_{Cor} is the estimated corrected correlation for the unreliability of measurements, r_{xy} is the observed correlation between variables x and y, and $r_{xx'}$ and $r_{yy'}$ are the estimated reliabilities of variables x and y, respectively. As there are usually more than two variables in MASEM, the above correction applies to all correlation coefficients. The correlation matrices corrected for unreliability are then used in MASEM.

There are several issues associated with this practice. First, not all studies report information on reliabilities. When values on the reliabilities are missing, some researchers may have to use the mean reliabilities from all studies. Second, the correlation matrices corrected for unreliability may be nonpositive definite. If this happens, the resultant correlation matrices cannot be used in MASEM. Third, the conditional sampling covariance matrix $\boldsymbol{V}_i$ of $\boldsymbol{r}_{\text{Cor}}$ is likely to have been underestimated. It is because the corrected correlation matrix is treated as an observed correlation matrix that does not take the uncertainty in estimating the unreliability into account.

By reviewing several published meta-analyses, Michel et al. (2011) recently claimed that substantive model conclusions in the psychological literature are generally unaffected by study artifacts and related statistical corrections. As their conclusions are based on real examples rather than on computer simulations, further research may address the effects of unreliability in MASEM.

7.6 Illustrations using R

This section illustrates how to conduct the fixed- and random-effects TSSEM using the `metaSEM` package. The examples include a CFA model, regression model, and path model. The examples also illustrate how to specify the structural equation models in the stage 2 analysis with the RAM formulation. These examples should be general enough so that the techniques can be generalized to other models. We consider the proposed models would fit the data reasonably well if the RMSEA is close to 0.05. If the RMSEA is larger than 0.10, the proposed model does not fit the data (Browne and Cudeck, 1993). Moreover, we also use the SRMR to evaluate the proposed models. The proposed models fit the data reasonably well if SRMR is smaller than 0.05. The statistics reported in the illustrations were captured by using the `Sweave` function in R. The numbers of decimal places may be slightly different for those reported in the selected output and in the text.

7.6.1 A higher-order confirmatory factor analytic model for the Big Five model

The objective of this example is to illustrate how to fit a CFA model with the TSSEM approach. Digman (1997) reported 14 correlation matrices among the five-factor model. He proposed that agreeableness (*A*), conscientiousness (*C*), and emotional stability (*ES*) were loaded under a higher-order factor, called *Alpha*, whereas extraversion (*E*) and intellect (*I*) were loaded under another higher-order factor, called *Beta*. The *Alpha* and *Beta* factors represent the general personality of socialization and personal growth. Figure 7.2 displays the higher-order model with the corresponding elements labeled in the RAM formulation.

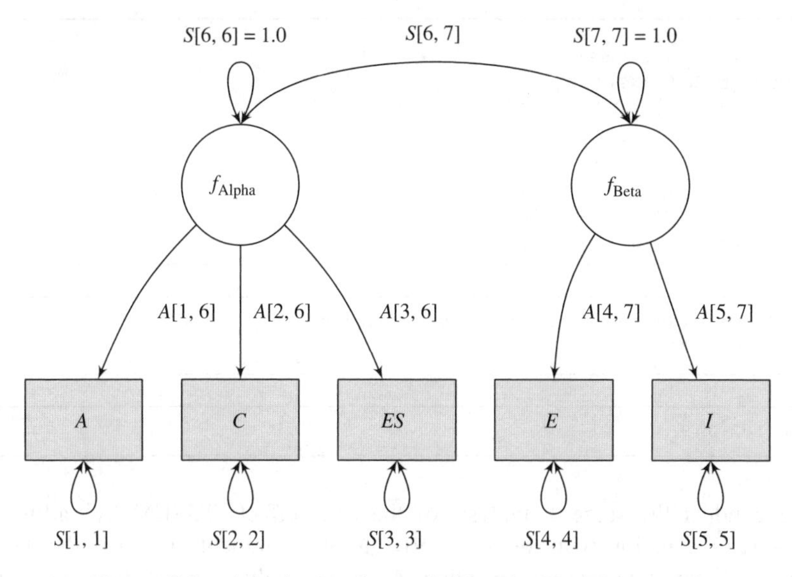

Figure 7.2 A higher-order model for the Big Five model.

The sample sizes of these studies vary from 70 to 1040 with a total sample size of 4496. The data set was stored as an R object called `Digman97` in the `metaSEM` package, where the correlation matrices and sample sizes were stored as `Digman97$data` and `Digman97$n`, respectively. Digman further grouped the studies under the samples of *Children*, *Young adults*, and *Mature adults*. They are stored as the `Digman97$cluster`. For ease of illustration, these groups are recoded to *younger* versus *older* participants in this demonstration. We begin the illustration by first fitting a fixed-effects TSSEM. Then, we refit the model by grouping the studies under *younger* versus *older* participants. Finally, we fit a random-effects TSSEM.

7.6.1.1 Fixed-effects model

We display the content of the first two studies by using the following command:

```
R> library("metaSEM")
R> ## Display the first two correlation matrices
R> Digman97$data[1:2]
```

```
$'Digman 1 (1994)'
       A     C   ES     E    I
A   1.00  0.62 0.41 -0.48 0.00
C   0.62  1.00 0.59 -0.10 0.35
ES  0.41  0.59 1.00  0.27 0.41
E  -0.48 -0.10 0.27  1.00 0.37
I   0.00  0.35 0.41  0.37 1.00
```

```
$'Digman 2 (1994)'
      A    C   ES     E     I
A   1.00 0.39 0.53 -0.30 -0.05
C   0.39 1.00 0.59  0.07  0.44
ES  0.53 0.59 1.00  0.09  0.22
E  -0.30 0.07 0.09  1.00  0.45
I  -0.05 0.44 0.22  0.45  1.00
```

```
R> ## Display the first two sample sizes
R> Digman97$n[1:2]
```

```
[1] 102 149
```

We conduct the stage 1 analysis of the fixed-effects TSSEM by calling the `tssem1()` function with the `method="FEM"` argument, which is the default value if this argument is not specified. After the analysis, we display the results using the `summary()` function.

```
R> fixed1 <- tssem1(Digman97$data, Digman97$n, method="FEM")
R> summary(fixed1)
R> ## Alternative approach if you do not want to save "fixed1"
R> ## summary( tssem1(Digman97$data, Digman97$n, method="FEM") )
```

```
----------------------- Selected output ----------------------

Coefficients:
      Estimate Std.Error z value Pr(>|z|)
S[1,2]  0.3631    0.0134   27.12  < 2e-16 ***
S[1,3]  0.3902    0.0129   30.24  < 2e-16 ***
S[1,4]  0.1038    0.0151    6.88  5.8e-12 ***
S[1,5]  0.0922    0.0151    6.12  9.3e-10 ***
S[2,3]  0.4160    0.0125   33.17  < 2e-16 ***
S[2,4]  0.1352    0.0148    9.14  < 2e-16 ***
S[2,5]  0.1412    0.0149    9.48  < 2e-16 ***
S[3,4]  0.2445    0.0142   17.25  < 2e-16 ***
S[3,5]  0.1382    0.0149    9.30  < 2e-16 ***
S[4,5]  0.4245    0.0124   34.25  < 2e-16 ***
---
Signif. codes:  0 '***' 0.001 '**' 0.01 '*' 0.05 '.' 0.1 ' ' 1

Goodness-of-fit indices:
                                 Value
Sample size                    4496.00
Chi-square of target model     1499.73
DF of target model              130.00
p value of target model           0.00
```

```
Chi-square of independence model 4454.60
DF of independence model          140.00
RMSEA                               0.18
SRMR                                0.16
TLI                                 0.66
CFI                                 0.68
AIC                              1239.73
BIC                               406.31
OpenMx status1: 0 ("0" or "1": The optimization is considered fine.
Other values indicate problems.)

-----------------------  Selected output  ----------------------
```

Before interpreting the results, we should check whether there are estimation problems in the analysis. We check the `OpenMx status1` at the end of the output. If the status is either 0 or 1, the optimization is fine; otherwise, the results are not trustworthy. The test statistic on testing the homogeneity test of the correlation matrix is $\chi^2(\text{df} = 130, N = 4496) = 1499.7340, p < 0.001$. On the basis of the test statistic, the null hypothesis of homogeneity of the correlation matrices was rejected. Moreover, both the RMSEA and SRMR are very large, indicating that the proposed model (homogeneity of correlation matrices) did not fit the data well. The estimates of the $S[i, j]$ represent the element of the pooled correlation (or covariance) matrix. They can be extracted by using the `coef()` function:

```
R> coef(fixed1)
```

```
         x1     x2     x3     x4      x5
x1 1.00000 0.3631 0.3902 0.1038 0.09225
x2 0.36312 1.0000 0.4160 0.1352 0.14121
x3 0.39018 0.4160 1.0000 0.2445 0.13817
x4 0.10375 0.1352 0.2445 1.0000 0.42451
x5 0.09225 0.1412 0.1382 0.4245 1.00000
```

As the fixed-effects model does not seem appropriate for this data set, we should consider either grouping the studies into clusters or using a random-effects model. As an illustration, we continue to fit the structural model on the pooled correlation matrix. The structural model in the stage 2 analysis is specified via the RAM formulation (see Section 2.2.3 for more details). The observed variables (*A*, *C*, *ES*, *E*, and *I*) and the latent variables (*f_Alpha* and *f_Beta*) are stacked together. Readers may refer to Figure 7.2 for the corresponding elements in the RAM formulation.

The three matrices (***F***, ***A***, and ***S***) should be `MxMatrix` objects. There are three equivalent approaches to create the `MxMatrix` objects. The first approach is to use the `mxMatrix()` function in the `OpenMx` package. We need to specify the starting values, free versus fixed parameters, and the labels for the parameters. As this approach may sound tedious to many new users, we use the other approaches here.

The second approach is to specify a character matrix with the `matrix()` function. After specifying the matrices, we use the `as.mxMatrix()` function to convert them into `MxMatrix` objects. If the elements are numeric, they are treated as fixed parameters with the values fixed at the given values. If the elements are characters, they are treated as free parameters. For example, if we specify `0.5*loading1`, it is treated as a free parameter with 0.5 as the starting value and `loading1` as the label of the parameter. If the labels are the same, these parameters will be constrained equally. The third approach is to specify these matrices via the `create.Fmatrix()` function to create the ***F*** matrix and the `create.mxMatrix()` to create the `MxMatrix` objects. We illustrate these two approaches in the following examples.

```
R> ## Define the S matrix that includes both Phi and Psi
R> ## Phi matrix: 2x2 correlation matrix between the latent factors
R> ( Phi <- matrix(c(1,".3*cor",".3*cor",1), nrow=2, ncol=2) )
```

```
     [,1]     [,2]
[1,] "1"      ".3*cor"
[2,] ".3*cor" "1"
```

```
R> ## Psi matrix: 5x5 diagonal matrix of the error variances
R> ( Psi <- Diag(c(".2*e1",".2*e2",".2*e3",".2*e4",".2*e5")) )
```

```
     [,1]    [,2]    [,3]    [,4]    [,5]
[1,] ".2*e1" "0"     "0"     "0"     "0"
[2,] "0"     ".2*e2" "0"     "0"     "0"
[3,] "0"     "0"     ".2*e3" "0"     "0"
[4,] "0"     "0"     "0"     ".2*e4" "0"
[5,] "0"     "0"     "0"     "0"     ".2*e5"
```

```
R> ## Create a block diagonal matrix as the S matrix
R> S1 <- bdiagMat(list(Psi, Phi))
R> ## This step is not necessary but useful
R> ## for inspecting the model.
R> dimnames(S1)[[1]] <- c("A","C","ES","E","I","f_Alpha","f_Beta")
R> dimnames(S1)[[2]] <- dimnames(S1)[[1]]
R> S1
```

```
        A       C       ES      E       I       f_Alpha  f_Beta
A       ".2*e1" "0"     "0"     "0"     "0"     "0"      "0"
C       "0"     ".2*e2" "0"     "0"     "0"     "0"      "0"
ES      "0"     "0"     ".2*e3" "0"     "0"     "0"      "0"
E       "0"     "0"     "0"     ".2*e4" "0"     "0"      "0"
I       "0"     "0"     "0"     "0"     ".2*e5" "0"      "0"
f_Alpha "0"     "0"     "0"     "0"     "0"     "1"      ".3*cor"
f_Beta  "0"     "0"     "0"     "0"     "0"     ".3*cor" "1"
```

```
R> ## Convert it into a MxMatrix class
R> S1 <- as.mxMatrix(S1)
R> ## Lambda matrix: 5x2 factor loadings
R> ## Arrange the data by row
R> ( Lambda <- matrix(c(".3*Alpha_A", 0,
                         ".3*Alpha_C", 0,
                         ".3*Alpha_ES", 0,
                         0, ".3*Beta_E",
                         0, ".3*Beta_I"),
                       nrow=5, ncol=2, byrow=TRUE) )
```

```
     [,1]          [,2]
[1,] ".3*Alpha_A"  "0"
[2,] ".3*Alpha_C"  "0"
[3,] ".3*Alpha_ES" "0"
[4,] "0"           ".3*Beta_E"
[5,] "0"           ".3*Beta_I"
```

```
R> ## Create a 5x5 of zeros
R> ( Zero5x5 <- matrix(0, nrow=5, ncol=5) )
```

```
     [,1] [,2] [,3] [,4] [,5]
[1,]    0    0    0    0    0
[2,]    0    0    0    0    0
[3,]    0    0    0    0    0
[4,]    0    0    0    0    0
[5,]    0    0    0    0    0
```

```
R> ## Create a 2x7 of zeros
R> ( Zero2x7 <- matrix(0, nrow=2, ncol=7) )
```

```
     [,1] [,2] [,3] [,4] [,5] [,6] [,7]
[1,]    0    0    0    0    0    0    0
[2,]    0    0    0    0    0    0    0
```

```
R> ## Define the A matrix
R> A1 <- rbind(cbind(Zero5x5, Lambda),
               Zero2x7)
R> ## This step is useful for inspecting the model.
R> dimnames(A1)[[1]] <- c("A","C","ES","E","I","f_Alpha","f_Beta")
R> dimnames(A1)[[2]] <- dimnames(A1)[[1]]
R> A1
```

```
         A   C   ES  E   I   f_Alpha      f_Beta
```

```
A        "0" "0" "0" "0" "0" ".3*Alpha_A"  "0"
C        "0" "0" "0" "0" "0" ".3*Alpha_C"  "0"
ES       "0" "0" "0" "0" "0" ".3*Alpha_ES" "0"
E        "0" "0" "0" "0" "0" "0"           ".3*Beta_E"
I        "0" "0" "0" "0" "0" "0"           ".3*Beta_I"
f_Alpha  "0" "0" "0" "0" "0" "0"           "0"
f_Beta   "0" "0" "0" "0" "0" "0"           "0"
```

```
R> ## Convert it into a MxMatrix class
R> A1 <- as.mxMatrix(A1)
R> ## F matrix to select the observed variables
R> ## First 5 elements are observed variables
R> ## Last 2 elements are latent variables
R> F1 <- create.Fmatrix(c(1,1,1,1,1,0,0), name="F1",
                        as.mxMatrix=FALSE)
R> ## This step is useful for inspecting the model.
R> dimnames(F1)[[1]]  <- c("A","C","ES","E","I")
R> dimnames(F1)[[2]]  <- c("A","C","ES","E","I","f_Alpha","f_Beta")
R> F1
```

```
   A C ES E I f_Alpha f_Beta
A  1 0  0 0 0       0      0
C  0 1  0 0 0       0      0
ES 0 0  1 0 0       0      0
E  0 0  0 1 0       0      0
I  0 0  0 0 1       0      0
```

```
R> ## Convert it into a MxMatrix class
R> F1 <- as.mxMatrix(F1)
```

After setting up the required matrices for the structural equation model, we call the `tssem2()` function to conduct the stage 2 analysis. The `tssem2()` function automatically extract the pooled correlation matrix and its asymptotic sampling covariance matrix from the `fixed1` object regardless of whether it is based on either a fixed-effects or a random-effects model. As there is no mediator in the model, we use the `diag.constraints=FALSE` argument to analyze the correlation structure.

```
R> fixed2 <- tssem2(fixed1, Amatrix=A1, Smatrix=S1, Fmatrix=F1,
                    diag.constraints=FALSE)
R> summary(fixed2)
```

```
------------------------ Selected output ------------------------

Goodness-of-fit indices:
```

```
                                                   Value
Sample size                                      4496.00
Chi-square of target model                         65.06
DF of target model                                  4.00
p value of target model                             0.00
Number of constraints imposed on "Smatrix"          0.00
DF manually adjusted                                0.00
Chi-square of independence model                 3100.24
DF of independence model                           10.00
RMSEA                                               0.06
SRMR                                                0.03
TLI                                                 0.95
CFI                                                 0.98
AIC                                                57.06
BIC                                                31.42
OpenMx status1: 0 ("0" or "1": The optimization is considered fine.
Other values indicate problems.)

-----------------------  Selected output  -----------------------
```

On the basis of the goodness-of-fit indices, the proposed model fits the data reasonably well with $\chi^2(\text{df} = 4, N = 4496) = 65.0609, p < 0.001$, CFI = 0.9802, RMSEA = 0.0583, and SRMR = 0.0284. Some readers may wonder why the proposed model fits the data well in the stage 2 analysis even though the homogeneity of correlation matrices was rejected in the stage 1 analysis. As the structural equation model in the stage 2 analysis is only fitted on the basis of the pooled correlation matrix and its asymptotic covariance matrix, whether or not the correlation matrices are homogeneous has little impact on the model fit of the structural models. The SEs of the parameter estimates are likely to have been underestimated in this case. Readers should be cautious when interpreting the results of the stage 2 analysis when the homogeneity of the correlation matrices is rejected in the stage 1 analysis.

7.6.1.2 Fixed-effects model with clusters

As the assumption of the homogeneity of the correlation matrices has not been met, we group the studies into clusters based on the study characteristics. `Digman97$cluster` stores the sample characteristics of these studies. For ease of illustration, we recode the data into younger and older participants.

```
R> ## Display the original study characteristics
R> Digman97$cluster
```

```
 [1] "Children"       "Children"       "Children"       "Children"
 [5] "Adolescents"    "Young adults"   "Young adults"   "Young adults"
 [9] "Mature adults"  "Mature adults"  "Mature adults"  "Mature adults"
[13] "Mature adults"  "Mature adults"
```

```
R> ## Convert 3 clusters into 2 clusters
R> ## Younger participants: "Children" and "Adolescents"
R> ## Older participants: "Mature adults"
R> sample <- ifelse(Digman97$cluster %in% c("Children",
                    "Adolescents"),
                    yes="Younger participants",
                    no="Older participants")
R> ## Show the recoded sample
R> sample
```

```
 [1] "Younger participants" "Younger participants"
 [3] "Younger participants" "Younger participants"
 [5] "Younger participants" "Older participants"
 [7] "Older participants"   "Older participants"
 [9] "Older participants"   "Older participants"
[11] "Older participants"   "Older participants"
[13] "Older participants"   "Older participants"
```

The syntax for the fixed-effects TSSEM with clusters is identical to that of the fixed-effects analysis except for specifying the `cluster` argument in the `tssem1()` function. It should be noted that the `cluster` argument is ignored when `method="REM"` is specified.

```
R> ## Fixed-effects TSSEM with two clusters
R> fixed1.cluster <- tssem1(Digman97$data, Digman97$n, method="FEM",
                            cluster=sample)
R> summary(fixed1.cluster)
```

```
-----------------------  Selected output  ----------------------

$'Older participants'
Goodness-of-fit indices:
                                      Value
Sample size                         3658.00
Chi-square of target model           823.88
DF of target model                    80.00
p value of target model                0.00
Chi-square of independence model 2992.93
DF of independence model              90.00
RMSEA                                  0.15
SRMR                                   0.15
TLI                                    0.71
CFI                                    0.74
AIC                                  663.88
BIC                                  167.50
OpenMx status1: 0 ("0" or "1": The optimization is considered fine.
Other values indicate problems.)

-----------------------  Selected output  ----------------------
```

```
------------------------ Selected output ----------------------

$'Younger participants'
Goodness-of-fit indices:
                                       Value
Sample size                           838.00
Chi-square of target model            344.18
DF of target model                     40.00
p value of target model                 0.00
Chi-square of independence model 1461.67
DF of independence model               50.00
RMSEA                                   0.21
SRMR                                    0.14
TLI                                     0.73
CFI                                     0.78
AIC                                   264.18
BIC                                    74.94
OpenMx status1: 0 ("0" or "1": The optimization is considered fine.
Other values indicate problems.)

------------------------ Selected output ----------------------
```

From the above test statistics and goodness-of-fit indices, the hypothesis of the homogeneity of the correlation matrices in these two samples was rejected. The `sample` type is not sufficient to explain the heterogeneity of the correlation matrices. We do not go on to fit the structural equation models. If a structural equation model is required to be fitted in the stage 2 analysis, the command is

```
R> fixed2.cluster <- tssem2(fixed1, Amatrix=A1, Smatrix=S1,
                            Fmatrix=F1, diag.constraints=FALSE)
R> summary(fixed2.cluster)
```

7.6.1.3 Random-effects model

We conduct a random-effects TSSEM by specifying the `method="REM"` argument in calling the `tssem1()` function to account for the heterogeneity of the correlation matrices. The R code is

```
R> ## Random-effects TSSEM
R> ## There were errors in the analysis.
R> random1 <- tssem1(Digman97$data, Digman97$n, method="REM")
R> summary(random1)
```

The above analysis encounters errors. As there are five variables in the model, this led to a total of 10 correlation coefficients in the analysis. If a random-effects model is fitted, there are 55 elements in the variance component matrix of the random effects. As there are not enough data to estimate the full variance component, a diagonal matrix was used in estimating the variance component. The default option in the `tssem1()` function is to estimate the full symmetric variance component (with the `RE.type="Symm"` argument). A diagonal variance component can be requested by specifying the `RE.type="Diag"` argument. If `RE.type="Zero"` is specified, it becomes the GLS approach discussed in Section 7.2.2.

```
R> ## Random-effects TSSEM with random effects on the diagonals
R> random1 <- tssem1(Digman97$data, Digman97$n, method="REM",
                     RE.type="Diag")
R> summary(random1)
```

```
-----------------------  Selected output  ----------------------

95% confidence intervals: z statistic approximation
Coefficients:
             Estimate Std.Error    lbound    ubound z value Pr(>|z|)
Intercept1   3.95e-01  5.42e-02  2.88e-01  5.01e-01    7.28  3.4e-13
Intercept2   4.40e-01  4.13e-02  3.59e-01  5.21e-01   10.67  < 2e-16
Intercept3   5.45e-02  6.17e-02 -6.64e-02  1.76e-01    0.88   0.3768
Intercept4   9.87e-02  4.62e-02  8.08e-03  1.89e-01    2.13   0.0328
Intercept5   4.30e-01  4.02e-02  3.51e-01  5.08e-01   10.70  < 2e-16
Intercept6   1.29e-01  4.08e-02  4.85e-02  2.09e-01    3.15   0.0016
Intercept7   2.05e-01  4.96e-02  1.08e-01  3.02e-01    4.14  3.5e-05
Intercept8   2.40e-01  3.19e-02  1.77e-01  3.03e-01    7.52  5.7e-14
Intercept9   1.89e-01  4.30e-02  1.05e-01  2.73e-01    4.40  1.1e-05
Intercept10  4.44e-01  3.25e-02  3.80e-01  5.08e-01   13.65  < 2e-16
Tau2_1_1     3.72e-02  1.50e-02  7.81e-03  6.66e-02    2.48   0.0131
Tau2_2_2     2.03e-02  8.43e-03  3.77e-03  3.68e-02    2.41   0.0161
Tau2_3_3     4.82e-02  1.97e-02  9.56e-03  8.69e-02    2.44   0.0145
Tau2_4_4     2.46e-02  1.06e-02  3.79e-03  4.54e-02    2.32   0.0205
Tau2_5_5     1.87e-02  8.25e-03  2.56e-03  3.49e-02    2.27   0.0232
Tau2_6_6     1.83e-02  8.79e-03  1.03e-03  3.55e-02    2.08   0.0378
Tau2_7_7     2.94e-02  1.23e-02  5.39e-03  5.35e-02    2.40   0.0164
Tau2_8_8     9.65e-03  4.88e-03  8.17e-05  1.92e-02    1.98   0.0481
Tau2_9_9     2.09e-02  9.13e-03  3.04e-03  3.88e-02    2.29   0.0218
Tau2_10_10   1.12e-02  5.05e-03  1.26e-03  2.10e-02    2.21   0.0271

Intercept1  ***
Intercept2  ***
Intercept3
Intercept4  *
Intercept5  ***
Intercept6  **
Intercept7  ***
Intercept8  ***
Intercept9  ***
Intercept10 ***
Tau2_1_1    *
Tau2_2_2    *
Tau2_3_3    *
```

```
Tau2_4_4     *
Tau2_5_5     *
Tau2_6_6     *
Tau2_7_7     *
Tau2_8_8     *
Tau2_9_9     *
Tau2_10_10   *
---
Signif. codes:  0 '***' 0.001 '**' 0.01 '*' 0.05 '.' 0.1 ' ' 1

Q statistic on the homogeneity of effect sizes: 2381
Degrees of freedom of the Q statistic: 130
P value of the Q statistic: 0

Heterogeneity indices (based on the estimated Tau2):
                              Estimate
Intercept1: I2 (Q statistic)      0.95
Intercept2: I2 (Q statistic)      0.91
Intercept3: I2 (Q statistic)      0.94
Intercept4: I2 (Q statistic)      0.89
Intercept5: I2 (Q statistic)      0.90
Intercept6: I2 (Q statistic)      0.85
Intercept7: I2 (Q statistic)      0.91
Intercept8: I2 (Q statistic)      0.77
Intercept9: I2 (Q statistic)      0.87
Intercept10: I2 (Q statistic)     0.84

Number of studies (or clusters): 14
Number of observed statistics: 140
Number of estimated parameters: 20
Degrees of freedom: 120
-2 log likelihood: -110.8
OpenMx status1: 0 ("0" or "1": The optimization is considered fine.
Other values indicate problems.)

-----------------------  Selected output  ----------------------
```

The Q statistic on testing the homogeneity test of the correlation matrix is $Q(\text{df} = 130) = 2381.0001, p < 0.001$. The test statistic is statistically significant in testing the null hypothesis of the homogeneity of the correlation matrices. The above output also displays the I^2 discussed in Section 4.3.3 that shows the between-study variation to the total variation. From the output, the minimum and the maximum I^2 are 0.7714 and 0.9487, respectively. These indicate that there is huge between-study heterogeneity on the correlation coefficients. It is more appropriate to use the random-effects model than the fixed-effects model on this data set. The pooled correlation matrix can be obtained using the following command:

```
R> ## Extract the fixed-effects estimates
R> (est_fixed <- coef(random1, select="fixed"))
```

```
 Intercept1  Intercept2  Intercept3  Intercept4  Intercept5
    0.39465     0.44009     0.05454     0.09867     0.42966
 Intercept6  Intercept7  Intercept8  Intercept9 Intercept10
    0.12851     0.20526     0.23994     0.18910     0.44413
```

```
R> ## Convert the estimated vector to a symmetrical matrix
R> ## where the diagonals are fixed at 1 (for a correlation matrix)
R> vec2symMat(est_fixed, diag=FALSE)
```

```
        [,1]    [,2]    [,3]     [,4]     [,5]
[1,] 1.00000 0.3946 0.4401 0.05454 0.09867
[2,] 0.39465 1.0000 0.4297 0.12851 0.20526
[3,] 0.44009 0.4297 1.0000 0.23994 0.18910
[4,] 0.05454 0.1285 0.2399 1.00000 0.44413
[5,] 0.09867 0.2053 0.1891 0.44413 1.00000
```

The stage 2 analysis for fitting the structural equation model can be conducted with the `tssem2()` function. The syntax for the stage 2 analysis is exactly the same regardless of whether a fixed-effects model, a fixed-effects model with clusters, or a random-effects model is fitted in the stage 1 analysis. This is because only the pooled correlation matrix and its asymptotic covariance matrix are used in fitting structural equation models in the stage 2 analysis. The `tssem2()` handles this issue automatically.

```
R> random2 <- tssem2(random1, Amatrix=A1, Smatrix=S1, Fmatrix=F1,
                     diag.constraints=FALSE)
R> summary(random2)
```

```
------------------------ Selected output -----------------------

95% confidence intervals: z statistic approximation
Coefficients:
          Estimate Std.Error lbound ubound z value Pr(>|z|)
Alpha_A     0.5726    0.0516 0.4714 0.6737   11.09  < 2e-16 ***
Alpha_C     0.5901    0.0518 0.4885 0.6917   11.38  < 2e-16 ***
Alpha_ES    0.7705    0.0610 0.6508 0.8901   12.62  < 2e-16 ***
Beta_E      0.6934    0.0748 0.5468 0.8400    9.27  < 2e-16 ***
Beta_I      0.6401    0.0689 0.5052 0.7751    9.30  < 2e-16 ***
cor         0.3937    0.0476 0.3004 0.4869    8.28  2.2e-16 ***
---
Signif. codes:  0 '***' 0.001 '**' 0.01 '*' 0.05 '.' 0.1 ' ' 1

Goodness-of-fit indices:
                                             Value
Sample size                                4496.00
Chi-square of target model                    8.51
DF of target model                            4.00
```

```
p value of target model                              0.07
Number of constraints imposed on "Smatrix"           0.00
DF manually adjusted                                 0.00
Chi-square of independence model                   514.56
DF of independence model                            10.00
RMSEA                                                0.02
SRMR                                                 0.05
TLI                                                  0.98
CFI                                                  0.99
AIC                                                  0.51
BIC                                                -25.13
OpenMx status1: 0 ("0" or "1": The optimization is considered fine.
Other values indicate problems.)

------------------------ Selected output -----------------------
```

On the basis of the goodness-of-fit indices, the proposed model fits the data well with $\chi^2(\text{df} = 4, N = 4496) = 8.5118, p = 0.0745$, CFI = 0.9911, RMSEA = 0.0158, and SRMR = 0.0463. The estimated correlation between the two factors (labeled `cor` in the output) is $\hat{S}[7,6] = 0.3937$, whereas the parameter estimates of the factor loadings ($\hat{A}[1,6]$ to $\hat{A}[5,7]$; labeled `Alpha_A` to `Beta_I` in the output) are 0.5726, 0.5901, 0.7705, 0.6934, and 0.6401, respectively. All of the parameter estimates are statistically significant. The lower bound (the column labeled `lbound`) and the upper bound (the column labeled `ubound`) of the 95% Wald CI are also listed. If an LBCI is required, we may specify `intervals.type="LB"` in the call.

As shown in the output, the error variances are not parameters in the model when the argument `diag.constraints=FALSE` is specified. Therefore, there is no estimate on them. The *computed* error variances are stored in a matrix called `Ematrix` in `random2$mx.fit`. We may request it by using the `mxEval()` function in the `OpenMx` package. Moreover, we can also calculate the $R^2 = 1 -$error variance, the percentage of the variance explained by the predictors, on the observed variables.

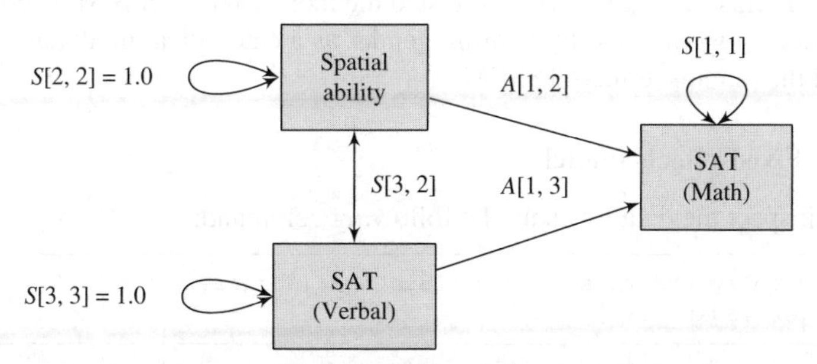

Figure 7.3 A regression model of math aptitude with spatial ability and verbal ability as predictors.

```
R> ## Extract and display the error variances
R> (Ematrix <- diag(mxEval(Ematrix, random2$mx.fit)))
```

```
[1] 0.6722 0.6518 0.4064 0.5192 0.5903
```

```
R> ## Calculate the R2 on the variables
R> 1 - Ematrix
```

```
[1] 0.3278 0.3482 0.5936 0.4808 0.4097
R> ## Load the library to plot the model
R> library("semPlot")

R> ## Convert the model to semPlotModel object
R> my.plot <- meta2semPlot(random2, latNames=c("Alpha","Beta"))

R> ## Plot the model with labels
R> semPaths(my.plot, whatLabels="path")

R> ## Plot the parameter estimates
R> semPaths(my.plot, whatLabels="est")
```

7.6.2 A regression model on SAT (Math)

The objective of this example is to demonstrate how to fit a regression model with the TSSEM approach. Becker and Schram (1994) reported 10 independent samples (five for males and five for females) from five studies with a total sample size of 538. Their study includes correlations among *SAT (Math)*, *SAT (Verbal)*, and *Spatial ability*. Becker and Schram (1994) synthesized the correlation matrices and fit a regression model by using *SAT (Math)* as the dependent variable and *SAT (Verbal)* and *Spatial ability* as the predictors. These 10 studies varied in sample sizes (18–153). The data set is stored as an R object named `Becker94` in the `metaSEM` package. Figure 7.3 depicts the model with the parameters labeled in the RAM formulation. In this illustration, we first tested the fixed-effects TSSEM by pooling all of the studies together and by treating gender as a categorical moderator. Finally, we fitted the random-effects TSSEM.

7.6.2.1 Fixed-effects model

We first inspect the data by using the following command:

```
R> ## Display the first two correlation matrices
R> Becker94$data[1:2]
```

```
$'Becker (1978) Females'
                 SAT (Math) Spatial SAT (Verbal)
```

```
SAT (Math)          1.00    0.47          -0.21
Spatial             0.47    1.00          -0.15
SAT (Verbal)       -0.21   -0.15           1.00

$'Becker (1978) Males'
             SAT (Math) Spatial SAT (Verbal)
SAT (Math)          1.00    0.28           0.19
Spatial             0.28    1.00           0.18
SAT (Verbal)        0.19    0.18           1.00
```

```
R> ## Display the first two sample sizes
R> Becker94$n[1:2]
```

```
[1]  74 153
```

```
R> ## Display the first two sample types
R> Becker94$gender[1:2]
```

```
[1] "Females" "Males"
```

We conduct the first stage of the analysis with a fixed-effects TSSEM by calling the `tssem1()` function specifying the `method="FEM"` argument.

```
R> fixed1 <- tssem1(Becker94$data, Becker94$n, method="FEM")
R> summary(fixed1)
```

```
------------------------ Selected output -----------------------

Goodness-of-fit indices:
                                       Value
Sample size                           538.00
Chi-square of target model             62.50
DF of target model                     27.00
p value of target model                 0.00
Chi-square of independence model      202.61
DF of independence model               30.00
RMSEA                                   0.16
SRMR                                    0.16
TLI                                     0.77
CFI                                     0.79
AIC                                     8.50
BIC                                  -107.27
OpenMx status1: 0 ("0" or "1": The optimization is considered fine.
Other values indicate problems.)

------------------------ Selected output -----------------------
```

The `OpenMx status1` looks fine. We then check the test of the homogeneity and the goodness-of-fit indices. The test statistic on testing the homogeneity test of the correlation matrix is $\chi^2(\mathrm{df} = 27, N = 538) = 62.4983, p < 0.001$. On the basis of the test statistic, the null hypothesis of the homogeneity of the correlation matrices was rejected. Both the RMSEA and the SRMR are very large, indicating that the proposed model (homogeneity of correlation matrices) does not fit the data. As the fixed-effects model is not appropriate for this data set, we do not fit the regression model on the pooled correlation matrix. We illustrate the stage 2 analysis in the later section of this illustration.

7.6.2.2 Fixed-effects model with clusters

We may test whether or not *gender* explains the heterogeneity of the correlation matrices by calling up the following syntax.

```
R> ## Fixed-effects TSSEM with two clusters
R> fixed.cluster <- tssem1(Becker94$data, Becker94$n, method="FEM",
                          cluster=Becker94$gender)
R> summary(fixed.cluster)
```

```
-----------------------  Selected output  ----------------------

$Females
                                   Value
Sample size                       235.00
Chi-square of target model         42.41
DF of target model                 12.00
p value of target model             0.00
Chi-square of independence model 120.42
DF of independence model           15.00
RMSEA                               0.23
SRMR                                0.21
TLI                                 0.64
CFI                                 0.71
AIC                                18.41
BIC                               -23.11
OpenMx status1: 0 ("0" or "1": The optimization is considered fine.
Other values indicate problems.)

-----------------------  Selected output  ----------------------
```

```
-----------------------  Selected output  ----------------------

$Males
                                   Value
Sample size                       303.00
Chi-square of target model         16.13
DF of target model                 12.00
```

```
p value of target model               0.19
Chi-square of independence model     82.19
DF of independence model             15.00
RMSEA                                 0.08
SRMR                                  0.10
TLI                                   0.92
CFI                                   0.94
AIC                                  -7.87
BIC                                 -52.43
OpenMx status1: 0 ("0" or "1": The optimization is considered fine.
Other values indicate problems.)

----------------------- Selected output ----------------------
```

From the above test statistics and goodness-of-fit indices, the null hypothesis of the homogeneity of the correlation matrices in these two samples was rejected in studies of female samples but not in male samples. As there were only five studies in each group, we should be cautious when interpreting nonsignificant results.

7.6.2.3 Random-effects model

We run a random-effects TSSEM by specifying the `method="REM"` argument to account for the heterogeneity on the correlation matrices.

```
R> ## Random-effects TSSEM
R> ## The OpenMx status was 6.
R> random1 <- tssem1(Becker94$data, Becker94$n, method="REM")
R> summary(random1)
```

The above analysis encounters errors. The main issue is that the estimated variance component is nonpositive definite. This problem may be attributed to the small number of studies. We use a diagonal matrix when estimating the variance component.

```
R> ## Random-effects TSSEM with random effects on the diagonals
R> random1 <- tssem1(Becker94$data, Becker94$n, method="REM",
                     RE.type="Diag")
R> summary(random1)
```

```
----------------------- Selected output ----------------------

95% confidence intervals: z statistic approximation
Coefficients:
            Estimate Std.Error    lbound    ubound z value Pr(>|z|)
Intercept1  3.71e-01  3.69e-02  2.98e-01  4.43e-01   10.05  < 2e-16
Intercept2  4.32e-01  7.73e-02  2.80e-01  5.83e-01    5.59  2.3e-08
Intercept3  2.03e-01  4.65e-02  1.12e-01  2.94e-01    4.36  1.3e-05
```

```
Tau2_1_1    1.00e-10  4.89e-03 -9.59e-03  9.59e-03     0.00      1.00
Tau2_2_2    4.75e-02  2.62e-02 -3.90e-03  9.88e-02     1.81      0.07
Tau2_3_3    5.16e-03  9.75e-03 -1.39e-02  2.43e-02     0.53      0.60

Intercept1 ***
Intercept2 ***
Intercept3 ***
Tau2_1_1
Tau2_2_2    .
Tau2_3_3
---
Signif. codes:  0 '***' 0.001 '**' 0.01 '*' 0.05 '.' 0.1 ' ' 1

Q statistic on the homogeneity of effect sizes: 81.66
Degrees of freedom of the Q statistic: 27
P value of the Q statistic: 2.112e-07

Heterogeneity indices (based on the estimated Tau2):
                             Estimate
Intercept1: I2 (Q statistic)     0.00
Intercept2: I2 (Q statistic)     0.81
Intercept3: I2 (Q statistic)     0.23

Number of studies (or clusters): 10
Number of observed statistics: 30
Number of estimated parameters: 6
Degrees of freedom: 24
-2 log likelihood: -23.64
OpenMx status1: 0 ("0" or "1": The optimization is considered fine.
Other values indicate problems.)

------------------------  Selected output  -----------------------
```

The Q statistic on testing the homogeneity test of the correlation matrix is $Q(\mathrm{df} = 27) = 81.6591, p < 0.001$. This indicates that the null hypothesis of the homogeneity of correlation matrices is rejected. From the output, the I^2 for the correlations between *SAT (Math)* and *Spatial ability*, between *SAT (Math)* and *SAT (Verbal)*, and between *Spatial ability* and *SAT (Verbal)* are 0.0000, 0.8052, and 0.2274, respectively. These indicate that there is not much between-study variation on the population correlation between *SAT (Math)* and *Spatial ability*, whereas there is heterogeneity on the other population correlation coefficients. A random-effects model is more appropriate than a fixed-effects model for this data set. The pooled correlation matrix can be obtained using the following command:

```
R> vec2symMat( coef(random1, select="fixed"), diag=FALSE )
```

```
       [,1]   [,2]   [,3]
[1,] 1.0000 0.3708 0.4316
[2,] 0.3708 1.0000 0.2029
[3,] 0.4316 0.2029 1.0000
```

We prepare the necessary matrices to fit the structural equation model. As all variables are observed, we may skip the ***F*** selection matrix. If it is skipped, the `tssem2()` function assumes that there are no latent variables. Besides creating the matrices with the method shown in the example of `Digman97`, we may also create the matrices with the `create.mxMatrix()` function.

```
R> ## Arrange the data by row
R> A2 <- create.mxMatrix(c(0, "0.2*Spatial2Math", "0.2*Verbal2Math",
                           0, 0, 0,
                           0, 0, 0),
                         type="Full", nrow=3, ncol=3,
                         byrow=TRUE, name="A2")
R> A2
```

```
FullMatrix 'A2'

$labels
     [,1] [,2]           [,3]
[1,] NA   "Spatial2Math" "Verbal2Math"
[2,] NA   NA             NA
[3,] NA   NA             NA

$values
     [,1] [,2] [,3]
[1,]    0  0.2  0.2
[2,]    0  0.0  0.0
[3,]    0  0.0  0.0

$free
      [,1]  [,2]  [,3]
[1,] FALSE  TRUE  TRUE
[2,] FALSE FALSE FALSE
[3,] FALSE FALSE FALSE

$lbound: No lower bounds assigned.

$ubound: No upper bounds assigned.
```

```
R> ## Only the elements in the lower triangle are required
R> S2 <- create.mxMatrix(c("0.2*ErrVarMath",
                           0,1,
                           0,"0.2*CorMathVerbal",1),
                         type="Symm", byrow=TRUE, name="S2")
R> S2
```

```
SymmMatrix 'S2'

$labels
     [,1]            [,2]                [,3]
```

```
[1,] "ErrVarMath" NA               NA
[2,] NA           NA               "CorMathVerbal"
[3,] NA           "CorMathVerbal"  NA

$values
     [,1] [,2] [,3]
[1,]  0.2  0.0  0.0
[2,]  0.0  1.0  0.2
[3,]  0.0  0.2  1.0

$free
      [,1]  [,2]  [,3]
[1,]  TRUE FALSE FALSE
[2,] FALSE FALSE  TRUE
[3,] FALSE  TRUE FALSE

$lbound: No lower bounds assigned.

$ubound: No upper bounds assigned.
```

The stage 2 analysis of fitting the structural model is conducted with the `tssem2()` function.

```
R> random2 <- tssem2(random1, Amatrix=A2, Smatrix=S2,
                     diag.constraints=FALSE)
R> summary(random2)
```

```
------------------------ Selected output -----------------------

95% confidence intervals: z statistic approximation
Coefficients:
              Estimate Std.Error lbound ubound z value Pr(>|z|)
Spatial2Math    0.2953    0.0404 0.2161 0.3746    7.30  2.8e-13 ***
Verbal2Math     0.3717    0.0806 0.2138 0.5295    4.61  3.9e-06 ***
CorMathVerbal   0.2029    0.0465 0.1118 0.2941    4.36  1.3e-05 ***
---
Signif. codes:  0 '***' 0.001 '**' 0.01 '*' 0.05 '.' 0.1 ' ' 1

Goodness-of-fit indices:
                                           Value
Sample size                                  538
Chi-square of target model                     0
DF of target model                             0
p value of target model                        0
Number of constraints imposed on "Smatrix"     0
DF manually adjusted                           0
Chi-square of independence model             130
DF of independence model                       3
RMSEA                                          0
```

```
SRMR                                                    0
TLI                                                  -Inf
CFI                                                     1
AIC                                                     0
BIC                                                     0
OpenMx status1: 0 ("0" or "1": The optimization is considered fine.
Other values indicate problems.)

----------------------- Selected output ----------------------
```

As the proposed model is the regression model, the model is just identified. Thus, the chi-square statistic on the model is always 0. The other goodness-of-fit indices are irrelevant. The estimated regression coefficients (with their 95% Wald CIs) from Spatial ability and SAT (Verbal) to SAT (Math) are 0.2953 $(0.2161, 0.3746)$ and 0.3717 $(0.2138, 0.5295)$, respectively. The estimated correlation between the predictors is 0.2029 $(0.1118, 0.2941)$. All the parameter estimates are statistically significant at $\alpha = 0.05$. We may compute the error variance and the R^2 on SAT (Math) by

```
R> ## Display the content of the Ematrix
R> mxEval(Ematrix, random2$mx.fit)
```

```
       [,1] [,2] [,3]
[1,] 0.7301    0    0
[2,] 0.0000    0    0
[3,] 0.0000    0    0
```

```
R> ## Error variance on SAT (Math)
R> ## Select the element [1,1] correspondent to SAT (Math)
R> (Ematrix <- mxEval(Ematrix, random2$mx.fit)[1,1])
```

```
[1] 0.7301
```

```
R> ## R2 on SAT (Math)
R> 1 - Ematrix
```

```
[1] 0.2699
```

The computed error variance on SAT (Math) is 0.7301, whereas the R^2 is 0.2699 by considering the joint effect of SAT (Verbal) and Spatial ability.

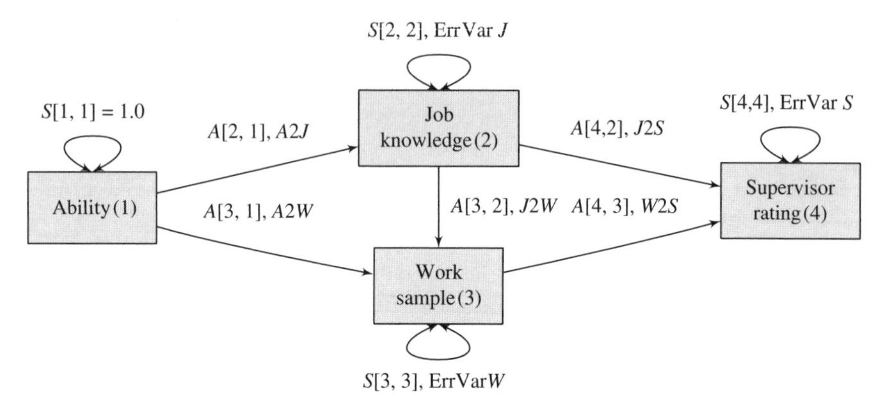

Figure 7.4 A path model of cognitive ability, job knowledge, work sample, and supervisor rating.

7.6.3 A path model for cognitive ability to supervisor rating

The purpose of this example is to illustrate how to fit a mediation model with the use of the `diag.constraints=TRUE` and `intervals.type="LB"`. Moreover, we will also illustrate how to create additional matrices to calculate functions of the parameters. This approach is used to calculate functions such as indirect effects and their LBCI. Hunter (1983) conducted an MASEM to test a mediator model from *Ability (A)* to *Supervisor rating (S)* on 14 studies with a total sample size of 3975. There are two intermediate mediators, *Job knowledge (J)* and *Work sample (W)*, in the model, whereas there is no direct effect from *Ability* to *Supervisor rating*. Figure 7.4 shows the proposed method with the labels of the parameters. For the ease of reference, the variables in the figure are also labeled in numbers according to the orders in the data set and the RAM specification.

7.6.3.1 Fixed-effects model

When there are missing data in the correlation or covariance matrices, it is of importance to check the patterns of the missing data. Users may also check whether the matrices are positive definite. We display some sample data for checking the patterns of missing (or present) data by using the `pattern.na()` function. Moreover, we may also calculate the accumulative sample sizes with the `pattern.n()` function. Users should make sure that there are enough data on each cell.

```
R> ## Display the correlation matrices
R> Hunter83$data[1:2]
```

```
$'Campbell et al. (1973)'
           Ability Knowledge Work sample Supervisor
Ability       1.00      0.65        0.48       0.33
```

```
Knowledge       0.65      1.00         0.52        0.40
Work sample     0.48      0.52         1.00        0.23
Supervisor      0.33      0.40         0.23        1.00

$'Corts et al. (1977)'
            Ability Knowledge Work sample Supervisor
Ability        1.00      0.53        0.50       0.03
Knowledge      0.53      1.00        0.49       0.04
Work sample    0.50      0.49        1.00       0.18
Supervisor     0.03      0.04        0.18       1.00
```

```
R> ## Display the sample sizes
R> Hunter83$n[1:2]
```

```
[1] 443 186
```

```
R> ## Show the missing data: show.na=TRUE
R> ## Show the present data: show.na=FALSE
R> pattern.na(Hunter83$data, show.na=FALSE)
```

```
            Ability Knowledge Work sample Supervisor
Ability          13        11          11         12
Knowledge        11        12          10         11
Work sample      11        10          12         11
Supervisor       12        11          11         13
```

```
R> ## Display the accumulative sample sizes for each correlation
R> pattern.n(Hunter83$data, Hunter83$n)
```

```
            Ability Knowledge Work sample Supervisor
Ability        3815      3372        3281       3605
Knowledge      3372      3532        2998       3322
Work sample    3281      2998        3441       3231
Supervisor     3605      3322        3231       3765
```

There are not too many missing data on the correlation coefficients. We check whether the matrices are positive definite by using the `is.pd()` function. It returns `TRUE` when the input matrix is positive definite. It returns `FALSE` and `NA` when the input matrix is not positive definite and some of the elements are `NA`, respectively. If one or some of the matrices are not positive definite, the analysis may return an error, especially for the fixed-effects TSSEM approach.

```
R> is.pd(Hunter83$data)
```

```
     Campbell et al. (1973)         Corts et al. (1977)
                       TRUE                        TRUE
O'Leary and Trattner (1977)      Trattner rt al. (1977)
                       TRUE                        TRUE
 Vineberg and Taylor (1972)  Vineberg and Taylor (1972)
                       TRUE                        TRUE
 Vineberg and Taylor (1972)  Vineberg and Taylor (1972)
                       TRUE                        TRUE
      Campell et al. (1973)              Drauden (1978)
                       TRUE                        TRUE
     Campbell et al. (1973)        Boyles et al. (19??)
                       TRUE                        TRUE
              Schoon (1979)   van Rijn and Payne (1980)
                       TRUE                        TRUE
```

We pool the correlation matrices based on the fixed-effects model.

```
R> fixed1 <- tssem1(Hunter83$data, Hunter83$n, method="FEM")
R> summary(fixed1)
```

```
----------------------- Selected output -----------------------

Coefficients:
       Estimate Std.Error z value Pr(>|z|)
S[1,2]   0.5105    0.0127    40.1   <2e-16 ***
S[1,3]   0.4270    0.0141    30.3   <2e-16 ***
S[1,4]   0.2077    0.0160    13.0   <2e-16 ***
S[2,3]   0.5229    0.0131    39.8   <2e-16 ***
S[2,4]   0.2846    0.0158    18.0   <2e-16 ***
S[3,4]   0.2432    0.0163    14.9   <2e-16 ***
---
Signif. codes:  0 '***' 0.001 '**' 0.01 '*' 0.05 '.' 0.1 ' ' 1

Goodness-of-fit indices:
                                   Value
Sample size                      3975.00
Chi-square of target model        263.34
DF of target model                 60.00
p value of target model             0.00
Chi-square of independence model 2767.41
DF of independence model           66.00
RMSEA                               0.11
SRMR                                0.09
TLI                                 0.92
CFI                                 0.92
AIC                               143.34
BIC                              -233.93
OpenMx status1: 0 ("0" or "1": The optimization is considered fine.
Other values indicate problems.)

----------------------- Selected output -----------------------
```

The OpenMx status1 looks fine. The test statistic on testing the homogeneity test of the correlation matrix is $\chi^2(\text{df} = 60, N = 3975) = 263.3389, p < 0.001$. On the basis of the test statistic, the null hypothesis on the homogeneity of the correlation matrices was rejected. Both the RMSEA and the SRMR are also very large, indicating that the proposed model of homogeneity of correlation matrices does not fit the data. We will fit the structural equation model based on the random-effects model later.

7.6.3.2 Random-effects model

As there was not enough data to estimate the full variance component, we estimate the variance component based on a diagonal matrix. Moreover, the OpenMx status1 was 6, indicating that there were estimation issues. We try to improve the estimation by rerunning the analysis with the rerun() function.

```
R> random1 <- tssem1(Hunter83$data, Hunter83$n, method="REM",
                     RE.type="Diag")
R> ## Rerun the analysis
R> random1 <- rerun(random1)
R> summary(random1)
```

```
-----------------------  Selected output  ----------------------

95% confidence intervals: z statistic approximation
Coefficients:
            Estimate Std.Error    lbound    ubound z value Pr(>|z|)
Intercept1  5.06e-01  2.79e-02  4.51e-01  5.61e-01   18.14  < 2e-16
Intercept2  4.47e-01  2.47e-02  3.99e-01  4.95e-01   18.09  < 2e-16
Intercept3  2.12e-01  2.52e-02  1.63e-01  2.62e-01    8.41  < 2e-16
Intercept4  5.31e-01  3.47e-02  4.63e-01  5.99e-01   15.28  < 2e-16
Intercept5  2.59e-01  3.95e-02  1.82e-01  3.37e-01    6.56  5.4e-11
Intercept6  2.41e-01  1.69e-02  2.08e-01  2.74e-01   14.27  < 2e-16
Tau2_1_1    6.37e-03  3.52e-03 -5.38e-04  1.33e-02    1.81    0.071
Tau2_2_2    4.41e-03  3.10e-03 -1.66e-03  1.05e-02    1.42    0.155
Tau2_3_3    4.31e-03  2.82e-03 -1.22e-03  9.84e-03    1.53    0.126
Tau2_4_4    1.02e-02  5.40e-03 -3.56e-04  2.08e-02    1.89    0.058
Tau2_5_5    1.39e-02  7.31e-03 -4.55e-04  2.82e-02    1.90    0.058
Tau2_6_6    1.32e-05  1.26e-03 -2.45e-03  2.48e-03    0.01    0.992

Intercept1 ***
Intercept2 ***
Intercept3 ***
Intercept4 ***
Intercept5 ***
Intercept6 ***
Tau2_1_1   .
Tau2_2_2
Tau2_3_3
Tau2_4_4   .
Tau2_5_5   .
```

```
Tau2_6_6
---
Signif. codes:  0 '***' 0.001 '**' 0.01 '*' 0.05 '.' 0.1 ' ' 1

Q statistic on the homogeneity of effect sizes: 314.4
Degrees of freedom of the Q statistic: 60
P value of the Q statistic: 0

Heterogeneity indices (based on the estimated Tau2):
                              Estimate
Intercept1: I2 (Q statistic)      0.76
Intercept2: I2 (Q statistic)      0.64
Intercept3: I2 (Q statistic)      0.57
Intercept4: I2 (Q statistic)      0.83
Intercept5: I2 (Q statistic)      0.82
Intercept6: I2 (Q statistic)      0.00
```

```
Number of studies (or clusters): 14
Number of observed statistics: 66
Number of estimated parameters: 12
Degrees of freedom: 54
-2 log likelihood: -126.3
OpenMx status1: 0 ("0" or "1": The optimization is considered fine.
Other values indicate problems.)

------------------------  Selected output  -----------------------
```

After rerunning the analysis, the `OpenMx status1` was fine. The Q statistic on testing the null hypothesis on the homogeneity test of the correlation matrix is $Q(\text{df} = 60) = 314.4232, p < 0.001$. This indicates that the null hypothesis on the homogeneity of correlation matrices is rejected. The range of I^2 on the correlation coefficients is from 0.0041 to 0.8349. A random-effects model is applied to this data set. The pooled correlation matrix can be obtained by the following command:

```
R> vec2symMat( coef(random1, select="fixed"), diag=FALSE )
```

```
       [,1]   [,2]   [,3]   [,4]
[1,] 1.0000 0.5059 0.4469 0.2122
[2,] 0.5059 1.0000 0.5307 0.2591
[3,] 0.4469 0.5307 1.0000 0.2410
[4,] 0.2122 0.2591 0.2410 1.0000
```

We prepare the necessary matrices to fit the structural model. As all variables are observed, we skip the $\boldsymbol{F}$ selection matrix.

```
R> ## Arrange the elements by row
```

```
R> ## Variables are arranged as:
R> ## Ability, Job knowledge, Work sample and Supervisor rating
R> (A3 <- create.mxMatrix(c(0, 0, 0, 0,
                           "0.1*A2J", 0, 0, 0,
                           "0.1*A2W", "0.1*J2W", 0, 0,
                           0, "0.1*J2S", "0.1*W2S", 0),
                         type="Full", nrow=4, ncol=4, byrow=TRUE))
```

```
FullMatrix 'untitled1'

$labels
     [,1]  [,2]  [,3]  [,4]
[1,] NA    NA    NA    NA
[2,] "A2J" NA    NA    NA
[3,] "A2W" "J2W" NA    NA
[4,] NA    "J2S" "W2S" NA

$values
     [,1] [,2] [,3] [,4]
[1,]  0.0  0.0  0.0    0
[2,]  0.1  0.0  0.0    0
[3,]  0.1  0.1  0.0    0
[4,]  0.0  0.1  0.1    0

$free
      [,1]  [,2]  [,3]  [,4]
[1,] FALSE FALSE FALSE FALSE
[2,]  TRUE FALSE FALSE FALSE
[3,]  TRUE  TRUE FALSE FALSE
[4,] FALSE  TRUE  TRUE FALSE

$lbound: No lower bounds assigned.

$ubound: No upper bounds assigned.
```

```
R> (S3 <- create.mxMatrix(c(1,"0.1*ErrVarJ",
                           "0.1*ErrVarW", "0.1*ErrVarS"),
                         type="Diag"))
```

```
DiagMatrix 'untitled1'

$labels
     [,1] [,2]      [,3]      [,4]
[1,] NA   NA        NA        NA
[2,] NA   "ErrVarJ" NA        NA
[3,] NA   NA        "ErrVarW" NA
[4,] NA   NA        NA        "ErrVarS"
```

```
$values
     [,1] [,2] [,3] [,4]
[1,]    1  0.0  0.0  0.0
[2,]    0  0.1  0.0  0.0
[3,]    0  0.0  0.1  0.0
[4,]    0  0.0  0.0  0.1

$free
      [,1]  [,2]  [,3]  [,4]
[1,] FALSE FALSE FALSE FALSE
[2,] FALSE  TRUE FALSE FALSE
[3,] FALSE FALSE  TRUE FALSE
[4,] FALSE FALSE FALSE  TRUE

$lbound: No lower bounds assigned.

$ubound: No upper bounds assigned.
```

The `tssem2()` function accepts newly defined matrices in the analysis. In this example, the indirect effect is $Ind = A2J * J2S + A2J * J2W * W2S + A2W * W2S$. We define a new matrix called `Ind` based on the formula. This object must be placed in the `mx.algebras` argument when calling up the `tssem2()` function. By specifying the `intervals.type="LB"` argument, the LBCI on the indirect effect will also be calculated. Moreover, it is necessary to specify the `diag.constraints=TRUE` because there are mediators in this model (see the discussion in Section 7.3.2).

```
R> random2 <- tssem2(random1, Amatrix=A3, Smatrix=S3,
                     intervals.type="LB",
                     diag.constraints=TRUE,
   mx.algebras=list(Ind=mxAlgebra(A2J*J2S+A2J*J2W*W2S+A2W*W2S,
                    name="Ind")))
R> summary(random2)
```

```
----------------------- Selected output -----------------------

95% confidence intervals: Likelihood-based statistic
Coefficients:
        Estimate Std.Error lbound ubound z value Pr(>|z|)
A2J       0.5164        NA 0.4629 0.5699      NA       NA
A2W       0.2461        NA 0.1629 0.3251      NA       NA
J2W       0.3966        NA 0.3002 0.4935      NA       NA
J2S       0.2314        NA 0.1382 0.3257      NA       NA
W2S       0.1214        NA 0.0544 0.1830      NA       NA
ErrVarJ   0.7334        NA 0.6752 0.7857      NA       NA
ErrVarW   0.6814        NA 0.6165 0.7381      NA       NA
ErrVarS   0.9023        NA 0.8631 0.9315      NA       NA
```

```
mxAlgebras objects (and their 95% likelihood-based CIs):
        lbound Estimate ubound
Ind[1,1] 0.1429   0.1742 0.2074

Goodness-of-fit indices:
                                              Value
Sample size                                 3975.00
Chi-square of target model                     3.81
DF of target model                             1.00
p value of target model                        0.05
Number of constraints imposed on "Smatrix"     3.00
DF manually adjusted                           0.00
Chi-square of independence model             938.23
DF of independence model                       6.00
RMSEA                                          0.03
SRMR                                           0.02
TLI                                            0.98
CFI                                            1.00
```

```
AIC                                            1.81
BIC                                           -4.47
OpenMx status1: 0 ("0" or "1": The optimization is considered fine.
Other values indicate problems.)

----------------------- Selected output ----------------------
```

The proposed model fits the data well with $\chi^2(\text{df} = 1, N = 3975) = 3.8132$, $p = 0.0509$, CFI $= 0.9970$, RMSEA $= 0.0266$, and SRMR $= 0.0220$. As this is a path model with df $= 1$, it is expected that the goodness-of-fit indices will be good. The estimated parameters for $\boldsymbol{A}$ and $\boldsymbol{S}$ are

$$\begin{array}{c} \\ A \\ J \\ W \\ S \end{array}\begin{array}{c}\begin{array}{cccc} A & J & W & S \end{array}\\ \begin{bmatrix} 0 & 0 & 0 & 0 \\ 0.5164 & 0 & 0 & 0 \\ 0.2461 & 0.3966 & 0 & 0 \\ 0 & 0.2314 & 0.1214 & 0 \end{bmatrix}\end{array}$$

and

$$\begin{array}{c} \\ A \\ J \\ W \\ S \end{array}\begin{array}{c}\begin{array}{cccc} A & J & W & S \end{array}\\ \begin{bmatrix} 1 & 0 & 0 & 0 \\ 0 & 0.7334 & 0 & 0 \\ 0 & 0 & 0.6814 & 0 \\ 0 & 0 & 0 & 0.9023 \end{bmatrix}\end{array}$$

, respectively. The indirect effect (and its 95% LBCI) from *Ability* to *Supervisor rating* is 0.1742 (0.1429, 0.2074).

7.7 Concluding remarks and further readings

This chapter introduced the key concepts and issues of MASEM. It briefly reviewed the conceptual issues related to MASEM. Becker (2009) and Viswesvaran and Ones (1995) provided more thorough treatment on the conceptual issues in MASEM. This chapter introduced the fixed- and random-effects TSSEM approach. The advantages and technical details of the TSSEM approach were also addressed. Several issues related to MASEM were also addressed. Three examples were used to illustrate the TSSEM approach in R.

References

Allison PD 1987. Estimation of linear models with incomplete data. *Sociological Methodology* **17**, 71–103.

Becker BJ 1992. Using results from replicated studies to estimate linear models. *Journal of Educational Statistics* **17**(4), 341–362.

Becker BJ 1995. Corrections to "using results from replicated studies to estimate linear models". *Journal of Educational and Behavioral Statistics* **20**(1), 100–102.

Becker G 1996. The meta-analysis of factor analyses: an illustration based on the cumulation of correlation matrices. *Psychological Methods* **1**(4), 341–353.

Becker BJ 2000. Multivariate meta-analysis In *Handbook of applied multivariate statistics and mathematical modeling* (ed. Tinsley HEA and Brown SD). Academic Press, San Diego, CA, pp. 499–525.

Becker BJ 2009. Model-based meta-analysis In *The handbook of research synthesis and meta-analysis* (ed. Cooper H, Hedges LV and Valentine JC), 2nd edn. Russell Sage Foundation, New York, pp. 377–395.

Becker BJ and Fahrbach K 1994. A comparison of approaches to the synthesis of correlation matrices. *Paper presented at the annual meeting of the American Educational Research Association*, New Orleans, LA.

Becker BJ and Schram CM 1994. Examining explanatory models through research synthesis In *The handbook of research synthesis* (ed. Cooper H and Hedges LV). Russell Sage Foundation, New York, pp. 357–381.

Becker BJ and Wu MJ 2007. The synthesis of regression slopes in meta-analysis. *Statistical Science* **22**(3), 414–429.

Bentler P 2007. Can scientifically useful hypotheses be tested with correlations? *American Psychologist* **62**(8), 772–782.

Bentler PM and Savalei V 2010. Analysis of correlation structures: current status and open problems In *Statistics in the social sciences* (ed. Kolenikov S, Steinley D and Thombs L). John Wiley & Sons, Inc., Hoboken, NJ, pp. 1–36.

Bergh DD, Aguinis H, Heavey C, Ketchen DJ, Boyd BK, Su P, Lau CLL and Joo H 2014. Using meta-analytic structural equation modeling to advance strategic management research: guidelines and an empirical illustration via the strategic leadership-performance relationship. Strategic Management Journal, Advance access. Available from: http://onlinelibrary.wiley.com/doi/10.1002/smj.2338/abstract doi:10.1002/smj.2338

Bollen KA 1990. Overall fit in covariance structure models: two types of sample size effects. *Psychological Bulletin* **107**(2), 256–259.

Brown SP and Peterson RA 1993. Antecedents and consequences of salesperson job satisfaction: meta-analysis and assessment of causal effects. *Journal of Marketing Research* **30**(1), 63–77.

Browne MW and Cudeck R 1993. Alternative ways of assessing model fit In *Testing structural equation models* (ed. Bollen KA and Long JS). Sage Publications, Inc., Newbury Park, CA, pp. 136–162.

Byrne BM, Shavelson RJ and Muthén B 1989. Testing for the equivalence of factor covariance and mean structures: the issue of partial measurement invariance. *Psychological Bulletin* **105**(3), 456–466.

Card NA 2012. *Applied meta-analysis for social science research*. The Guilford Press, New York.

Cheung SF 2000. *Examining solutions to two practical issues in meta-analysis: dependent correlations and missing data in correlation matrices* PhD thesis. The Chinese University of Hong Kong, Hong Kong.

Cheung MWL 2002. *Meta-analysis for structural equation modeling: a two-stage approach* PhD thesis. The Chinese University of Hong Kong, Hong Kong.

Cheung MWL 2007. Comparison of approaches to constructing confidence intervals for mediating effects using structural equation models. *Structural Equation Modeling: A Multidisciplinary Journal* **14**(2), 227–246.

Cheung MWL 2009. TSSEM: a LISREL syntax generator for two-stage structural equation modeling (version 1.11) [Computer software and manual].

Cheung MWL 2010. Fixed-effects meta-analyses as multiple-group structural equation models. *Structural Equation Modeling: A Multidisciplinary Journal* **17**(3), 481–509.

Cheung MWL 2014a. Fixed- and random-effects meta-analytic structural equation modeling: examples and analyses in R. *Behavior Research Methods* **46**(1), 29–40.

Cheung MWL 2014b. metaSEM: meta-analysis using structural equation modeling. Frontiers in Psychology 5(1521).

Cheung MWL and Chan W 2004. Testing dependent correlation coefficients via structural equation modeling. *Organizational Research Methods* **7**(2), 206–223.

Cheung MWL and Chan W 2005a. Classifying correlation matrices into relatively homogeneous subgroups: a cluster analytic approach. *Educational and Psychological Measurement* **65**(6), 954–979.

Cheung MWL and Chan W 2005b. Meta-analytic structural equation modeling: a two-stage approach. *Psychological Methods* **10**(1), 40–64.

Cheung MWL and Chan W 2009. A two-stage approach to synthesizing covariance matrices in meta-analytic structural equation modeling. *Structural Equation Modeling: A Multidisciplinary Journal* **16**(1), 28–53.

Cheung SF and Cheung MWL 2010. Random effects models for meta-analytic structural equation modeling. Paper presented at the 7th Conference of the International Test Commission, Hong Kong SAR, China.

Cheung MWL, Leung K and Au K 2006. Evaluating multilevel models in cross-cultural research an illustration with social axioms. *Journal of Cross-Cultural Psychology* **37**(5), 522–541.

Colquitt JA, LePine JA and Noe RA 2000. Toward an integrative theory of training motivation: a meta-analytic path analysis of 20 years of research. *Journal of Applied Psychology* **85**(5), 678–707.

Conway JM 1999. Distinguishing contextual performance from task performance for managerial jobs. *Journal of Applied Psychology* **84**(1), 3–13.

Cudeck R 1989. Analysis of correlation-matrices using covariance structure models. *Psychological Bulletin* **105**(2), 317–327.

Demidenko E 2013. *Mixed models: theory and applications with R*, 2nd edn. Wiley-Interscience, Hoboken, NJ.

Digman JM 1997. Higher-order factors of the Big Five. *Journal of Personality and Social Psychology* **73**(6), 1246–1256.

Dunst CJ and Trivette CM 2009. Meta-analytic structural equation modeling of the influences of family-centered care on parent and child psychological health. *International Journal of Pediatrics* **2009**, 1–9.

Eby LT, Freeman DM, Rush MC and Lance CE 1999. Motivational bases of affective organizational commitment: a partial test of an integrative theoretical model. *Journal of Occupational and Organizational Psychology* **72**(4), 463–483.

Enders CK 2010. *Applied missing data analysis*. Guilford Press, New York.

Fouladi RT 2000. Performance of modified test statistics in covariance and correlation structure analysis under conditions of multivariate nonnormality. *Structural Equation Modeling: A Multidisciplinary Journal* **7**(3), 356–410.

Furlow CF and Beretvas NS 2005. Meta-analytic methods of pooling correlation matrices for structural equation modeling under different patterns of missing data. *Psychological Methods* **10**(2), 227–254.

Gasparrini A, Armstrong B and Kenward MG 2012. Multivariate meta-analysis for non-linear and other multi-parameter associations. *Statistics in Medicine* **31**(29), 3821–3839.

Geyskens I, Steenkamp JBE and Kumar N 1998. Generalizations about trust in marketing channel relationships using meta-analysis. *International Journal of Research in Marketing* **15**(3), 223–248.

Greenwald AG, Pratkanis AR, Leippe MR and Baumgardner MH 1986. Under what conditions does theory obstruct research progress? *Psychological Review* **93**(2), 216–229.

Hafdahl AR 2001. *Multivariate meta-analysis for exploratory factor analytic research* PhD thesis. The University of North Carolina at Chapel Hill United States, North Carolina.

Hafdahl AR 2007. Combining correlation matrices: simulation analysis of improved fixed-effects methods. *Journal of Educational and Behavioral Statistics* **32**(2), 180–205.

Hafdahl AR 2008. Combining heterogeneous correlation matrices: simulation analysis of fixed-effects methods. *Journal of Educational and Behavioral Statistics* **33**(4), 507–533.

Hafdahl AR 2009. *Meta-analysis for functions of heterogeneous multivariate effect sizes* Master of arts. Washington University in St. Louis United States, Washington, DC.

Hedges LV and Olkin I 1985. *Statistical methods for meta-analysis*. Academic Press, Orlando, FL.

Hedges LV and Pigott TD 2001. The power of statistical tests in meta-analysis. *Psychological Methods* **6**(3), 203–217.

Hedges LV and Pigott TD 2004. The power of statistical tests for moderators in meta-analysis. *Psychological Methods* **9**(4), 426–445.

Hedges LV and Vevea JL 1998. Fixed- and random-effects models in meta-analysis. *Psychological Methods* **3**(4), 486–504.

Hom PW, Caranikas-Walker F, Prussia GE and Griffeth RW 1992. A meta-analytical structural equations analysis of a model of employee turnover. *Journal of Applied Psychology* **77**(6), 890–909.

Hunter JE 1983. A causal analysis of cognitive ability, job knowledge, job performance, and supervisor ratings In *Performance Measurement and Theory* (ed. Landy F, Zedeck S and Cleveland J). Erlbaum, Hillsdale, NJ, pp. 257–266.

Hunter JE and Schmidt FL 1990. *Methods of meta-analysis: correcting error and bias in research findings*. Sage Publications, Inc., Newbury Park, CA.

Hunter JE and Schmidt FL 2004. *Methods of meta-analysis: correcting error and bias in research findings*, 2nd edn. Sage Publications, Inc., Thousand Oaks, CA.

Jak S, Oort FJ, Roorda DL and Koomen HMY 2013. Meta-analytic structural equation modelling with missing correlations. *Netherlands Journal of Psychology* **67**(4), 132–139.

Jöreskog KG 1978. Structural analysis of covariance and correlation matrices. *Psychometrika* **43**(4), 443–477.

Jöreskog KG and Sörbom D 1996. *LISREL 8: a user's reference guide*. Scientific Software International, Inc., Chicago, IL.

Jöreskog KG, Sörbom D, Du Toit S and Du Toit M 1999. *LISREL 8: new statistical features*. Scientific Software International, Inc., Chicago, IL.

Kaplan D 1990. Evaluating and modifying covariance structure models: a review and recommendation. *Multivariate Behavioral Research* **25**(2), 137–155.

Landis RS 2013. Successfully combining meta-analysis and structural equation modeling: recommendations and strategies. *Journal of Business and Psychology* **28**(3), 251–261.

MacCallum RC and Austin JT 2000. Applications of structural equation modeling in psychological research. *Annual Review of Psychology* **51**(1), 201–226.

Michel JS, Viswesvaran C and Thomas J 2011. Conclusions from meta-analytic structural equation models generally do not change due to corrections for study artifacts. *Research Synthesis Methods* **2**(3), 174–187.

Murayama K and Elliot AJ 2012. The competition-performance relation: a meta-analytic review and test of the opposing processes model of competition and performance. *Psychological Bulletin* **138**(6), 1035–1070.

Muthén B, Kaplan D and Hollis M 1987. On structural equation modeling with data that are not missing completely at random. *Psychometrika* **52**(3), 431–462.

National Research Council 1992. *Combining information: statistical issues and opportunities for research*. National Academies Press, Washington, DC.

Norton S, Cosco T, Doyle F, Done J and Sacker A 2013. The hospital anxiety and depression scale: a meta confirmatory factor analysis. *Journal of Psychosomatic Research* **74**(1), 74–81.

Paul PA, Lipps PE and Madden LV 2006. Meta-analysis of regression coefficients for the relationship between Fusarium head blight and deoxynivalenol content of wheat. *Phytopathology* **96**(9), 951–961.

Premack SL and Hunter JE 1988. Individual unionization decisions. *Psychological Bulletin* **103**(2), 223–234.

Rosenthal R 1991. *Meta-analytic procedures for social research*, revised edn. Sage Publications, Inc., Newbury Park, CA.

Schmidt FL and Hunter JE 2015. *Methods of meta-analysis: correcting error and bias in research findings*, 3rd edn. Sage Publications, Inc., Thousand Oaks, CA.

Schmidt FL, Oh IS and Hayes TL 2009. Fixed- versus random-effects models in meta-analysis: model properties and an empirical comparison of differences in results. *British Journal of Mathematical and Statistical Psychology* **62**(1), 97–128.

Smith KW, Avis NE and Assmann SF 1999. Distinguishing between quality of life and health status in quality of life research: a meta-analysis. *Quality of Life Research* **8**(5), 447–459.

Steiger JH 1980. Tests for comparing elements of a correlation matrix. *Psychological Bulletin* **87**(2), 245–251.

Steiger JH 2002. When constraints interact: a caution about reference variables, identification constraints, and scale dependencies in structural equation modeling. *Psychological Methods* **7**(2), 210–227.

Steinmetz H, Isidor R and Baeuerle N 2012. Testing the circular structure of human values: a meta-analytical structural equation modelling approach. *Survey Research Methods* **6**(1), 61–75.

Tett RP and Meyer JP 1993. Job satisfaction, organizational commitment, turnover intention, and turnover: path analyses based on meta-analytic findings. *Personnel Psychology* **46**(2), 259–293.

Vandenberg RJ and Lance CE 2000. A review and synthesis of the measurement invariance literature: suggestions, practices, and recommendations for organizational research. *Organizational Research Methods* **3**(1), 4–70.

Viswesvaran C and Ones DS 1995. Theory testing: combining psychometric meta-analysis and structural equations modeling. *Personnel Psychology* **48**(4), 865–885.

World Values Study Group 1994. *World Values Survey, 1981–1984 and 1990–1993 [Computer file]*. Inter-university Consortium for Political and Social Research, Ann Arbor, MI.

Wothke W 1993. Nonpositive definite matrices in structural modeling In *Testing structural equation models* (ed. Bollen KA and Long JS). Lawrence Erlbaum Associates, Newbury Park, CA, pp. 256–293.

Wothke W 2000. Longitudinal and multigroup modeling with missing data In *Modeling longitudinal and multilevel data: practical issues, applied approaches, and specific examples* (ed. Little TD, Schnable KU and Baumert J). Lawrence Erlbaum Associates, Mahwah, NJ, pp. 219–240.

Yuan KH and Chan W 2005. On nonequivalence of several procedures of structural equation modeling. *Psychometrika* **70**(4), 791–798.

Yufik T and Simms LJ 2010. A meta-analytic investigation of the structure of posttraumatic stress disorder symptoms. *Journal of Abnormal Psychology* **119**(4), 764–776.

8

Advanced topics in SEM-based meta-analysis

This chapter discusses two advanced topics in the structural equation modeling (SEM)-based meta-analysis. The first topic is the restricted (or residual) maximum likelihood (REML) estimation. We compare the pros and cons of the maximum likelihood (ML) estimation against the REML estimation. A graphical model is proposed to represent the transformation of the REML estimation. How to implement the REML estimation in SEM to conduct the SEM-based meta-analyses is introduced. The next topic is how to handle missing values in the moderators in a mixed-effects meta-analysis. Problems of and common methods on handling missing values in the moderators in meta-analysis are reviewed. ML estimation is proposed as a preferred method to handle the missing values. Examples are used to illustrate these procedures in the R statistical environment.

8.1 Restricted (or residual) maximum likelihood estimation

There are several estimation methods available in SEM, for example, two-stage least squares, unweighted least squares, generalized least squares, ML estimation, and weighted least squares (see Bentler, 2006; Jöreskog and Sörbom, 1996; Muthén and Muthén, 2012). In the previous chapters, we mainly focus on the ML estimation method. Under some regularity conditions (e.g., Millar, 2011), ML estimators have many desirable properties. For instance, they are consistent, asymptotically unbiased, asymptotically efficient, and asymptotically normally distributed. Several estimators are also available in meta-analysis (see Section 4.3.1). One of

Meta-Analysis: A Structural Equation Modeling Approach, First Edition. Mike W. -L. Cheung.

Companion Website: www.wiley.com/go/cheung/meta_analysis

the most popular estimator in meta-analysis is the weighted method of moments proposed by DerSimonian and Laird (1986). Another popular choice is the REML estimation, which was found to have a good balance between the unbiasedness and efficiency (Viechtbauer, 2005). This section discusses how REML estimation can be implemented in univariate, multivariate, and three-level meta-analyses in the SEM-based meta-analysis. In the following subsections, we first discuss the problems of the ML estimation method and how the REML estimation method can be used to address these problems. Then we introduce how to implement the REML estimation in SEM. This section is primarily based on Cheung (2013).

8.1.1 Reasons for and against the maximum likelihood estimation

There are two components in a mixed-effects model—the fixed effects and the variance components of the random effects. One problem of the ML estimator is that the estimated variance components are negatively biased. The reason of the bias is due to the fact that the variance components are estimated based on the estimated fixed effects. Let us consider the ML estimator of the variance in normally distributed data as an example. If we knew the population mean μ, an unbiased estimator on the variance is $\sum_{i=1}^{n}(x_i - \mu)^2/n$, where n is the sample size. As we rarely know the population mean in practice, the ML estimator based on the sample data is $\sum_{i=1}^{n}(x_i - \bar{x})^2/n$, where $\bar{x} = \sum_{i=1}^{n} x_i/n$ is the sample mean. It is always true that $\sum_{i=1}^{n}(x_i - \bar{x})^2/n \leq \sum_{i=1}^{n}(x_i - \mu)^2/n$. The equality sign holds only when $\bar{x} = \mu$. Therefore, the estimated variance based on the ML estimation is negatively biased. A well-known unbiased estimator on the variance is $\sum_{i=1}^{n}(x_i - \bar{x})^2/(n-1)$ that adjusts the uncertainty in estimating $\bar{x}$ by replacing n with $(n-1)$. The situation is more complicated in multilevel modeling or meta-analysis because the negative bias on the variance component cannot be corrected by a simple scalar adjustment.

The REML estimation was proposed to minimize the bias on estimating the variance components. The basic idea of the REML estimation is to break the analysis into two steps. In the first step, we estimate the variance components of the random effects on the transformed data (the residuals) so that the fixed-effects parameters have been removed from the estimation. As the fixed-effects parameters are not in the model, the variance components can be estimated without bias. In the second step, we estimate the fixed effects by treating the estimated variance components as known.

However, REML is not without its own limitations. As the fixed-effects parameters are removed before estimating the variance components, there is no fixed-effects estimates in REML estimation. Ad hoc calculations are required to compute the estimates of the fixed-effects parameters. Therefore, likelihood ratio (LR) test cannot be applied to compare models involving the fixed-effects parameters. LR statistic can only be used to compare nested models that are different on the variance components. Researchers have to switch back to the ML estimation when the analyses involve comparisons on the fixed effects.

Although the estimated variance components are negatively biased with the ML estimation, it should be noted that the estimated fixed effects with the ML estimation are unbiased even in small samples (Demidenko, 2013). Moreover, many new statistical developments are based on the ML estimation. It is possible to combine several techniques using the ML estimation. For example, we may handle missing data and conduct robust statistics with the ML estimation. "As to the question 'ML' or 'REML?'," (Searle et al. 1992, p.255) succinctly summarized that "there is probably no hard and fast answer." If the estimated variance components play a critical role in research hypotheses and the sample sizes are small, ML estimation is not recommended; otherwise, ML estimation can be a reasonable choice for the analysis.

8.1.2 Applying the restricted (or residual) maximum likelihood estimation in SEM-based meta-analysis

The rationale behind the REML estimation is to remove the fixed-effects parameters before estimating the variance components. A contrast matrix is chosen in such a way that the fixed-effects parameters are not estimated. As the fixed-effects parameters are not part of the model, the estimated variance components will not be biased by treating the fixed-effects estimates as known. Let us consider a typical mixed-effects meta-analytic model in the ith study,

$$\boldsymbol{y}_i = \boldsymbol{X}_i\boldsymbol{\beta} + \boldsymbol{Z}_i\boldsymbol{u}_i + \boldsymbol{e}_i, \tag{8.1}$$

where $\boldsymbol{y}_i$ is a vector of effect sizes, $\boldsymbol{X}_i$ is a design matrix for the fixed effects including a column of one in the first column, $\boldsymbol{\beta}$ is a vector of regression coefficients including the intercept, $\boldsymbol{Z}_i$ is a selection matrix of 1 and 0 to select the random effects, $\boldsymbol{T}^2 = \mathsf{Var}(\boldsymbol{u}_i)$ is the variance component of the random effects, and $\boldsymbol{V}_i = \mathsf{Var}(\boldsymbol{e}_i)$ is the known variance–covariance matrix of the sampling error.

The above model is general enough to represent both univariate and multivariate meta-analyses. The model can be slightly modified to include another random effect for the three-level meta-analysis. If there is only one effect size per study, $\boldsymbol{T}^2$ and $\boldsymbol{V}_i$ are scalars of the heterogeneity variance and the known sampling variance, respectively. $\boldsymbol{T}^2$ and $\boldsymbol{V}_i$ are matrices of the heterogeneity variance components and the known covariance matrix of the sampling errors, respectively, in a multivariate meta-analysis.

As there are k studies, we may stack all studies together. Suppose that the total number of the stacked effect sizes is $\tilde{k}$ ($\tilde{k} = k$ for a univariate meta-analysis); the model of $\underset{\tilde{k}\times 1}{\boldsymbol{y}}$ is

$$\boldsymbol{y} = \boldsymbol{X}\boldsymbol{\beta} + \boldsymbol{Z}\boldsymbol{u} + \boldsymbol{e}, \quad \text{or}$$

$$\begin{bmatrix} \boldsymbol{y}_1 \\ \boldsymbol{y}_2 \\ \vdots \\ \boldsymbol{y}_k \end{bmatrix} = \begin{bmatrix} \boldsymbol{X}_1 \\ \boldsymbol{X}_2 \\ \vdots \\ \boldsymbol{X}_k \end{bmatrix} \boldsymbol{\beta} + \begin{bmatrix} \boldsymbol{Z}_1 & & & \\ 0 & \boldsymbol{Z}_2 & & \\ \vdots & \ddots & \ddots & \\ 0 & \cdots & 0 & \boldsymbol{Z}_k \end{bmatrix} \begin{bmatrix} \boldsymbol{u}_1 \\ \boldsymbol{u}_2 \\ \vdots \\ \boldsymbol{u}_k \end{bmatrix} + \begin{bmatrix} \boldsymbol{e}_1 \\ \boldsymbol{e}_2 \\ \vdots \\ \boldsymbol{e}_k \end{bmatrix} \tag{8.2}$$

Let $G = \mathrm{Var}(u) = \mathrm{Diag}(T^2, \ldots, T^2)$ and $V = \mathrm{Var}(e) = \mathrm{Diag}(V_1, V_2, \ldots, V_k)$. Instead of analyzing y with the ML estimation, we analyze its residuals $\tilde{y}$,

$$\begin{aligned} \underset{(\tilde{k}-p)\times 1}{\tilde{y}} &= \underset{(\tilde{k}-p)\times\tilde{k}}{A} \underset{\tilde{k}\times 1}{y}, \\ &= AX\beta + A(Zu + e), \end{aligned} \tag{8.3}$$

where $A = I - X(X^{\mathrm{T}}X)^{-1}X^{\mathrm{T}}$ with p arbitrary rows removed and I is a $\tilde{k} \times \tilde{k}$ identity matrix, and p is the number of columns in X (Harville, 1977; Patterson and Thompson, 1971). Several characteristics bear explaining. After the calculations, the contrast A (without deleting the p arbitrary rows) is a $\tilde{k} \times \tilde{k}$ matrix. As the rank of this matrix is only $(\tilde{k} - p)$, it is not of full rank. Thus, p redundant rows have to be deleted. Harville (1977) showed that these p rows can be arbitrarily selected without affecting the results. The common practice is to delete the last p rows. Therefore, A becomes a $(\tilde{k} - p) \times \tilde{k}$ matrix after deleting the last p rows. It is of importance to note that $\tilde{y}$ is now a column vector with $(\tilde{k} - p)$ rows. In practice, the transpose of it is used in the analysis using the SEM approach. That is, the final data is a row vector with $(\tilde{k} - p)$ columns of variables. With a slight abuse of notation, it is assumed that the last p rows of the contrast matrix A have been removed in the following discussion.

$AX\beta$ and $A(Zu + e)$ are the fixed and the random effects, respectively. After the transformation, the expected value of $\tilde{y}$, the fixed effects, is

$$\begin{aligned} \mathrm{E}(\tilde{y}) &= (I - X(X^{\mathrm{T}}X)^{-1}X^{\mathrm{T}})X\beta, \\ &= X\beta - X(X^{\mathrm{T}}X)^{-1}X^{\mathrm{T}}X\beta, \\ &= \mathbf{0}. \end{aligned} \tag{8.4}$$

The population means of $\tilde{y}$ is always zero regardless of what β is. Moreover, β is not estimable by analyzing $\tilde{y}$. Effectively, the variance components can be estimated without estimating β. The expected covariance matrix of $\tilde{y}$, the variance component of the random effects, is

$$\begin{aligned} \mathrm{Cov}(\tilde{y}) &= A\mathrm{Cov}(Zu + e)A^{\mathrm{T}}, \\ &= A\Omega A^{\mathrm{T}}, \end{aligned} \tag{8.5}$$

where $\Omega = ZGZ^{\mathrm{T}} + V$. Before the transformation, the between clusters of y are independent; the between clusters become systematically dependent after the transformation.

Another approach to obtain the REML estimates is to directly analyze the −2*log-likelihood (LL) of the transformed data. The −2LL function without the constant term on $\tilde{y}$ is

$$-2\mathrm{LL}_{\mathrm{REML}} = \log|\Omega| + (y - X\alpha)^{\mathrm{T}}\Omega^{-1}(y - X\alpha) + \log|X^{\mathrm{T}}\Omega^{-1}X|), \tag{8.6}$$

where $\alpha = (X\Omega^{-1}X)^{-1}X^{\mathrm{T}}\Omega^{-1}y$ (e.g., Muller and Stewart, 2006, p. 286). There are two main differences between the −2LL functions in the ML and the REML estimations. First, the fixed effects $\boldsymbol{\beta}$ are parameters that have to be estimated in the $-2LL$ function in the ML estimation (e.g., Equation 2.42), whereas $\boldsymbol{\alpha}$ is not a parameter in the $-2LL$ in the REML estimation. Second, an additional term $\log|\boldsymbol{X}^{\mathrm{T}}\boldsymbol{\Omega}^{-1}\boldsymbol{X}|$ is in the $-2LL$ in the REML estimation. It should be noted that there was a typo in Equation 7 in Cheung (2013). The log operation was missing in $\log|\boldsymbol{X}^{\mathrm{T}}\boldsymbol{\Omega}^{-1}\boldsymbol{X}|$ in the equation. The `OpenMx` package (Boker et al., 2011) can be used to fit the model with the $-2LL$ (see Cheung, 2013).

Once the variance components $\hat{\boldsymbol{G}}$ have been estimated, we may compute the fixed-effects by

$$\hat{\boldsymbol{\beta}}_{\mathrm{REML}} = (\boldsymbol{X}^{\mathrm{T}}\hat{\boldsymbol{\Omega}}_{\mathrm{REML}}^{-1}\boldsymbol{X})^{-1}\boldsymbol{X}^{\mathrm{T}}\hat{\boldsymbol{\Omega}}_{\mathrm{REML}}^{-1}\boldsymbol{y}, \tag{8.7}$$

where $\hat{\boldsymbol{\Omega}}_{\mathrm{REML}} = \boldsymbol{Z}\hat{\boldsymbol{G}}\boldsymbol{Z}^{\mathrm{T}} + \boldsymbol{V}$. Therefore, two steps are required to estimate the variance components and the fixed effects separately.

8.1.3 Implementation in structural equation modeling

Cheung (2013) showed how the above two approaches to obtain the REML estimates could be implemented in SEM. In this part, we show how the model of $\tilde{\boldsymbol{y}}$ can be fitted by using the so-called *wide* format using the first approach in SEM. Suppose that the model-implied mean vector and model-implied covariance matrix using the ML estimation are $\underset{\tilde{k}\times 1}{\mu(\boldsymbol{\theta})}$ and $\underset{\tilde{k}\times\tilde{k}}{\Sigma(\boldsymbol{\theta})}$, respectively. It should be noted that we are using the *wide* format here (see Section 6.3.1 for details). There is only one *subject* with $\tilde{k}$ *variables* in the stacked effect size $\underset{\tilde{k}\times 1}{\boldsymbol{y}}$.

The model-implied mean vector $\mu(\boldsymbol{\theta})$ and the model-implied covariance matrix $\Sigma(\boldsymbol{\theta})$ represent the fixed effects and the variance components of the random effects, respectively. We do not specify the specific structure here because the structure depends on the types of meta-analysis. Concrete examples on these models will be illustrated later.

From the above discussion, we fit a structural equation model using the REML estimation with the following model-implied mean vector and model-implied covariance matrix:

$$\begin{aligned}
\underset{(\tilde{k}-p)\times 1}{\mu(\tilde{\boldsymbol{\theta}})} &= \underset{(\tilde{k}-p)\times\tilde{k}}{\boldsymbol{A}}\ \underset{\tilde{k}\times 1}{\mu(\boldsymbol{\theta})} = \underset{(\tilde{k}-p)\times 1}{\boldsymbol{0}} \quad \text{and} \\
\underset{(\tilde{k}-p)\times(\tilde{k}-p)}{\Sigma(\tilde{\boldsymbol{\theta}})} &= \underset{(\tilde{k}-p)\times\tilde{k}}{\boldsymbol{A}}\ \underset{\tilde{k}\times\tilde{k}}{\Sigma(\boldsymbol{\theta})}\ \underset{\tilde{k}\times(\tilde{k}-p)}{\boldsymbol{A}^{\mathrm{T}}}.
\end{aligned} \tag{8.8}$$

Several issues are worthy to be mentioned here. First, all effect sizes are stacked together into a single vector of $\boldsymbol{y}$ regardless of whether it is a univariate meta-analysis or a multivariate meta-analysis. Second, the contrast matrix $\underset{(\tilde{k}-p)\times\tilde{k}}{\boldsymbol{A}}$ transforms $\underset{\tilde{k}\times 1}{\boldsymbol{y}}$ to $\underset{(\tilde{k}-p)\times 1}{\tilde{\boldsymbol{y}}}$. Third, the means for all *variables* are zero, while the model-implied covariance matrix is structured in a particular pattern. Fourth, the only parameters in this model are the variance components.

After obtaining the estimated variance components, we may estimate $\hat{\boldsymbol{\beta}}_{\text{REML}}$ by fitting another model with the variance component estimated from REML as known values,

$$\underset{\tilde{k}\times 1}{\mu_i(\boldsymbol{\theta})} = \boldsymbol{X\beta} \quad \text{and} \\ \underset{\tilde{k}\times\tilde{k}}{\Sigma_i(\boldsymbol{\theta})} = \boldsymbol{Z}\tilde{\boldsymbol{G}}\boldsymbol{Z}^{\text{T}} + \tilde{\boldsymbol{V}}, \tag{8.9}$$

where $\tilde{\boldsymbol{G}}$ and $\tilde{\boldsymbol{V}}$ are obtained from REML estimation. In this model, the only parameters are the fixed effects $\boldsymbol{\beta}$.

8.1.3.1 Univariate meta-analysis

Let us illustrate the univariate random-effects meta-analysis using the SEM approach. Suppose that there are k studies in a meta-analysis; the model-implied means and model-implied covariance matrix are

$$\begin{aligned}
\underset{k\times 1}{\boldsymbol{\mu}(\boldsymbol{\theta})} &= \underset{k\times 1}{\mathbf{1}}\,\beta_{\text{R}} = \begin{bmatrix} \beta_{\text{R}} \\ \vdots \\ \beta_{\text{R}} \end{bmatrix} \quad \text{and} \\
\underset{k\times k}{\boldsymbol{\Sigma}(\boldsymbol{\theta})} &= \underset{k\times k}{\boldsymbol{I}}\,\tau^2 + \underset{k\times k}{\text{Diag}(v_1, v_2, \ldots, v_k)} \\
&= \begin{bmatrix} \tau^2 & & & \\ 0 & \tau^2 & & \\ \vdots & \ddots & \ddots & \\ 0 & \cdots & 0 & \tau^2 \end{bmatrix} + \begin{bmatrix} v_1 & & & \\ 0 & v_2 & & \\ \vdots & \ddots & \ddots & \\ 0 & \cdots & 0 & v_k \end{bmatrix},
\end{aligned} \tag{8.10}$$

where $\underset{k\times 1}{\mathbf{1}}$ is a vector of ones and $\underset{k\times k}{\boldsymbol{I}}$ is an identity matrix. There are only two parameters, the fixed effects β_{R} and the random effects τ^2, in this model (see Section 6.3.1).

Figure 8.1 displays a univariate random-effects meta-analysis with three studies. Figure 8.1a shows the three studies arranged in the wide format. There are three *variables* with only one *subject* in the analysis. Figure 8.1b displays the same model by introducing three latent variables η_1 to η_3. The factor loadings are all one meaning that η_1 to η_3 equal y_1 to y_3. Although the introduction of η_1 to η_3 seems to be redundant, these latent variables are useful in explaining what the transformation means in SEM.

When we apply an REML estimation on $\boldsymbol{y}$, the model-implied mean vector and model-implied covariance matrix of $\tilde{\boldsymbol{y}} = \boldsymbol{Ay}$ are

$$\begin{aligned}
\boldsymbol{\mu}(\tilde{\boldsymbol{\theta}}) &= \underset{(k-1)\times 1}{\mathbf{0}} \quad \text{and} \\
\boldsymbol{\Sigma}(\tilde{\boldsymbol{\theta}}) &= \underset{(k-1)\times k}{\boldsymbol{A}}\,\underset{k\times k}{(\boldsymbol{I}\tau^2 + \text{Diag}(v_1, v_2, \ldots, v_k))}\,\underset{k\times(k-1)}{\boldsymbol{A}^{\text{T}}}.
\end{aligned} \tag{8.11}$$

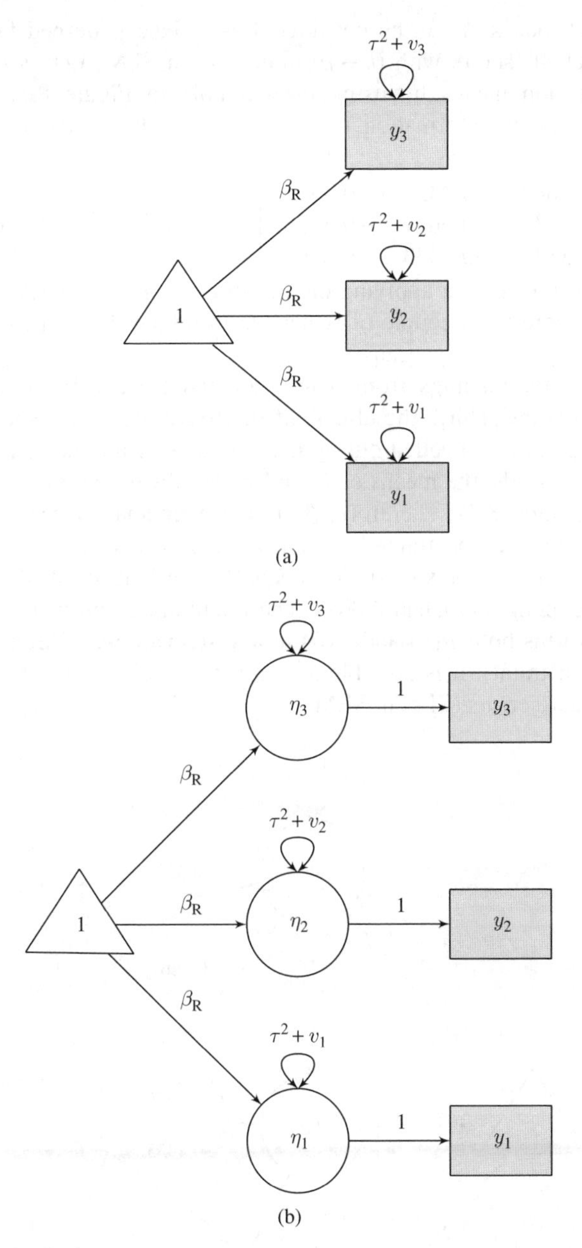

Figure 8.1 A univariate random-effects meta-analysis with three studies.

The mean structure is simply fixed at 0 because the fixed-effects parameters are not involved in the model. After the transformation, there are only $k - 1$ *subjects*, because one row has been removed from $\boldsymbol{A}$. There is only one parameter, the heterogeneity variance τ^2, because v_i is known in this model.

The contrast matrix $\boldsymbol{A}$ can be considered as a fixed patterned factor loadings matrix of k latent factors with $(k - p)$ indicators in SEM. Let us illustrate what the transformation means by using the example in Figure 8.1. As there are three studies, the transformation matrix without deleting the last row ($\tilde{y}_3$) is

$$\boldsymbol{A} = \begin{array}{c} \\ \tilde{y}_1 \\ \tilde{y}_2 \\ \tilde{y}_3 \end{array} \begin{array}{c} \begin{array}{ccc} \eta_1 & \eta_2 & \eta_3 \end{array} \\ \begin{bmatrix} 0.6667 & -0.3333 & -0.3333 \\ -0.3333 & 0.6667 & -0.3333 \\ -0.3333 & -0.3333 & 0.6667 \end{bmatrix} \end{array}$$

(in four decimal places). Figure 8.2 shows the model of $\tilde{\boldsymbol{y}}$ after applying the transformation with the $\boldsymbol{A}$ matrix deleted the last row. There are a couple of issues worth noting. First, there are only two studies $\tilde{y}_1$ and $\tilde{y}_2$ after the transformation because the last row $\tilde{y}_3$ is deleted. Second, the factor loadings from η to $\tilde{y}$ are fixed according to the $\boldsymbol{A}$ matrix without the last row. Third, the effects of the fixed effects β_{R} (shown in dashed lines in the figure) is cancelled out by the $\boldsymbol{A}$ matrix or the factor loadings in the figure. In other words, the means of $\tilde{y}_1$ and $\tilde{y}_2$ are always zero regardless of what β_{R} is (see Equation 8.4). Therefore, β_{R} is not estimable in this model. Similar approaches using the constrained optimization have also been used to analyze ipsative data, sum of scores of subjects equal a constant, in SEM (e.g., Cheung, 2004, 2006, Cheung and Chan, 2002). These authors found that the ipsatization process transforms both $\boldsymbol{\mu}(\boldsymbol{\theta})$ and $\boldsymbol{\Sigma}(\boldsymbol{\theta})$ in a systematic way. After imposing the appropriate constraints, it is possible to recover the original parameter estimates and their standard errors (SEs) in SEM.

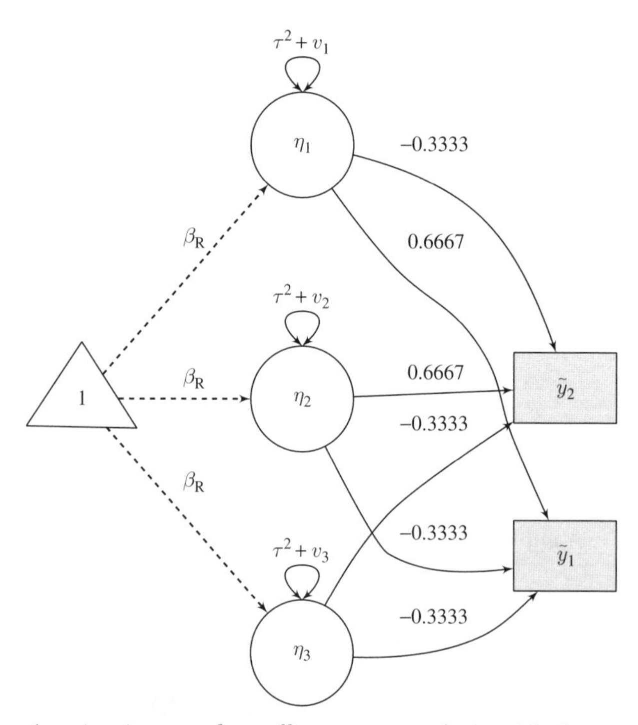

Figure 8.2 A univariate random-effects meta-analysis with three studies using REML estimation.

8.1.3.2 Multivariate meta-analysis

The same idea applies to the univariate meta-analysis can be extended to multivariate meta-analysis. We stack all multivariate effect sizes of the k studies into a single column vector with Equation 8.2. The model for $\tilde{\boldsymbol{y}}$ is

$$\begin{aligned}\boldsymbol{\mu}(\boldsymbol{\theta}) &= \boldsymbol{X}\boldsymbol{\beta} \quad \text{and} \\ \boldsymbol{\Sigma}(\boldsymbol{\theta}) &= \boldsymbol{Z}\boldsymbol{G}\boldsymbol{Z}^{\mathrm{T}} + \mathrm{Diag}(\boldsymbol{V}_1, \boldsymbol{V}_2, \dots, \boldsymbol{V}_k) \\ &= \begin{bmatrix} \boldsymbol{Z}_1\boldsymbol{T}^2\boldsymbol{Z}_1^{\mathrm{T}} & & & \\ 0 & \boldsymbol{Z}_2\boldsymbol{T}^2\boldsymbol{Z}_2^{\mathrm{T}} & & \\ \vdots & \ddots & \ddots & \\ 0 & \cdots & 0 & \boldsymbol{Z}_k\boldsymbol{T}^2\boldsymbol{Z}_k^{\mathrm{T}} \end{bmatrix} + \begin{bmatrix} \boldsymbol{V}_1 & & & \\ 0 & \boldsymbol{V}_2 & & \\ \vdots & \ddots & \ddots & \\ 0 & \cdots & 0 & \boldsymbol{V}_k \end{bmatrix},\end{aligned} \tag{8.12}$$

where $\boldsymbol{\beta}$ is the vector of the fixed effects including the intercept and $\boldsymbol{T}^2$ is the variance components of the random effects. It should also be noted that $\boldsymbol{T}^2$ is duplicated across the k studies. Thus, this model representation is not very efficient for large data set.

Now, we fit a model on the transformed residuals $\tilde{\boldsymbol{y}}$ using the REML estimation. The model-implied mean vector and the covariance matrix are

$$\begin{aligned}\underset{(\tilde{k}-p)\times 1}{\boldsymbol{\mu}(\tilde{\boldsymbol{\theta}})} &= \underset{(\tilde{k}-p)\times 1}{\boldsymbol{0}} \quad \text{and} \\ \underset{(\tilde{k}-p)\times(\tilde{k}-p)}{\boldsymbol{\Sigma}(\tilde{\boldsymbol{\theta}})} &= \underset{(\tilde{k}-p)\times\tilde{k}}{\boldsymbol{A}} \underset{k\times k}{(\boldsymbol{Z}\boldsymbol{G}\boldsymbol{Z}^{\mathrm{T}} + \mathrm{Diag}(\boldsymbol{V}_1, \boldsymbol{V}_2, \dots, \boldsymbol{V}_k))} \underset{\tilde{k}\times(\tilde{k}-p)}{\boldsymbol{A}^{\mathrm{T}}},\end{aligned} \tag{8.13}$$

where $\tilde{k}$ is the number of rows in $\tilde{\boldsymbol{y}}$ and p is the number of rows deleted. As $\boldsymbol{V}_i$ is known, the only parameter in this model is $\boldsymbol{G} = \mathrm{Diag}(\boldsymbol{T}^2, \dots, \boldsymbol{T}^2)$.

8.1.3.3 Three-level meta-analysis

As a three-level meta-analysis also consists of fixed effects and random effects, the above approach can be extended to the three-level meta-analysis (Cheung, 2014b). Suppose that there are k clusters with a maximum of m effect sizes per study; the model-implied means and model-implied covariance matrix for the jth cluster are

$$\begin{aligned}\underset{m\times 1}{\boldsymbol{\mu}_j(\boldsymbol{\theta})} &= \underset{m\times 1}{\boldsymbol{1}}\beta_{\mathrm{R}} \quad \text{and} \\ \underset{m\times m}{\boldsymbol{\Sigma}_j(\boldsymbol{\theta})} &= \underset{m\times m}{\boldsymbol{1}}\tau^2_{(3)} + \underset{m\times m}{\boldsymbol{I}}\tau^2_{(2)} + \underset{m\times m}{\boldsymbol{V}_j} \\ &= \begin{bmatrix} \tau^2_{(3)} & & & \\ \tau^2_{(3)} & \tau^2_{(3)} & & \\ \vdots & \ddots & \ddots & \\ \tau^2_{(3)} & \cdots & \tau^2_{(3)} & \tau^2_{(3)} \end{bmatrix} + \begin{bmatrix} \tau^2_{(2)} & & & \\ 0 & \tau^2_{(2)} & & \\ \vdots & \ddots & \ddots & \\ 0 & \cdots & 0 & \tau^2_{(2)} \end{bmatrix} \\ &\quad + \begin{bmatrix} v_{1j} & & & \\ 0 & v_{2j} & & \\ \vdots & \ddots & \ddots & \\ 0 & \cdots & 0 & v_{mj} \end{bmatrix}.\end{aligned} \tag{8.14}$$

As there are k clusters, we need to stack all effect sizes together before applying the transformation. The model-implied mean vector and model-implied covariance matrix for all studies are

$$\boldsymbol{\mu}(\boldsymbol{\theta}) = \begin{bmatrix} \boldsymbol{\mu}_1(\boldsymbol{\theta}) \\ \boldsymbol{\mu}_2(\boldsymbol{\theta}) \\ \vdots \\ \boldsymbol{\mu}_k(\boldsymbol{\theta}) \end{bmatrix} \quad \text{and}$$
$$\boldsymbol{\Sigma}(\boldsymbol{\theta}) = \begin{bmatrix} \boldsymbol{\Sigma}_1(\boldsymbol{\theta}) & & & \\ 0 & \boldsymbol{\Sigma}_2(\boldsymbol{\theta}) & & \\ \vdots & \ddots & \ddots & \\ 0 & \cdots & 0 & \boldsymbol{\Sigma}_k(\boldsymbol{\theta}) \end{bmatrix}. \tag{8.15}$$

Now, we apply the contrast matrix $\boldsymbol{A}$ on $\boldsymbol{y}$. The model of the residuals $\tilde{\boldsymbol{y}}$ is

$$\underset{(\tilde{k}-p)\times 1}{\boldsymbol{\mu}(\tilde{\boldsymbol{\theta}})} = \underset{(\tilde{k}-p)\times \tilde{k}}{\boldsymbol{A}} \; \underset{\tilde{k}\times 1}{\boldsymbol{\mu}(\boldsymbol{\theta})} = \underset{(\tilde{k}-p)\times 1}{\mathbf{0}} \quad \text{and}$$
$$\underset{(\tilde{k}-p)\times(\tilde{k}-p)}{\boldsymbol{\Sigma}(\tilde{\boldsymbol{\theta}})} = \underset{(\tilde{k}-p)\times \tilde{k}}{\boldsymbol{A}} \; \underset{\tilde{k}\times\tilde{k}}{\boldsymbol{\Sigma}(\boldsymbol{\theta})} \; \underset{\tilde{k}\times(\tilde{k}-p)}{\boldsymbol{A}^{\mathrm{T}}}. \tag{8.16}$$

Although the above model looks complicated, there are only two parameters, $\tau^2_{(2)}$ and $\tau^2_{(3)}$, in the model.

8.1.3.4 Issues related to implementing the restricted maximum likelihood estimation

In the above discussion, we mainly focused on the models without any study characteristics. It is easy to extend the above models to the mixed-effects meta-analyses by including the study characteristics in the design matrix $\boldsymbol{X}$. The estimated variance components are then the estimated variance component of the residual heterogeneity. The fixed effects including the regression coefficients can be estimated by Equation 8.9 .

As discussed in the above subsections, there are two equivalent approaches to implementing the REML estimation in SEM. One approach is to analyze the transformed data by removing the fixed effects. The other approach is to directly fit the −2LL. This chapter mainly focuses on analyzing the transformed data because only the implied mean and implied variance–covariance matrix are required. The models still look like structural equation models on the transformed data. The REML estimation is basically the same as the ML estimation by imposing the transformation matrix $\boldsymbol{A}$ as the fixed pattern of factor loadings. There is one main limitation of analyzing the transformed data, however. The input data is a $\tilde{k} - p$ columns of variables with one row (or *subject*) in SEM. Mehta and Neale (2005) has also warned that this approach is not very efficient when *subjects* are treated as *variables* in handling nested data. The situation may become even worse in implementing REML estimation because there are constraints in the model-implied covariance matrix. Nonconvergent solutions may occur. Similar issues also occur in the constrained optimization of ipsative data (e.g., Cheung, 2004, 2006). On the other hand, the approach based on the −2LL is more stable. The main limitation is that most SEM packages cannot implement arbitrary fitting functions. Moreover, analysis using the −2LL does not look like a structural equation model any more. Future research may explore how to better implement the REML estimation in SEM.

8.2 Missing values in the moderators

No matter how good the design of an experiment is, it is the rule rather than the exception to have missing or incomplete data. The presence of missing data introduces a considerable challenge to researchers and methodologists in various disciplines. This applies equally well to meta-analysis. Cooper and Hedges (2009, p. 565) have called missing data "perhaps the most pervasive practical problem in research synthesis, which obviously influences any statistical analyses." The main problem is that most statistical analyses were developed to handle complete data. When there are missing data, the missing data have to be handled before the analyses.

There are several types of missing data in a meta-analysis, for example, incomplete effect sizes in a multivariate meta-analysis (see Section 5.1.2) and publication bias. This section focuses on another type of missing data—missing values in the moderators in a mixed-effects meta-analysis. The first part of this section introduces different types of missing mechanisms and how they can be applied in a meta-analysis. The second part discusses some common methods and *modern* methods to handle missing data. Finally, we introduce how the SEM-based meta-analysis can be used to handle missing values in the moderators.

8.2.1 Types of missing mechanisms

The modern theory of missing data was developed by Rubin (1976). Rubin (1976) defined three types of missingness mechanism. The missingness on a variable, Y, is said to be missing completely at random (MCAR) if the missingness is unrelated to the value of Y itself or to the values of any other variables in the model. For example, a participant forgets to report his or her salary by mistake. Another example is that there are three long questionnaires: A, B, and C, where A is essential. If participants are required to complete all three versions, the data quality may be compromised because of participant fatigue. We may improve the data quality by using the so-called planned missingness (Graham et al., 1996). All participants are required to complete questionnaire A. Half of the participants are randomly chosen to complete questionnaire B, while the other half are required to complete questionnaire C. Therefore, there are systematically missing data in either questionnaires B or C. A third example is in longitudinal studies. After conducting a cross-sectional study, we randomly select 20% of the participants for a follow-up study because of the limited resource. In these three examples, the missingness of the data is not related to neither the values of the missing data nor other variables. Therefore, it is called MCAR.

The assumption of MCAR is rather strong in applied research. A considerably weaker assumption is missing at random (MAR). MAR means that the missingness on Y is unrelated to the value of Y after controlling for other variables in the analysis. For example, a participant fails to report his or her salary (a missing datum on Y) because he or she has a low (or high) educational level (another variable). In a longitudinal study, participants are selected for a follow-up study only if their pretest scores exceed a cutoff value. Therefore, the missing values at posttest scores are related to the pretest scores. Whether the missingness is MAR or not depends on the presence of other variables. When there are more and more variables in the model, it is more likely that the missingness is MAR.

If the missing mechanism is neither MCAR nor MAR, it is not missing at random (NMAR). The missingness on Y is said to be NMAR if the missingness on Y is related to the value of Y itself even after controlling for other variables in the model. For example, a participant fails to report his or her salary because his or her income is very high (or low). In a longitudinal study, a participant drops out from a follow-up study on the effectiveness of a training program because he or she finds that the training is not effective. Therefore, the missing data on the follow-up are those with low training effectiveness.

Rubin's definitions have been directly adopted for meta-analysis (e.g., Cheung, 2008; Pigott, 2001, 2009, 2012; Sutton, 2000; Sutton and Pigott, 2005). When we are applying these definitions in the context of missing values in the moderators, MCAR means that the missing values in a moderator is unrelated to the value of that moderator or other variables. MAR means that the missing values of a moderator can be related to other variables, for instance, the effect size or other moderator in the model, while NMAR means that the missing values in a moderator is related to the value of that moderator. As researchers are not very likely to fail to report a study characteristic, for example, mean age of the participants, because of the value of that moderator, Sutton and Pigott (2005, p. 235) stated that "the assumption that [study-level characteristics] are MCAR or MAR may be reasonable and standard missing-data methods may suffice in some situations." It should be noted that the missingness on the effect sizes, for example, a correlation coefficient between job satisfaction and performance, is likely to be NMAR. If the effect sizes are nonsignificant, they are less likely to be reported or published. This is known as publication bias (Rothstein et al., 2005), which is beyond the scope of this chapter.

8.2.2 Common methods to handling missing data

Enders (2010) provided a thorough treatment on the conventional and modern methods to handling missing data in general. Pigott (2009, 2012) also provided a summary in the context of meta-analysis. The common methods include listwise deletion, pairwise deletion, and mean substitution; the modern methods include multiple imputation (MI) and full information maximum likelihood (FIML or ML) estimation.

Listwise deletion (also known as complete-case analysis) excludes cases with missing values. Suppose that we are conducting a mixed-effects meta-analysis with three study characteristics as moderators; any study with missing values is deleted before the analysis. The main advantage is its ease to apply. Therefore, it is usually the default option in most statistical packages. There are two main disadvantages of this method. The first problem is that the sample size (or number of studies) is sharply reduced. This is more noticeable when there are lots of moderators in the analysis. The second problem is that it relies on the assumption of MCAR of the missing data. If the data are not MCAR, the parameter estimates can be biased.

Another conventional method is pairwise deletion (also known as available-case analysis). It attempts to keep as many data as possible by deleting the missing data on an analysis-by-analysis basis. Using the previous example as an illustration, we may conduct three separate mixed-effect meta-analyses by using one moderator at a time. As the patterns of missing values may be different in the moderators, the studies used in the analyses may also be different. Although the number of studies is retained as many as possible, there are a few drawbacks. The main issue is that it limits the potential of including several moderators in the same analysis. As the moderators are included one-by-one in the analyses, it is not possible to test and compare their relative contributions.

A third method is mean substitution or similar techniques. Researchers may replace the missing values with their correspondent means or predicted values from other variables. Although these methods may utilize all available cases in the analysis, there is one main problem—the means or predicted values are treated as observed data. The estimated SEs are likely underestimated. Thus, the empirical Type I error is inflated and the confidence intervals (CIs) are too short.

8.2.3 Maximum likelihood estimation

As full information (in contrast to limited information) is usually assumed in the ML estimation, some researchers argue that we should call it ML estimation rather than FIML estimation. In the SEM community, ML estimation may be used to refer to analysis of summary statistics (means and covariance matrix). Therefore, SEM users usually call the applications of ML estimation to individuals as FIML estimation (see Section 2.4.1). We use the term FIML here to emphasize that raw data, not the summary statistics, are used in the analysis.

Nearly all the empirical studies so far support the contention that both MI and FIML perform better than conventional methods such as listwise deletion, pairwise deletion, and mean imputation in handling missing data when the missingness is either MCAR or MAR (see, e.g., Cheung (2007) and Enders (2010) for some empirical findings on a comparison of methods of handling missing covariates in the context of a latent growth model and a regression analysis). When the missingness is NMAR, none of the above methods is unbiased (Schafer, 1997). However, the bias of the FIML is still less than that resulting from listwise deletion, pairwise deletion, and mean substitution (e.g., Jamshidian and Bentler, 1999; Muthén et al., 1987). Therefore, the MI and FIML are usually recommended for handling missing data (e.g., Enders, 2010; Graham, 2009; Little, 1992; Schafer and Graham, 2002). We introduce these two methods in more details.

Besides the above advantages, there are also other benefits of using the ML (FIML or MI) estimation to handling missing values in the moderators in a meta-analysis. In order to calculate the percentage of variance explained or R^2, the studies must be the same in calculating $\hat{\tau}_0^2$ and $\hat{\tau}_1^2$ (see Section 4.5.2). Studies with missing values in the moderators have to be deleted in calculating $\hat{\tau}_0^2$. When we are testing multiple moderators, it is likely that more studies have to be deleted because of the missing values in the moderators. It is difficult to test several moderators simultaneously. If we use the ML estimation, we can keep *all* studies in the analysis. This makes it feasible to compare the relative contributions of the moderators by testing all moderators simultaneously.

Both the MI and FIML are based on the principle of ML estimation. The main difference is on the implementations. MI generates multiple samples of imputation to obtain the ML estimates, while the FIML obtain the ML estimates by direct optimization of the LL. Compared with MI, FIML has at least three advantages (Cheung, 2014b). First, the results based on MI are asymptotically equivalent to those based on FIML. That is, they are equal when the number of imputations in MI approaches infinity. It is generally advised that three to five imputations are considered sufficient to obtain excellent results using MI (e.g., Schafer and Olsen, 1998). However, Graham et al. (2007) showed that many more imputations are required. They suggested requiring m imputations based on the fraction of missing information (γ). They recommended that researchers use $m = 20$, 20, 40, 100, and >100 imputations for data with $\gamma = 0.1$, 0.3, 0.5, 0.7, and 0.9, respectively. On the basis of their simulations results, they concluded that FIML is superior to MI in terms of power for testing small effect sizes (unless one has a sufficient number of imputations).

Second, conducting MI for multilevel data is still challenging (e.g., van Buuren and Groothuis-Oudshoorn, 2011). More importantly, how to apply MI on univariate, multivariate, and three-level meta-analytic data remains unclear. Lastly, FIML is more robust than MI when the data are nonnormal (Yuan et al., 2012). Specifically, the performance of FIML is less biased and more efficient than that of MI. Therefore, we focus on the FIML in this chapter.

The basic idea of FIML estimation is to analyze the data at the subject level. A filter or selection matrix is used to select the variables observed in each subject. Therefore, the numbers of variables used for each subject can be different. More importantly, no cases with missing values are excluded (see Enders (2010) for more details). FIML handles missing values on the dependent variables. For example, Section 5.1.2 discusses how multivariate meta-analysis handles missing effect sizes with the ML estimation. The model in Figure 5.2a explicitly shows that the multiple effect sizes $y_{1,j}$ and $y_{2,j}$ are considered as dependent variables. Therefore, missing values are handled by FIML in the SEM framework.

On the other hand, cases with missing values on the independent variables (or moderators in a meta-analysis) will be deleted before the analysis. We may "trick" the SEM program to treat the predictors as dependent variables with latent variables (e.g., Enders, 2010). This is similar to the model in Figure 5.2a. Some SEM packages, for example, M*plus*, do this automatically by estimating the means and variances of independent variables with missing values (see Section 9.2.4 for an illustration in M*plus*).

Figure 8.3 shows a univariate mixed-effects meta-analysis with a moderator x_i in the ith study. As x_i is an independent variable, we create a latent variable η_i to trick the SEM program that x_i is a dependent variable. The factor loading from η_i to x_i is fixed at 1, while the error variance of x_i is set at 0. As η_i is an independent variable, we have to estimate its mean and variance. By using this setup, the mean and variance of η_i are exactly the same as those for x_i. In other words, η_i is equivalent to x_i exception that x_i is a dependent variable and η_i is an independent variable. β_1 is the regression coefficient by regressing the *true* effect size f_i on η_i.

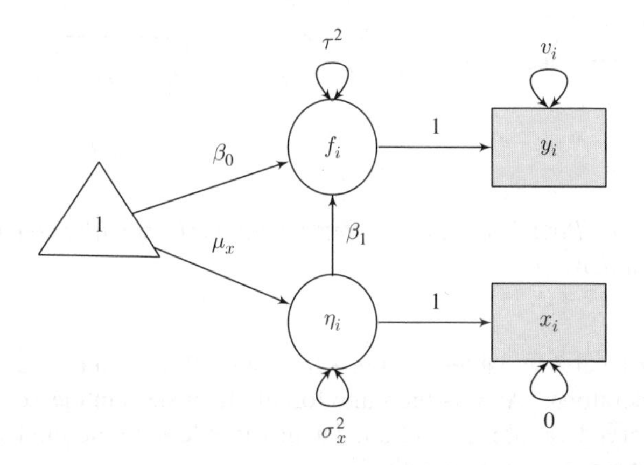

Figure 8.3 A univariate random-effects meta-analysis with a moderator.

When there is no missing values in x_i, this model is equivalent to the mixed-effects meta-analysis shown in Figure 4.6. As the model in Figure 8.3 requires additional latent variables to represent the moderators, it makes the model specification more complicated. Researchers seldom use this model unless there are lots of missing values in the predictors. When there are missing values in the moderators, however, the results between the models in Figures 8.3 and 4.6 can be different depending on the amount of missing values in the moderators.

We can extend the univariate model to multivariate model. Figure 8.4 displays a multivariate meta-analysis with two effect sizes per study ($y_{1,i}$ and $y_{2,i}$ in the ith study); x_i is the moderator with missing values. A latent variable η_i is created to represent the moderator x_i. As η_i is a variable, we estimate its mean and variance. $\beta_{1,1}$ and $\beta_{2,1}$ are the regression coefficients by regressing the *true* effect sizes $f_{1,i}$ and $f_{2,i}$ on η_i, respectively.

The above approach can be extended to handle missing values in moderators in a three-level meta-analysis. As the moderators are treated as variables rather than as a design matrix, the models are different depending on whether the moderators

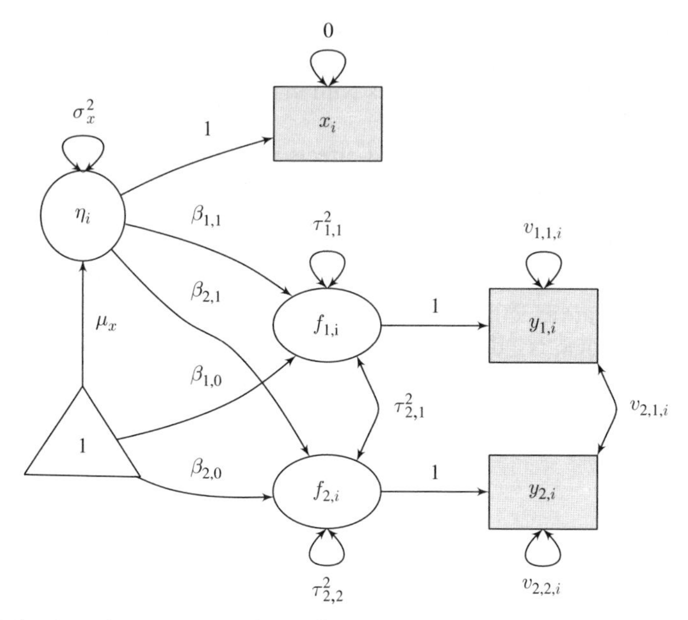

Figure 8.4 A multivariate random-effects meta-analysis with two effect sizes per study and a moderator.

are level-2 or level-3 variables. Figure 8.5 shows a three-level meta-analysis with a level-3 moderator x_j. As x_j is the same for all effect sizes in the j cluster, there is only one observed variable x_j and one latent variable η_j in the model. We regress the *true* effect size f_j on η_j at the jth cluster.

Figure 8.6 displays a three-level meta-analysis with a level-2 moderator. As the value of the moderator varies across the effect sizes in the jth cluster, $x_{1,j}$ and $x_{2,j}$ are created to store the level-2 moderator. Moreover, $\eta_{1,j}$ and $\eta_{2,j}$ are also created to represent the latent variables for $x_{1,j}$ and $x_{2,j}$, respectively. It should be noted that $\eta_{1,j}$ only predicts its correspondent effect size $y_{1,j}$.

8.3 Illustrations using R

This section demonstrates how to use R to implement the techniques discussed in this chapter. The first part illustrates how to analyze meta-analytic models with the REML estimation, while the second part shows how to handle missing values in the moderators in a mixed-effects meta-analysis. As the data sets have been introduced in previous chapters, they are not repeated here. The statistics reported in the illustrations were captured by using the `Sweave` function in R. The numbers of decimal places may be slightly different from those reported in the selected output and in the text.

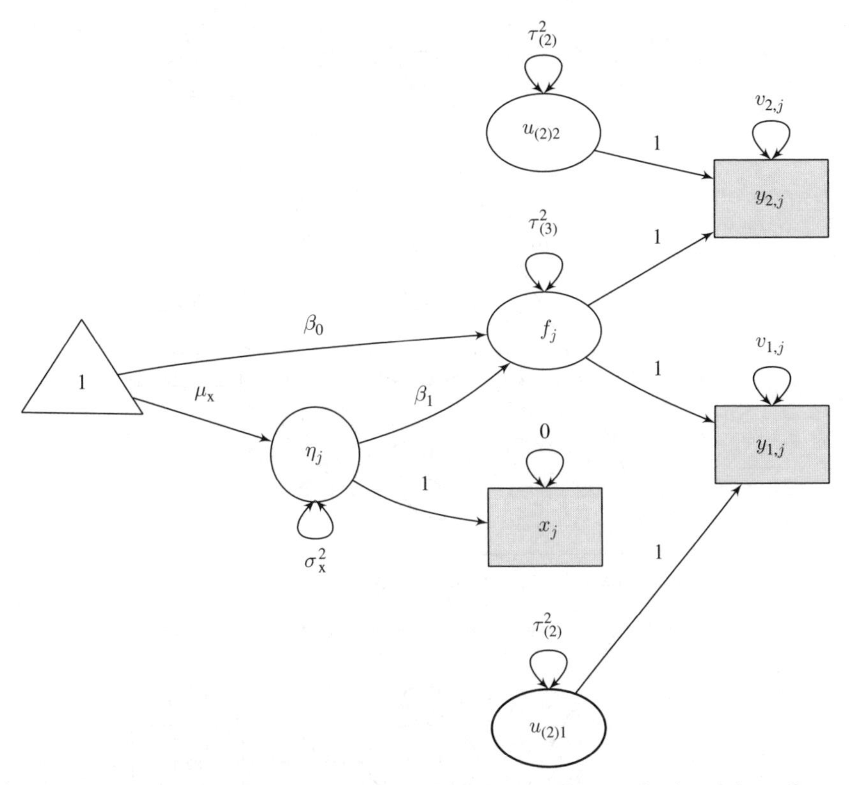

Figure 8.5 A three-level meta-analysis with two studies and a level-3 moderator in the jth cluster.

8.3.1 Restricted (or residual) maximum likelihood estimation

The `metaSEM` package (Cheung, 2014a) has implemented functions to conduct meta-analysis using the REML estimation. The `reml()` function can be used to conduct univariate and multivariate meta-analyses, while the `reml3()` function conducts three-level meta-analysis using the REML estimation. The syntax of `reml()` and `reml3()` are similar to those of `meta()` and `meta3()` functions using the ML estimation. For comparisons, we also report the parameter estimates using the ML estimation. It should be noted that both `reml()` and `reml3()` only estimates the variance components of the random effects. If we want to estimate the fixed effects, we need to use `meta()` or `meta3()` functions by treating the estimated variance components as known values.

8.3.1.1 Univariate meta-analysis

We illustrate the analyses using a data set from Jaramillo et al. (2005). The data set includes 61 studies on the relationship between organizational commitment

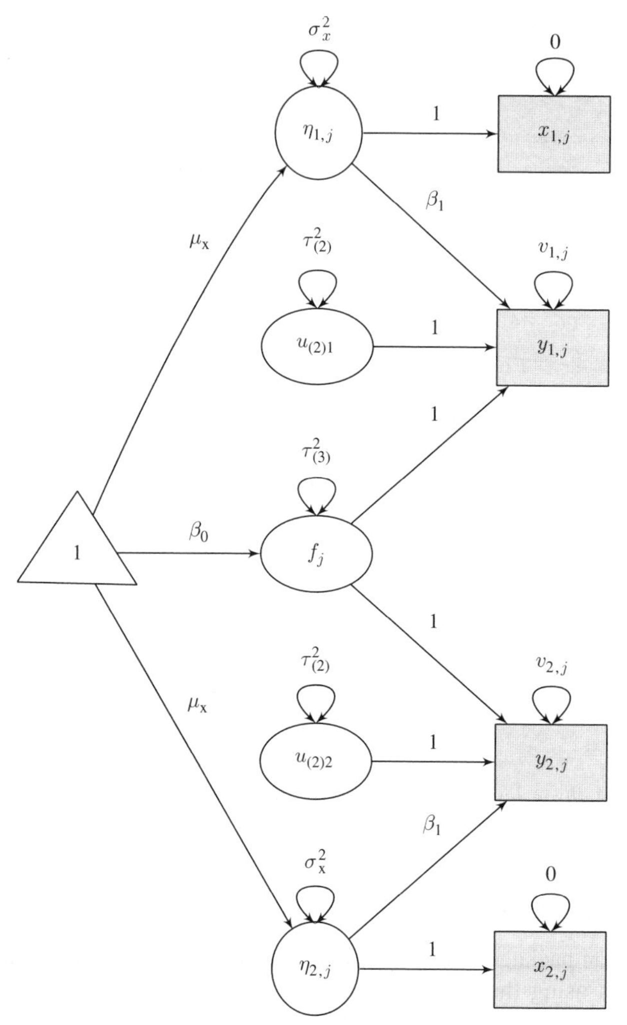

Figure 8.6 A three-level meta-analysis with two studies and a level-2 moderator in the jth cluster.

and salesperson job performance. The effect size was a correlation coefficient (see Section 4.7.2 for the analyses using the ML estimation). As the sampling distribution of the heterogeneity variance is hardly normally distributed unless the number of studies is huge, it is preferred to use a likelihood-based confidence interval (LBCI) to a Wald CI when investigating the heterogeneity variance.

```
R> ## Request LBCI: intervals.type="LB"
R> summary( jaramillo1 <- reml(y=r, v=r_v, intervals.type="LB",
                              data=Jaramillo05) )
```

```
------------------------ Selected output ------------------------

95% confidence intervals: Likelihood-based statistic
Coefficients:
         Estimate Std.Error lbound ubound z value Pr(>|z|)
Tau2_1_1   0.0174        NA 0.0108 0.0283      NA       NA

Number of studies (or clusters): 61
Number of observed statistics: 60
Number of estimated parameters: 1
Degrees of freedom: 59
-2 log likelihood: -159.7
OpenMx status1: 0 ("0" or "1": The optimization is considered fine.
Other values indicate problems.)

------------------------ Selected output ------------------------
```

The $\hat{\tau}^2_{\text{REML}}$ with its 95% LBCI is 0.0174 (0.0108, 0.0283), while the $\hat{\tau}^2_{\text{ML}}$ with its 95% LBCI is 0.0170 (0.0106, 0.0276). The $\hat{\tau}^2_{\text{REML}}$ is only slightly larger than that of $\hat{\tau}^2_{\text{ML}}$.

To calculate the estimate of the fixed effects, we extract the heterogeneity variance and treat it as a known value in calling the `meta()` function with the `RE.constraints` argument.

```
R> summary( meta(y=r, v=r_v, data=Jaramillo05,
                 RE.constraints=coef(jaramillo1)) )
```

```
------------------------ Selected output ------------------------

95% confidence intervals: z statistic approximation
Coefficients:
           Estimate Std.Error lbound ubound z value Pr(>|z|)
Intercept1   0.1866    0.0195 0.1484 0.2248    9.57   <2e-16 ***
---
Signif. codes:  0 '***' 0.001 '**' 0.01 '*' 0.05 '.' 0.1 ' ' 1

------------------------ Selected output ------------------------
```

The estimated average correlation with its 95% Wald CI based on the REML estimation is 0.1866 (0.1484, 0.2248), while the estimated average correlation using ML estimation is 0.1866 (0.1487, 0.2245). The estimated fixed effects and their Wald CIs are nearly identical under both REML and ML estimation methods.

8.3.1.2 Multivariate meta-analysis

We illustrate the multivariate meta-analysis using a data set from the World Values Survey II (World Values Study Group, 1994). It included standardized mean difference (SMD) between males and females on *life satisfaction* (SMD_{life_sat}) and *life control* (SMD_{life_con}) in 42 nations. Positive values on the effect sizes indicate that males have higher scores than females do (see Section 5.7.2 for the analyses using ML estimation).

```
R> summary( wvs1 <- reml(y=cbind(lifesat, lifecon),
                          v=cbind(lifesat_var, inter_cov, lifecon_var),
                          data=wvs94a) )
```

```
------------------------  Selected output  -----------------------

95% confidence intervals: z statistic approximation
Coefficients:
         Estimate Std.Error   lbound   ubound z value Pr(>|z|)
Tau2_1_1 0.004936  0.001831 0.001346 0.008525    2.70  0.00704 **
Tau2_2_1 0.004060  0.001751 0.000628 0.007492    2.32  0.02041 *
Tau2_2_2 0.008694  0.002631 0.003538 0.013851    3.30  0.00095 ***
---
Signif. codes:  0 '***' 0.001 '**' 0.01 '*' 0.05 '.' 0.1 ' ' 1

Number of studies (or clusters): 42
Number of observed statistics: 82
Number of estimated parameters: 3
Degrees of freedom: 79
-2 log likelihood: -299.3
OpenMx status1: 0 ("0" or "1": The optimization is considered fine.
Other values indicate problems.)

------------------------  Selected output  -----------------------
```

The estimated variance component $\hat{\boldsymbol{T}}^2$ using the REML and the ML estimation are $\begin{bmatrix} 0.0049 & \\ 0.0041 & 0.0087 \end{bmatrix}$ and $\begin{bmatrix} 0.0047 & \\ 0.0039 & 0.0084 \end{bmatrix}$, respectively. The estimated variance component using the REML is again slightly larger than that using the ML estimation.

To estimate the fixed effects, we need to extract the variance component and convert it into a matrix. The estimated variance component is treated as known values in calling the `meta()` function.

```
R> ## Extract the variance component of the random effects
R> ( RE <- vec2symMat(coef(wvs1)) )
```

```
         [,1]     [,2]
[1,] 0.004936 0.004060
[2,] 0.004060 0.008694
```

```
R> summary( meta(y=cbind(lifesat, lifecon),
                 v=cbind(lifesat_var, inter_cov, lifecon_var),
                 RE.constraints=RE, data=wvs94a) )
```

```
------------------------  Selected output  ----------------------

95% confidence intervals: z statistic approximation
Coefficients:
           Estimate Std.Error   lbound   ubound z value Pr(>|z|)
Intercept1  0.00124   0.01401 -0.02623  0.02870    0.09     0.93
Intercept2  0.06885   0.01701  0.03551  0.10218    4.05  5.2e-05 ***
---
Signif. codes:  0 '***' 0.001 '**' 0.01 '*' 0.05 '.' 0.1 ' ' 1

------------------------  Selected output  ----------------------
```

The estimated average effect sizes for SMD_{life_sat} and SMD_{life_con} (and their 95% Wald CIs) using the REML estimation are 0.0012 (−0.0262, 0.0287) and 0.0688 (0.0355, 0.1022), respectively, while the average effect sizes for SMD_{life_sat} and SMD_{life_con} (and their 95% Wald CIs) using the ML estimation are 0.0013 (−0.0258, 0.0285) and 0.0688 (0.0359, 0.1018), respectively. The results are nearly identical for both estimation methods.

8.3.1.3 Three-level meta-analysis

We illustrate the analyses using a data set from Bornmann et al. (2007) and Marsh et al. (2009). It consisted of 66 effect sizes from 21 studies of gender differences in the peer reviews of grant and fellowship applications. The effect size was a (log) odds ratio that measured the odds of being approved among the female applicants divided by the odds of being approved among the male applicants. If the effect size was positive, a female applicant was favored to receive the grant or fellowship, whereas if it was negative a male applicant was favored to obtain grant or fellowship. When we ran the analysis, the `OpenMx status1` was "6," indicating that the optimality conditions could not be reached. We reran the analysis with the `rerun()` function. It was then fine.

```
R> Bornmann1 <- reml3(y=logOR, v=v, cluster=Cluster, data=Bornmann07)
R> ## The OpenMx status1 is 6. We rerun the model.
R> summary( Bornmann2 <- rerun(Bornmann1) )
```

```
------------------------  Selected output  ----------------------

95% confidence intervals: z statistic approximation
Coefficients:
       Estimate Std.Error   lbound   ubound z value Pr(>|z|)
Tau2_2  0.00375   0.00268 -0.00151  0.00901    1.40     0.16
Tau2_3  0.01608   0.01040 -0.00430  0.03647    1.55     0.12

Number of studies (or clusters): 66
Number of observed statistics: 65
Number of estimated parameters: 2
Degrees of freedom: 63
-2 log likelihood: -89.1
```

```
OpenMx status1: 0 ("0" or "1": The optimization is considered fine.
Other values indicate problems.)

------------------------ Selected output ----------------------
```

The estimated level-2 heterogeneity variance $\hat{\tau}^2_{(2)}$ using the REML and the ML estimation are 0.0038 and 0.0038, respectively, while the estimated level-3 heterogeneity variance $\hat{\tau}^2_{(3)}$ using the REML and the ML estimation are 0.0161 and 0.0141, respectively. The estimated heterogeneity variances are similar under these two estimation methods.

We extract the heterogeneity variances as an R object RE that stores $\hat{\tau}^2_{(2)}$ and $\hat{\tau}^2_{(3)}$ as a vector. We then estimate the fixed effects by treating the estimated heterogeneity variances as known values.

```
R> ## Extract the level-2 and level-3 heterogeneity variances
R> ( RE <- coef(Bornmann2) )
```

```
  Tau2_2   Tau2_3
0.003753 0.016085
```

```
R> summary(meta3(y=logOR, v=v, cluster=Cluster, data=Bornmann07,
                 RE2.constraints=RE[1], RE3.constraints=RE[2]))
```

```
------------------------ Selected output ----------------------

95% confidence intervals: z statistic approximation
Coefficients:
          Estimate Std.Error  lbound  ubound z value Pr(>|z|)
Intercept  -0.1010    0.0417 -0.1828 -0.0192   -2.42    0.016 *
---
Signif. codes:  0 '***' 0.001 '**' 0.01 '*' 0.05 '.' 0.1 ' ' 1

------------------------ Selected output ----------------------
```

The estimated average effect sizes (and their 95% Wald CIs) using the REML and ML estimation are −0.1010(−0.1828, −0.0192) and −0.1008(−0.1794, −0.0221), respectively. Both estimates are very close to each other.

8.3.2 Missing values in the moderators

The `metaSEM` package has only implemented FIML estimation to handle missing values in the moderators for the three-level meta-analysis via the `meta3X()` function. When there are missing values in the moderators for univariate and multivariate meta-analyses, studies with missing values will be deleted before the analysis. Future versions of the `metaSEM` package will implement the FIML estimation for the univariate and multivariate meta-analyses. In this section, we demonstrate how the FIML estimation can be implemented in the `OpenMx` package. We first illustrate how to handle missing values in the moderators in the univariate and multivariate meta-analyses. Then we show how to use the `meta3X()` function to handle missing values in the moderators in three-level meta-analysis. Readers may refer to Enders (2010) for the empirical findings on various approaches in handling missing data.

8.3.2.1 Univariate meta-analysis

In this example, we conduct a mixed-effects meta-analysis by using the individualism (`IDV` in the data set `Jaramillo05`) as a moderator. We artificially create 20 missing values out of 61 studies under the assumption of MCAR. Then we analyze the data by using listwise deletion and FIML estimation.

```
R> library(metaSEM)
R> ## Set seed for replication
R> set.seed(1000000)
R> ## Create a copy of data
R> my.df1 <- Jaramillo05[, c("r", "r_v", "IDV")]
R> ## Create 20 missing data out of 61 studies with MCAR
R> my.df1$IDV[sample(1:61, 20)] <- NA
R> my.df1$IDV
```

```
 [1] 48 NA 91 67 91 91 NA 53 91 NA 91 NA 46 18 91 91 NA 89 NA NA 80 NA
[23] 91 91 NA 91 91 NA NA NA 20 NA 91 90 91 NA 91 NA 80 91 91 89 91 91
[45] 91 91 91 91 91 91 NA 91 91 NA 91 NA 38 91 NA 91 NA
```

```
R> ## Center the moderator
R> my.df1$IDV <- scale(my.df1$IDV, scale=FALSE)
```

```
R> ## Run the analysis by using the listwise deletion (the default)
R> summary( meta(y=r, v=r_v, x=IDV, data=my.df1) )
```

```
-----------------------  Selected output  ----------------------

95% confidence intervals: z statistic approximation
Coefficients:
            Estimate Std.Error    lbound    ubound z value Pr(>|z|)
Intercept1  0.180848  0.024316  0.133190  0.228506    7.44    1e-13
Slope1_1   -0.001702  0.001201 -0.004056  0.000651   -1.42  0.15632
Tau2_1_1    0.017789  0.005373  0.007257  0.028320    3.31  0.00093

Intercept1 ***
Slope1_1
Tau2_1_1   ***
```

```
---
Signif. codes:  0 '***' 0.001 '**' 0.01 '*' 0.05 '.' 0.1 ' ' 1

Q statistic on the homogeneity of effect sizes: 245.1
Degrees of freedom of the Q statistic: 40
P value of the Q statistic: 0

Explained variances (R2):
                          y1
Tau2 (no predictor)     0.02
Tau2 (with predictors) 0.02
R2                      0.00

Number of studies (or clusters): 41
Number of observed statistics: 41
Number of estimated parameters: 3
Degrees of freedom: 38
-2 log likelihood: -34.79
OpenMx status1: 0 ("0" or "1": The optimization is considered fine.

------------------------ Selected output -----------------------
```

As shown in the above output, there were only 41 studies in the analysis with the listwise deletion. The estimated slope and its SE and p value for the listwise deletion are −0.0017 (SE = 0.0012) and $p = 0.1563$, respectively.

We fit the model in Figure 8.3 that handles missing values in the moderator. The following `OpenMx` code may be used to conduct the analysis. The variables are arranged as `r` (the observed effect size), `IDV` (the moderator), `f` (the *true* effect size), and `eta` (the latent variable for `IDV`). The `A1` matrix specifies the fixed factor loadings and the regression coefficient (β_1 in Figure 8.3 and `Slope1_1` in the R code). The `S1` matrix specified the symmetric covariance matrix of the variables. The known sampling variance v_i is fixed via the definition variable with the label `data.r_v`. The heterogeneity variance of the *true* effect size and the variance of `IDV` are `Tau2_1_1` and `VarIDV`, respectively. The intercept of the *true* effect size and the mean of `IDV` are `Intercept1` and `MeanIDV`, respectively. A selection matrix `F1` is required to select the observed variables `r` and `IDV`.

```
R> ## Create an A matrix for the asymmetric paths
R> A1 <- matrix(c(0,0,1,0,
                  0,0,0,1,
                  0,0,0,"0*Slope1_1",
                  0,0,0,0), byrow=TRUE, ncol=4)
R> dimnames(A1) <- list(c("r","IDV","f","eta"),
                        c("r","IDV","f", "eta"))
R> A1
```

```
    r   IDV f   eta
r   "0" "0" "1" "0"
```

```
IDV "0" "0" "0" "1"
f   "0" "0" "0" "0*Slope1_1"
eta "0" "0" "0" "0"
```

```
R> ## Convert the A matrix into mxMatrix class
R> A1 <- as.mxMatrix(A1)
R> ## Create an S matrix for the covariance matrix among variables
R> S1 <- mxMatrix("Symm", nrow=4, ncol=4, values=0, byrow=TRUE,
                  free=c(FALSE,
                         FALSE, FALSE,
                         FALSE, FALSE, TRUE,
                         FALSE, FALSE, FALSE, TRUE),
                  labels=c("data.r_v",
                           NA, NA,
                           NA, NA, "Tau2_1_1",
                           NA, NA, NA, "VarIDV"),
                  name="S1")
R> S1@labels
```

```
     [,1]       [,2] [,3]       [,4]
[1,] "data.r_v" NA   NA         NA
[2,] NA         NA   NA         NA
[3,] NA         NA   "Tau2_1_1" NA
[4,] NA         NA   NA         "VarIDV"
```

```
R> ## Create an M matrix for the means
R> M1 <- matrix(c(0,0,"0*Intercept1","300*MeanIDV"), nrow=1)
R> dimnames(M1)[[2]] <- c("r","IDV","f","eta")
R> M1
```

```
     r   IDV f              eta
[1,] "0" "0" "0*Intercept1" "300*MeanIDV"
```

```
R> M1 <- as.mxMatrix(M1)
R> ## Create an F matrix to selecting the observed variables
R> F1 <- create.Fmatrix(c(1,1,0,0), name="F", as.mxMatrix=FALSE)
R> dimnames(F1) <- list(c("r","IDV"), c("r","IDV","f","eta"))
R> F1
```

```
    r IDV f eta
r   1   0 0   0
IDV 0   1 0   0
```

```
R> F1 <- as.mxMatrix(F1)
R> ## Create an mx model
R> uni.MCAR <- mxModel("MCAR",
```

```
                            mxData(my.df1, type="raw"),
                            A1, S1, F1, M1,
                            mxExpectationRAM("A1","S1","F1","M1",
                                     dimnames=c("r","IDV","f","eta")),
                            mxFitFunctionML())
```

```
R> ## Run the analysis
R> summary(mxRun(uni.MCAR))
```

```
------------------------ Selected output -----------------------

free parameters:
        name matrix row col   Estimate Std.Error lbound ubound
1   Slope1_1     A1   3   4  -0.001578  0.001103
2   Tau2_1_1     S1   3   3   0.016009  0.004029
3     VarIDV     S1   4   4 386.805744 85.148006
4 Intercept1     M1   1   f   0.185984  0.019106
5    MeanIDV     M1   1 eta  -0.134300  3.048705

observed statistics:  102
estimated parameters:  5
degrees of freedom:  97
-2 log likelihood:  303.4
number of observations:  61

------------------------ Selected output -----------------------
```

The estimated slope and its SE and p value for the FIML estimation is -0.0016 (SE $= 0.0011$) and $p = 0.1525$, respectively. The results are similar to those with listwise deletion in this example.

8.3.2.2 Multivariate meta-analysis

This example extends the univariate meta-analysis to the multivariate meta-analysis with missing values in the moderator. We fit the model in Figure 8.4 with the sample data set `wvs94a`. As there are already a few missing values in `GNP`, we do not need to introduce new missing values to the data. The variables in the model are arranged as `lifesat`, `lifecon`, `gnp`, `f1` (the *true* effect size of `lifesat`), `f2` (the *true* effect size of `lifecon`), and `eta` (the latent variable for `gnp`).

The `A2` matrix specifies the fixed factor loadings and the regression coefficients `Slope1_1` and `Slope2_1`. The `S2` matrix specifies the symmetric covariance among the variables. The known conditional sampling covariance matrix between `lifesat` and `lifecon` are imposed via the definition variables with the labels `data.lifesat_var`, `data.inter_cov`, and `data.lifecon_var`, while the covariance matrix of the random effects are `Tau2_1_1`, `Tau2_2_1`, and `Tau2_2_2`.

```
R> ## Create a copy of the original data by excluding the country
R> my.df2 <- wvs94a[, -1]
R> ## Center the predictor
R> my.df2$gnp <- scale(my.df2$gnp/10000, scale=FALSE)
R> ## Create the matrix for regression coefficients
R> A2 <- matrix(c(0,0,0,1,0,0,
                  0,0,0,0,1,0,
                  0,0,0,0,0,1,
                  0,0,0,0,0,"0*Slope1_1",
                  0,0,0,0,0,"0*Slope2_1",
                  0,0,0,0,0,0), byrow=TRUE, ncol=6)
R> dimnames(A2) <- list(c("lifesat","lifecon","gnp","f1",
                         "f2","eta"),
                       c("lifesat","lifecon","gnp","f1","f2","eta"))
R> ## Display the content of A2
R> A2
```

```
        lifesat lifecon gnp f1  f2  eta
lifesat "0"     "0"     "0" "1" "0" "0"
lifecon "0"     "0"     "0" "0" "1" "0"
gnp     "0"     "0"     "0" "0" "0" "1"
f1      "0"     "0"     "0" "0" "0" "0*Slope1_1"
f2      "0"     "0"     "0" "0" "0" "0*Slope2_1"
eta     "0"     "0"     "0" "0" "0" "0"
```

```
R> ## Convert A2 into mxMatrix class
R> A2 <- as.mxMatrix(A2)
R> ## Symmetric matrix for the variables
R> S2 <- mxMatrix("Symm", nrow=6, ncol=6, byrow=TRUE, values=0,
                 free=c(FALSE,
                        FALSE,FALSE,
                        FALSE,FALSE,FALSE,
                        FALSE,FALSE,FALSE,TRUE,
                        FALSE,FALSE,FALSE,TRUE,TRUE,
                        FALSE,FALSE,FALSE,FALSE,FALSE,TRUE),
                 labels=c("data.lifesat_var",
                          "data.inter_cov","data.lifecon_var",
                          NA,NA,NA,
                          NA,NA,NA,"Tau2_1_1",
                          NA,NA,NA,"Tau2_2_1","Tau2_2_2",
                          NA,NA,NA,NA,NA,"VarGNP"),
                 name="S2")
R> S2@labels
```

```
     [,1]               [,2]               [,3] [,4]       [,5]
[1,] "data.lifesat_var" "data.inter_cov"   NA   NA         NA
[2,] "data.inter_cov"   "data.lifecon_var" NA   NA         NA
[3,] NA                 NA                 NA   NA         NA
[4,] NA                 NA                 NA   "Tau2_1_1" "Tau2_2_1"
[5,] NA                 NA                 NA   "Tau2_2_1" "Tau2_2_2"
```

```
[6,] NA                  NA                  NA   NA          NA
     [,6]
[1,] NA
[2,] NA
[3,] NA
[4,] NA
[5,] NA
[6,] "VarGNP"
```

```
R> ## Create the vector for means
R> M2 <- matrix(c(0,0,0,"0*Intercept1","0*Intercept2","0*MeanGNP"),
                nrow=1)
R> dimnames(M2)[[2]] <- c("lifesat","lifecon","gnp","f1","f2","eta")
R> M2
```

```
     lifesat lifecon gnp f1             f2             eta
[1,] "0"     "0"     "0" "0*Intercept1" "0*Intercept2" "0*MeanGNP"
```

```
R> M2 <- as.mxMatrix(M2)
R> ## Create a selection matrix
R> F2 <- create.Fmatrix(c(1,1,1,0,0,0), name="F2",
                         as.mxMatrix=FALSE)
R> dimnames(F2) <- list(c("lifesat","lifecon","gnp"),
                        c("lifesat","lifecon","gnp","f1","f2","eta"))
R> F2
```

```
        lifesat lifecon gnp f1 f2 eta
lifesat       1       0   0  0  0   0
lifecon       0       1   0  0  0   0
gnp           0       0   1  0  0   0
```

```
R> F2 <- as.mxMatrix(F2)
R> ## Create a model
R> multi.MCAR <- mxModel("MCAR",
                         mxData(my.df2, type="raw"),
                         A2, S2, F2, M2,
                         mxExpectationRAM("A2","S2","F2","M2",
                         dimnames=c("lifesat","lifecon","gnp",
                                    "f1","f2","eta")),
                         mxFitFunctionML())
```

```
R> summary(mxRun(multi.MCAR))
```

```
------------------------  Selected output  -----------------------

free parameters:
```

```
         name matrix row col  Estimate Std.Error lbound ubound
1    Slope1_1     A2   4   6 -0.022600  0.014734
2    Slope2_1     A2   5   6 -0.035986  0.017494
3    Tau2_1_1     S2   4   4  0.004311  0.001654
4    Tau2_2_1     S2   4   5  0.003244  0.001536
5    Tau2_2_2     S2   5   5  0.007247  0.002298
6      VarGNP     S2   6   6  0.915137  0.211776
7 Intercept1     M2   1  f1  0.001467  0.013532
8 Intercept2     M2   1  f2  0.068946  0.016058
9     MeanGNP     M2   1 eta  0.004477  0.156339

observed statistics:  121
estimated parameters:  9
degrees of freedom:  112
-2 log likelihood:  -64.4
number of observations:  42

-----------------------  Selected output  ----------------------
```

The estimated regression coefficients and their SEs in predicting the average $f_{\text{life\_sat}}$ and $f_{\text{life\_con}}$ are -0.0226 (SE $= 0.0147$) and -0.0360 (SE $= 0.0175$), respectively. Both the parameter estimates and their SEs are similar to those listed in Section 5.7.2; however, the SEs appear to be slightly smaller for those based on the FIML estimation.

8.3.2.3 Three-level meta-analysis

This example demonstrates how to use the `meta3X()` function to handle missing values in the moderators in a three-level meta-analysis. We used the data set from Bornmann et al. (2007) and Marsh et al. (2009) that has also been used in Section 6.5. We first create some missing values in `Type` with the assumption of MCAR.

```
R> ## Set seed for replication
R> set.seed(1000000)
R> ## Create a copy of Type
R> ## "Fellowship": 1; "Grant": 0
R> Type_MCAR <- ifelse(Bornmann07$Type=="Fellowship", yes=1, no=0)
R> ## Create 17 missing values out of 66 studies with MCAR
R> Type_MCAR[sample(1:66, 17)] <- NA
R> ## Display the content
R> Type_MCAR
```

```
 [1] 1 NA  1  1  1  1 NA  0  0  0 NA  0 NA  0  0  1  1 NA NA  1 NA  1
[23] 0 NA  1  0  1  1  1  1 NA  1 NA  1 NA NA  0  0 NA  0  0 NA  0  0
[45] 0  1  1  1  0  0  1  0  1  0  0 NA  0 NA  0  0  0  0  0  0  0 NA
```

We conduct a mixed-effects meta-analysis with `Type_MCAR` as the moderator. When there are missing values in the moderators, these data will be excluded from the analysis.

```
R> summary( meta3(y=logOR, v=v, cluster=Cluster, x=Type_MCAR,
                 data=Bornmann07) )
```

```
------------------------  Selected output  ----------------------

Coefficients:
            Estimate Std.Error    lbound     ubound z value Pr(>|z|)
Intercept -0.004845  0.039344 -0.081959  0.072268   -0.12     0.90
Slope_1   -0.210901  0.053462 -0.315685 -0.106117   -3.94    8e-05
Tau2_2     0.004468  0.005493 -0.006298  0.015234    0.81     0.42
Tau2_3     0.000929  0.003365 -0.005666  0.007524    0.28     0.78

Intercept
Slope_1   ***
Tau2_2
Tau2_3
---
Signif. codes:  0 '***' 0.001 '**' 0.01 '*' 0.05 '.' 0.1 ' ' 1

Q statistic on the homogeneity of effect sizes: 151.6
Degrees of freedom of the Q statistic: 48
P value of the Q statistic: 1.116e-12

Explained variances (R2):
                       Level 2 Level 3
Tau2 (no predictor)    0.00427    0.01
Tau2 (with predictors) 0.00447    0.00
R2                     0.00000    0.94

Number of studies (or clusters): 20
Number of observed statistics: 49
Number of estimated parameters: 4
Degrees of freedom: 45
-2 log likelihood: 13.14
OpenMx status1: 0 ("0" or "1": The optimization is considered fine.

------------------------  Selected output  ----------------------
```

The syntax of the `meta3X()` function is similar to that of the `meta3X()` function except for two differences. As the moderators are treated as variables rather than as a design matrix in `meta3X()`, there are separate arguments for the level-2 moderators (`x2`) and the level-3 moderator (`x3`). Moreover, there are also two arguments `av2` and `av3` for the auxiliary variables at level 2 and level 3, respectively. Auxiliary variables are variables that are predictive to the missing values or correlated with the variables with missing values. The estimation will be more efficient by including the auxiliary variables (see Enders (2010) for details). As `Type_MCAR` is a level-2 moderator, we specify the argument `x2=Type_MCAR`.

```
R> summary( meta3X(y=logOR, v=v, cluster=Cluster, x2=Type_MCAR,
                   data=Bornmann07) )
```

```
------------------------ Selected output ----------------------

95% confidence intervals: z statistic approximation
Coefficients:
           Estimate Std.Error   lbound    ubound z value Pr(>|z|)
Intercept -0.01063   0.03977 -0.08858  0.06731   -0.27   0.7892
SlopeX2_1 -0.17532   0.05826 -0.28952 -0.06113   -3.01   0.0026 **
Tau2_2     0.00303   0.00268 -0.00223  0.00829    1.13   0.2583
Tau2_3     0.00368   0.00428 -0.00471  0.01208    0.86   0.3896
---
Signif. codes:  0 '***' 0.001 '**' 0.01 '*' 0.05 '.' 0.1 ' ' 1

Explained variances (R2):
                        Level 2 Level 3
Tau2 (no predictor)     0.00380    0.01
Tau2 (with predictors)  0.00303    0.00
R2                      0.20091    0.74

Number of studies (or clusters): 21
Number of observed statistics: 115
Number of estimated parameters: 7
Degrees of freedom: 108
-2 log likelihood: 49.76
OpenMx status1: 0 ("0" or "1": The optimization is considered fine.
Other values indicate problems.)

------------------------ Selected output ----------------------
```

The estimated regression coefficients and their SEs for the listwise deletion and the FIML estimation are −0.2109 (SE = 0.0535) and −0.1753 (SE = 0.0583), respectively. The estimated $R^2_{(2)}$ for the listwise deletion and the FIML estimation are 0.0000 and 0.2009, respectively, while the estimated $R^2_{(3)}$ for the listwise deletion and the FIML estimation are 0.9361 and 0.7394, respectively. The results are similar for these two approaches in this empirical example.

8.4 Concluding remarks and further readings

This chapter discussed the pros and cons of the ML and REML estimation and how the REML estimation could be implemented in SEM. A graphical model was proposed to show how the transformation matrix in the REML estimation could be considered as the fixed patterned factor loadings in SEM. This helps to formulate the REML estimation as a constrained confirmatory factor analytic model in SEM. This chapter also introduced the FIML estimation to handle missing value in the moderators in the SEM-based meta-analysis. The `metaSEM` and `OpenMx` packages were demonstrated to handle missing data in univariate, multivariate, and three-level meta-analyses. Although FIML and MI are generally preferred to handle missing data in applied research, limited studies have addressed the empirical performance of these techniques in meta-analysis. Future research should fill this research gap by comparing the performance of FIML and other methods in meta-analysis.

References

Bentler PM 2006. *EQS 6 structural equations program manual*. Multivariate Software, Encino, CA.

Boker S, Neale M, Maes H, Wilde M, Spiegel M, Brick T, Spies J, Estabrook R, Kenny S, Bates T, Mehta P and Fox J 2011. OpenMx: an open source extended structural equation modeling framework. *Psychometrika* **76**(2), 306–317.

Bornmann L, Mutz R and Daniel HD 2007. Gender differences in grant peer review: a meta-analysis. *Journal of Informetrics* **1**(3), 226–238.

Cheung MWL 2004. A direct estimation method on analyzing ipsative data with Chan and Bentler's (1993) method. *Structural Equation Modeling: A Multidisciplinary Journal* **11**(2), 217–243.

Cheung MWL 2006. Recovering preipsative information from additive ipsatized data a factor score approach. *Educational and Psychological Measurement* **66**(4), 565–588.

Cheung MWL 2007. Comparison of methods of handling missing time-invariant covariates in latent growth models under the assumption of missing completely at random. *Organizational Research Methods* **10**(4), 609–634.

Cheung MWL 2008. A model for integrating fixed-, random-, and mixed-effects meta-analyses into structural equation modeling. *Psychological Methods* **13**(3), 182–202.

Cheung MWL 2013. Implementing restricted maximum likelihood estimation in structural equation models. *Structural Equation Modeling: A Multidisciplinary Journal* **20**(1), 157–167.

Cheung MWL 2014a. metaSEM: An R package for meta-analysis using structural equation modeling. Frontiers in Psychology 5(1521).

Cheung MWL 2014b. Modeling dependent effect sizes with three-level meta-analyses: a structural equation modeling approach. *Psychological Methods* **19**(2), 211–229.

Cheung MWL and Chan W 2002. Reducing uniform response bias with ipsative measurement in multiple-group confirmatory factor analysis. *Structural Equation Modeling: A Multidisciplinary Journal* **9**(1), 55–77.

Cooper H and Hedges LV 2009. Potentials and limitations In *The handbook of research synthesis and meta-analysis* (ed. Cooper H, Hedges LV and Valentine JC), 2nd edn. Russell Sage Foundation, New York, pp. 561–572.

Demidenko E 2013. *Mixed models: theory and applications with R*, 2nd edn. Wiley-Interscience, Hoboken, NJ.

DerSimonian R and Laird N 1986. Meta-analysis in clinical trials. *Controlled Clinical Trials* **7**(3), 177–188.

Enders CK 2010. *Applied missing data analysis*. Guilford Press, New York.

Graham J 2009. Missing data analysis: making it work in the real world. *Annual Review of Psychology* **60**, 549–576.

Graham J, Hofer S and MacKinnon D 1996. Maximizing the usefulness of data obtained with planned missing value patterns: an application of maximum likelihood procedures. *Multivariate Behavioral Research* **31**(2), 197–218.

Graham J, Olchowski A and Gilreath T 2007. How many imputations are really needed? Some practical clarifications of multiple imputation theory. *Prevention Science* **8**(3), 206–213.

Harville DA 1977. Maximum likelihood approaches to variance component estimation and to related problems. *Journal of the American Statistical Association* **72**(358), 320–338.

Jamshidian M and Bentler PM 1999. ML estimation of mean and covariance structures with missing data using complete data routines. *Journal of Educational and Behavioral Statistics* **24**(1), 21–24.

Jaramillo F, Mulki JP and Marshall GW 2005. A meta-analysis of the relationship between organizational commitment and salesperson job performance: 25 years of research. *Journal of Business Research* **58**(6), 705–714.

Jöreskog KG and Sörbom D 1996. *LISREL 8: a user's reference guide*. Scientific Software International, Inc., Chicago, IL.

Little RJA 1992. Regression with missing x's: a review. *Journal of the American Statistical Association* **87**(420), 1227–1237.

Marsh HW, Bornmann L, Mutz R, Daniel HD and O'Mara A 2009. Gender effects in the peer reviews of grant proposals: a comprehensive meta-analysis comparing traditional and multilevel approaches. *Review of Educational Research* **79**(3), 1290–1326.

Mehta PD and Neale MC 2005. People are variables too: multilevel structural equations modeling. *Psychological Methods* **10**(3), 259–284.

Millar RB 2011. *Maximum likelihood estimation and inference: with examples in R, SAS, and ADMB*. John Wiley & Sons, Inc., Hoboken, NJ.

Muller KE and Stewart PW 2006. *Linear model theory: univariate, multivariate, and mixed models*. Wiley-Interscience, Hoboken, NJ.

Muthén B, Kaplan D and Hollis M 1987. On structural equation modeling with data that are not missing completely at random. *Psychometrika* **52**(3), 431–462.

Muthén BO and Muthén LK 2012. *Mplus user's guide*, 7th edn. Muthén & Muthén, Los Angeles, CA.

Patterson HD and Thompson R 1971. Recovery of inter-block information when block sizes are unequal. *Biometrika* **58**(3), 545–554.

Pigott T 2001. Missing predictors in models of effect size. *Evaluation & the Health Professions* **24**(3), 277–307.

Pigott TD 2009. Handling missing data In *The handbook of research synthesis and meta-analysis* (ed. Cooper H, Hedges LV and Valentine JC), 2nd edn. Russell Sage Foundation New York, pp. 399–416.

Pigott TD 2012. *Advances in meta-analysis*. Springer-Verlag, New York.

Rothstein HR, Sutton AJ and Borenstein M 2005. *Publication bias in meta-analysis: prevention, assessment and adjustments*. John Wiley & Sons, Ltd, Chichester.

Rubin DB 1976. Inference and missing data. *Biometrika* **63**(3), 581–592.

Schafer JL 1997. *Analysis of incomplete multivariate data*. Chapman & Hall, London.

Schafer JL and Graham JW 2002. Missing data: our view of the state of the art. *Psychological Methods* **7**(2), 147–177.

Schafer JL and Olsen MK 1998. Multiple imputation for multivariate missing-data problems: a data analyst's perspective. *Multivariate Behavioral Research* **33**(4), 545–571.

Searle SR, Casella G and McCulloch CE 1992. *Variance components*. Wiley-Interscience, New York.

Sutton AJ 2000. *Methods for meta-analysis in medical research*. John Wiley & Sons, Ltd, Chichester.

Sutton AJ and Pigott TD 2005. Bias in meta-analysis: induced by incompletely reported studies In *Publication bias in meta-analysis: prevention, assessment and adjustments* (ed. Rothstein HR, Sutton AJ and Borenstein M). John Wiley & Sons, Ltd, Chichester, pp. 223–239.

van Buuren S and Groothuis-Oudshoorn K 2011. MICE: multivariate imputation by chained equations in R. *Journal of Statistical Software* **45**(3), 1–67.

Viechtbauer W 2005. Bias and efficiency of meta-analytic variance estimators in the random-effects model. *Journal of Educational and Behavioral Statistics* **30**(3), 261–293.

World Values Study Group 1994. *World Values Survey, 1981–1984 and 1990–1993 [Computer file]*. Inter-university Consortium for Political and Social Research, Ann Arbor, MI.

Yuan KH, Yang-Wallentin F and Bentler PM 2012. ML versus MI for missing data with violation of distribution conditions. *Sociological Methods & Research* **41**(4), 598–629.

9

Conducting meta-analysis with M*plus*

Most users of structural equation modeling (SEM) are familiar with at least one popular SEM package, such as M*plus*, `LISREL`, or `EQS`. This chapter illustrates how to analyze the meta-analytic models introduced in previous chapters with M*plus*. We show how M*plus* can be used to conduct univariate, multivariate, and three-level meta-analyses using a transformed variables approach. Although we use M*plus* in this chapter, the proposed transformed variables approach can also be applied to other SEM packages to conduct some of the SEM-based meta-analysis.

9.1 Introduction

M*plus* (Muthén and Muthén, 2012) provides a unified framework to conduct data analyses using a latent variable modeling approach. It combines SEM, multilevel models, mixture modeling, survival analysis, latent class models, categorical variables, missing data analysis, item response theory models, robust test statistics, and Bayesian analysis into a single statistical modeling framework. The main strength of a unified framework to conduct data analysis is that some of these techniques can be combined together to address the research questions. For example, researchers may handle the missing data with maximum likelihood (ML) estimation method and the nonnormal data with robust statistics in the same analysis. It will be beneficial to methodologists and applied researchers if meta-analytic models can also be integrated as part of the SEM framework. The combination of meta-analysis and SEM allows researchers to easily apply advanced statistical methods (see Cheung (2008) for a discussion). For example, researchers may apply ML estimation to handle missing values in the moderators in a mixed-effects meta-analysis (see Section 8.2.3); Bayesian approach may also be used in the M*plus* framework

Meta-Analysis: A Structural Equation Modeling Approach, First Edition. Mike W. -L. Cheung.

Companion Website: www.wiley.com/go/cheung/meta_analysis

(Muthén and Muthén, 2012). M*plus* was chosen for the illustrations in this chapter because of its capability to handle random slopes and multilevel data, whereas other SEM packages without the random slopes or definition variables can only be used to conduct fixed-effects meta-analyses. We first introduce the univariate meta-analysis in the next section. Then we extend the methods to multivariate meta-analysis and three-level meta-analysis.

9.2 Univariate meta-analysis

Chapter 4 shows how a univariate meta-analysis can be analyzed as a structural equation model using the definition variables. Specifically, we fix the known sampling variance in a meta-analysis as the measurement error variance using the definition variable implemented in the `OpenMx` package (Boker et al., 2011). As most SEM packages do not have the definition variables, this approach cannot be used in other SEM packages. Cheung (2008) proposed an alternative model to conduct meta-analyses in SEM. In this section, we first show how a fixed-effects meta-analysis can be analyzed in M*plus*.

9.2.1 Fixed-effects model

A fixed-effects model in Equation 4.5 is repeated here for the ease of reference,

$$y_i = \beta_{\mathrm{F}} + e_i, \tag{9.1}$$

where y_i is the effect size, $\mathsf{Var}(e_i) = v_i$ is the known sampling variance in the ith study, and β_{F} is the common effect size under the fixed-effects model. As most SEM packages do not allow fixing values at the subject level, it is not trivial to implement the fixed-effects model in SEM. Cheung (2008) noticed that we may transform the known sampling variances to a common value of 1 by multiplying the above model by $\sqrt{w_i} = 1/\sqrt{v_i}$. This approach is termed the *transformed variables approach* here.

The fixed-effects model in Equation 9.1 based on the transformed variables approach is

$$\begin{aligned} \sqrt{w_i}y_i &= \sqrt{w_i}\beta_{\mathrm{F}} + \sqrt{w_i}e_i, \\ \tilde{y}_i &= \sqrt{w_i}\beta_{\mathrm{F}} + \tilde{e}_i, \end{aligned} \tag{9.2}$$

where $\tilde{y}_i = \sqrt{w_i}y_i$ and $\tilde{e}_i = \sqrt{w_i}e_i$. Under this model parameterization, $\sqrt{w_i}$ is a predictor without the intercept. It is easy to show that the sampling variance of $\tilde{e}_i$ is distributed with a known variance of 1:

$$\begin{aligned} \mathsf{Var}(\tilde{e}_i) &= \sqrt{w_i}\mathsf{Var}(e_i)\sqrt{w_i}, \\ &= w_i v_i, \\ &= 1 \quad \text{as} \quad w_i = 1/v_i. \end{aligned} \tag{9.3}$$

Now, the transformed error $\tilde{e}_i$ is independent and identically distributed (i.i.d.) with a unit variance. Ordinary least squares (OLS) and SEM can be directly applied to analyze the data. The weighted least squares (WLS) fit function in Equation 4.6 can be rewritten as an OLS fit function,

$$F_{\text{OLS}} = \sum_{i=1}^{k} \left(\tilde{y}_i - \sqrt{w_i}\beta_{\text{F}}\right)^2, \tag{9.4}$$

where k is the number of studies. Several researchers, for example, Kalaian and Raudenbush (1996) and Raudenbush et al. (1988), suggested this "trick" to conduct a meta-analysis with conventional regression software.

The parameter estimates based on an OLS regression are equivalent to those in a meta-analysis with the WLS fit function. However, the standard errors (SEs) reported by the OLS regression are incorrect. The correct SE ($\text{SE}_{\text{correct}}$) can be obtained by $\text{SE}_{\text{correct}} = \text{SE}_{\text{reported}}/\sqrt{\text{MS}_e}$, where $\text{SE}_{\text{reported}}$ is the reported SE and MS_e is the mean square error (see Hedges and Olkin, 1985). The reason for the correction is that the error variance is assumed to be fixed at 1. However, the error variance is still estimated in a WLS or an OLS regression. It is rarely to be exactly 1. By using the above correction factor, we adjust the SE by fixing the error variance to the known value. The above model can be implemented in M*plus* or other SEM packages to conduct a fixed-effects model. As we can fix $\text{Var}(\tilde{e}_i) = 1$ in the SEM packages, the reported SEs in SEM are already correct without the correction.

In a conventional fixed-effects meta-analysis represented by a graphical model (see Figure 4.4), there is only one parameter for the intercept, which is usually represented by an triangle. Figure 9.1 shows a fixed-effects meta-analysis based on the transformed variables approach in M*plus*. There are some issues that required further elaborations. First, the error variance of $\tilde{y}_i$ is fixed at 1. This ensures that the reported SE is correct without any correction. Second, $\sqrt{w_i}$ is considered as an independent variable in the model. Therefore, the mean and variance of $\sqrt{w_i}$ are estimated in the analysis. To simplify the figures, the means, variances, and covariances of the independent variables, for example, $\sqrt{w_i}$ of the transformed constants, are not shown in this chapter. Third, the estimated common effect β_{F} is represented by the regression coefficient from $\sqrt{w_i}$ to $\tilde{y}_i$. As the intercept of $\tilde{y}_i$ is explicitly fixed at 0, there is no intercept represented by the triangle in the figure.

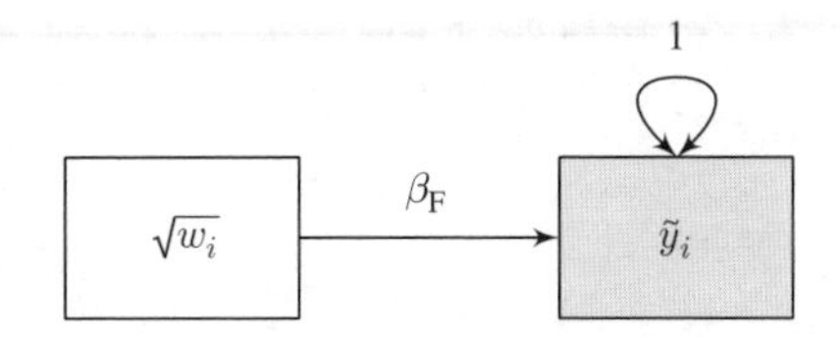

*Figure 9.1 A univariate fixed-effects meta-analytic model in M*plus.

We illustrate the analysis with the example of Jaramillo et al. (2005), which has been used in Section 4.7.2. The data file was exported as a plain text file

`Jaramillo05.dat`. The effect size was a correlation coefficient indicating the relationship between organizational commitment and salesperson job performance. The following box shows the first few cases in `Jaramillo05.dat`. The variables are `id`, `r`, `r_v`, `IDV`, `OC_alpha`, and `JP_alpha` (see the original data in Section 4.7.2). The missing values are represented by the `*` symbol in the data set.

```
1 0.02 0.00558212379888268 48 0.87 0.89
2 0.12 0.00418710068965517 91 0.82 *
3 0.09 0.001756902875 91 0.83 0.76
4 0.2 0.00509171270718232 67 * 1
5 0.08 0.00632846769230769 91 0.83 *
6 0.04 0.005537792 91 0.83 *
```

The following M*plus* code was used to conduct a fixed-effects meta-analysis. As $\sqrt{w_i}$ is considered as a variable, we need to declare it in `USEVARIABLES ARE w2;`. We calculate $\sqrt{w_i} = 1/\sqrt{v_i}$ and $\tilde{y}_i = \sqrt{w_i}y_i$ using `w2=SQRT (r_v**(-1));` and `r=w2*r;`, respectively. The mean and variance of the transformed effect size `r` are fixed at 0 and 1 using `[r@0.0];` and `r@1.0;`, respectively. The common effect size is estimated by regressing the transformed effect size on $\sqrt{w_i}$ with `r ON w2;`.

```
TITLE:    Fixed-effects model
DATA:     FILE IS Jaramillo05.dat;
VARIABLE: NAMES id r r_v IDV OC_alpha JP_alpha;
          USEVARIABLES ARE r w2;   ! Use both r and w2 in the analysis
          MISSING ARE *;           ! Define missing values

DEFINE: w2 = SQRT(r_v**(-1));      ! Weight for transformation
          r = w2*r;                ! Transformed r

MODEL:
          [r@0.0];                 ! Intercept fixed at 0
          r@1.0;                   ! Error variance fixed at 1
          r ON w2;                 ! Common effect estimate beta_F

OUTPUT: SAMPSTAT;
          CINTERVAL(symmetric);    ! Wald CI
```

```
------------------------  Selected output  ------------------------
MODEL FIT INFORMATION

Number of Free Parameters                          1

Loglikelihood

          H0 Value                          -225.750
          H1 Value                          -138.581
```

```
Information Criteria

          Akaike (AIC)                             453.499
          Bayesian (BIC)                           455.610
          Sample-Size Adjusted BIC                 452.464
            (n* = (n + 2) / 24)

MODEL RESULTS
                                                              Two-Tailed
                         Estimate       S.E.  Est./S.E.    P-Value

 R        ON
    W2                      0.194      0.008     24.428      0.000

 Intercepts
    R                       0.000      0.000    999.000    999.000

 Residual Variances
    R                       1.000      0.000    999.000    999.000
------------------------ Selected output ----------------------
```

The estimated common effect $\hat{\beta}_F$ is 0.194 (SE= 0.008), $p < 0.001$. This seems to suggest that there is a weak correlation coefficient between organizational commitment and salesperson job performance. It should be noted that the SE is underestimated when the effect sizes are heterogeneous. As shown in the next analysis, a random-effects model is more appropriate for this data set.

9.2.2 Random-effects model

We extend the above fixed-effects model to a random-effects model. The random-effects model in Equation 4.15 based on the transformed variables approach is

$$\begin{array}{ll} \text{Level 1:} & \tilde{y}_i = f_i\sqrt{w_i} + \tilde{e}_i, \\ \text{Level 2:} & f_i = \beta_R + u_i, \end{array} \tag{9.5}$$

where f_i is the *true* effect size in the ith study, β_R is the average population effect size under a random-effects model, and $u_i \sim \mathcal{N}(0, \tau^2)$ is the heterogeneity variance that has to be estimated.

Figure 9.2 shows the random-effects meta-analysis in M*plus*. In this model specification, f_i is a *random slope* by regressing $\tilde{y}_i$ on $\sqrt{w_i}$, meaning that f_i may vary across subjects (studies in a meta-analysis). f_i is a latent variable representing the *true* effect size in a meta-analysis. As f_i is a random variable, it has its own mean (the average effect β_R under a random-effects model) and variance (the heterogeneity variance τ^2). Random slopes analysis, as that implemented in M*plus*, is required to conduct the random-effects meta-analysis using the transformed variables approach.

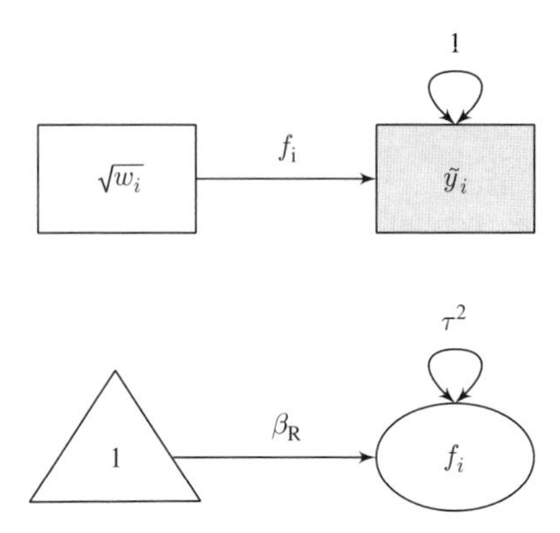

*Figure 9.2 A univariate random-effects meta-analytic model in M*plus.

The following M*plus* code shows how to analyze a random-effects meta-analysis. `w2` is similarly defined as that in the fixed-effects meta-analysis. We specify `ANALYSIS: TYPE=RANDOM;` to request the random slopes analysis. Moreover, we specify `f | r ON w2;` to define the random slopes `f` that is the regression slope from `w2` to `r` (the transformed effect size). Then, we specify the mean `[f*]` and the variance `f*` of `f`.

```
TITLE:   Random-effects model: Single level approach
DATA:    FILE IS Jaramillo05.dat;
VARIABLE: NAMES id r r_v IDV OC_alpha JP_alpha;
         USEVARIABLES ARE r w2; ! Use both r and w2 in the analysis
         MISSING ARE *;

DEFINE: w2 = SQRT(r_v**(-1));   ! Weight for transformation
        r = w2*r;               ! Transformed r

ANALYSIS: TYPE=RANDOM;          ! Use random slopes analysis
          ESTIMATOR=ML;         ! Use ML estimation

MODEL:
        [r@0.0];                ! Intercept fixed at 0
        r@1.0;                  ! Error variance fixed at 1
        f | r ON w2;            ! f: Study specific random effects
        f*;                     ! var(f): tau^2
        [f*];                   ! mean(f): Average effect size beta_R

OUTPUT: SAMPSTAT;
        CINTERVAL(symmetric);    ! Wald CI
```

Instead of using a single-level model as shown in Cheung (2008), Bengt Muthén (personal communication, May 2, 2011) also suggested an equivalent model specification using the two-level model. The following M*plus* code may also be used to analyze the above random-effects model. The syntax `ANALYSIS:`

TYPE=TWOLEVEL RANDOM; requests a two-level model with random slopes. The model in Equation 9.5 can be directly implemented as a two-level model.

```
TITLE:  Random-effects model: Two-level approach
DATA:   FILE IS Jaramillo05.dat;
VARIABLE: NAMES id r r_v IDV OC_alpha JP_alpha;
        USEVARIABLES ARE r w2;  ! Use both r and w2 in the analysis
        MISSING ARE *;
        WITHIN=ALL;             ! All variables are within
        CLUSTER=id;             ! id is the cluster label

DEFINE: w2 = SQRT(r_v**(-1));   ! Weight for transformation
        r = w2*r;               ! Transformed r

ANALYSIS: TYPE=TWOLEVEL RANDOM; ! Use random slopes analysis
                                ! Use two-level model
          ESTIMATOR=ML;         ! Use ML estimation

MODEL:
        %WITHIN%                ! Within model
        [r@0.0];                ! Intercept fixed at 0
        r@1.0;                  ! Error variance fixed at 1
        f | r ON w2;            ! f: Study specific random effects

        %BETWEEN%               ! Between model
        f*;                     ! var(f): tau^2
        [f*];             ! mean(f): Average effect size beta_R

OUTPUT: SAMPSTAT;
        CINTERVAL(symmetric);   ! Wald CI
```

```
----------------------- Selected output -----------------------
MODEL FIT INFORMATION

Number of Free Parameters                          2

Loglikelihood

         H0 Value                           -133.494

Information Criteria

         Akaike (AIC)                        270.987
         Bayesian (BIC)                      275.209
         Sample-Size Adjusted BIC            268.917
           (n* = (n + 2) / 24)

MODEL RESULTS
                                                    Two-Tailed
                    Estimate       S.E.  Est./S.E.    P-Value

 Means
```

```
    F                    0.187        0.019        9.654        0.000

 Intercepts
    R                    0.000        0.000      999.000      999.000

 Variances
    F                    0.017        0.004        4.119        0.000

 Residual Variances
    R                    1.000        0.000      999.000      999.000
------------------------ Selected output -----------------------
```

The estimated average effect $\hat{\beta}_R$ under a random-effects model is 0.187 (SE = 0.019), $p < 0.001$. The estimated heterogeneity variance $\hat{\tau}^2$ is 0.017. It is usually difficult to interpret the degree of heterogeneity based on $\hat{\tau}^2$. The following example illustrates how to calculate the Q statistic and I^2 to quantify the heterogeneity of effect sizes (Cheung, 2008).

In the above analyses, we fixed $\text{Var}(\tilde{e}_i) = 1$ in order to avoid the need of adjusting the SE. We may free this constraint in order to calculate the Q statistic and I^2. Let us explain the rationale. The Q statistic in Equation 4.14 can be rewritten as

$$\begin{aligned} Q &= \sum_{i=1}^{k} w_i \left(y_i - \hat{\beta}_F\right)^2, \\ &= \sum_{i=1}^{k} \left(\sqrt{w_i} y_i - \sqrt{w_i} \hat{\beta}_F\right)^2, \\ &= \sum_{i=1}^{k} \left(\tilde{y}_i - \sqrt{w_i} \hat{\beta}_F\right)^2. \end{aligned} \tag{9.6}$$

Therefore, the Q statistic is

$$Q = k\hat{\sigma}^2_{\tilde{e}_i}, \tag{9.7}$$

where $\hat{\sigma}^2_{\tilde{e}_i} = \left(\sum_{i=1}^{k} \left(\tilde{y}_i - \sqrt{w_i}\hat{\beta}_F\right)^2\right)/k$ is the estimated error variance of $\tilde{e}_i$. It should be noted that k instead of $(k - 1)$ is used because the ML estimation is used as the estimation method in SEM.

Once we have estimated the Q statistic, we can also calculate the I^2 (Higgins and Thompson, 2002) (see Section 4.3.3 for details),

$$I^2 = 1 - \frac{k-1}{Q}. \tag{9.8}$$

As I^2 is a function of $\hat{\sigma}^2_{\tilde{e}_i}$, an approximate Wald confidence interval (CI) can be constructed on I^2 using the delta method (see Section 3.4.1). However, the constructed CI is not accurate unless the number of studies is very large. Readers should be cautious when interpreting the Wald CI. The following M*plus* code shows the analysis. The estimated error variance of $\tilde{e}_i$ is labeled as `var` in `r (var);`. It can be used as a variable for further calculations. We define two new functions of parameters with `NEW(Q I2)`.

```
TITLE: Calculate the Q statistic and the heterogeneity indices
DATA:  FILE IS Jaramillo05.dat;
VARIABLE: NAMES id r r_v IDV OC_alpha JP_alpha;
        USEVARIABLES ARE r w2;  ! Use both r and w2 in the analysis
        MISSING ARE *;

DEFINE: w2 = SQRT(r_v**(-1));    ! Weight for transformation
        r = w2*r;                ! Transformed r

ANALYSIS: ESTIMATOR=ML;          ! Use ML estimation

Model:
        r ON w2;
        [r@0.0];                 ! Intercept fixed at 0
        r (var);                 ! Estimated error variance

MODEL CONSTRAINT:
        NEW(Q I2);               ! Define functions of parameters
                                 ! There are 61 studies.
        Q = 61*var;              ! Q statistic
        I2 = 1-60/Q;             ! I2 index
OUTPUT: SAMPSTAT;
        CINTERVAL(symmetric);  ! Wald CI
```

```
----------------------- Selected output ----------------------
MODEL RESULTS
                                                   Two-Tailed
                     Estimate       S.E.  Est./S.E.    P-Value

 R        ON
    W2                  0.194      0.019     10.356      0.000

 Intercepts
    R                   0.000      0.000    999.000    999.000

 Residual Variances
    R                   5.564      1.007      5.523      0.000

 New/Additional Parameters
    Q                 339.389     61.454      5.523      0.000
    I2                  0.823      0.032     25.716      0.000
----------------------- Selected output ----------------------
```

The Q statistic to test the homogeneity of effect sizes is $Q(df = 60) = 339.389$, $p < 0.001$. The I^2 index is 0.823, indicating that 82% of the variation is due to the between-study variation, whereas 18% of it is due to the within-study sampling variance.

9.2.3 Mixed-effects model

A random-effects model can be extended to a mixed-effects model by including moderators in the analysis. Suppose that there is a study characteristic z_i in the ith study; the mixed-effects model on the transformed effect size is

$$\begin{aligned} &\text{Level 1:} \quad \tilde{y}_i = f_i\sqrt{w_i} + \tilde{e}_i, \\ &\text{Level 2:} \quad f_i = \beta_0 + \beta_1 z_i + u_i, \end{aligned} \tag{9.9}$$

where β_0 is the intercept when $z_i = 0$, β_1 is the regression coefficient, and $\text{Var}(u_i) = \tau^2$ is the residual heterogeneity after controlling for z_i. Figure 9.3 shows the model with one moderator. f_i is regressed on z_i in Figure 9.3. The trick of the transformed variables approach is to treat the random slope as the *true* effect size f_i. Once the *true* effect size is available, we may use it as either a predictor or a dependent variable in further statistical modeling. It should be noted that the predictors, $\sqrt{w_i}$ and z_i, are correlated; however, this correlation is not shown in the figure.

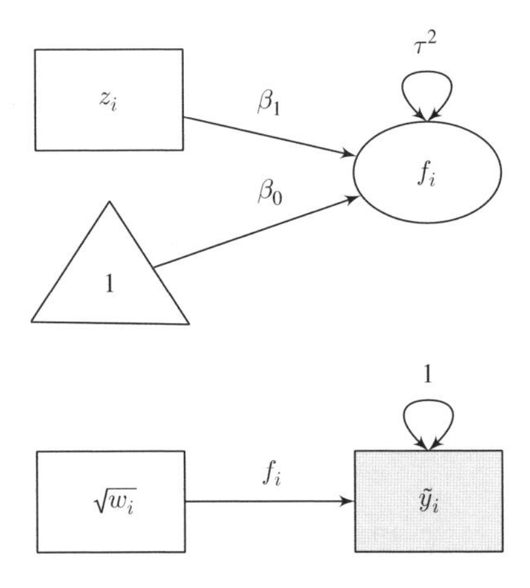

*Figure 9.3 A univariate mixed-effects meta-analytic model in M*plus.

The following M*plus* code shows how to conduct a mixed-effects meta-analysis with two moderators, `OC_alpha` and `JP_alpha`. The regression coefficients are specified via `f ON OC_alpha JP_alpha;`.

```
TITLE:   Mixed-effects model: OC_alpha and JP_alpha as the predictors
DATA:    FILE IS Jaramillo05.dat;
VARIABLE: NAMES id r r_v IDV OC_alpha JP_alpha;
         USEVARIABLES ARE r OC_alpha JP_alpha w2;
         MISSING ARE *;

DEFINE: w2 = SQRT(r_v**(-1));     ! Weight for transformation
         r = w2*r;                  ! Transformed r
```

```
ANALYSIS: TYPE=RANDOM;            ! Use random slope analysis
        ESTIMATOR=ML;             ! Use ML estimation

MODEL:
        [r@0.0];                  ! Intercept fixed at 0
        r@1.0;                    ! Error variance fixed at 1
        f | r ON w2;              ! f: Study specific random effects
        f*;                       ! var(f): tau^2
        [f*];                     ! beta_0
        f ON OC_alpha JP_alpha; ! beta_1 and beta_2

OUTPUT: SAMPSTAT;
        CINTERVAL(symmetric);    ! Wald CI
```

```
----------------------- Selected output -----------------------
*** WARNING
  Data set contains cases with missing on x-variables.
  These cases were not included in the analysis.
  Number of cases with missing on x-variables:  26
   1 WARNING(S) FOUND IN THE INPUT INSTRUCTIONS

SUMMARY OF ANALYSIS

Number of groups                                                    1
Number of observations                                             35

Number of dependent variables                                       1
Number of independent variables                                     3
Number of continuous latent variables                               1

MODEL FIT INFORMATION

Number of Free Parameters                       4

Loglikelihood

          H0 Value                          -80.082

Information Criteria

          Akaike (AIC)                      168.164
          Bayesian (BIC)                    174.385
          Sample-Size Adjusted BIC          161.895
            (n* = (n + 2) / 24)

MODEL RESULTS

                                                        Two-Tailed
                    Estimate       S.E.  Est./S.E.    P-Value

 F          ON
    OC_ALPHA           0.131      0.459      0.286      0.775
```

```
   JP_ALPHA            0.804      0.430      1.869      0.062

 Intercepts
   R                   0.000      0.000    999.000    999.000
   F                  -0.576      0.502     -1.148      0.251

 Residual Variances
   R                   1.000      0.000    999.000    999.000
   F                   0.019      0.006      3.313      0.001
----------------------- Selected output -----------------------
```

The estimated regression coefficients with their SEs and p values for `OC_Alpha` and `JP_Alpha` are 0.131 (SE= 0.459), $p = 0.775$ and 0.804 (SE= 0.430), $p = 0.062$, respectively. The estimated regression coefficient for `JP_Alpha` is marginally significant. Although there were 61 studies in the data set, 26 studies were excluded in the analysis because of the missing values in the predictors (`OC_Alpha` and `JP_Alpha`). We will demonstrate how to handle the missing values in the predictors using the full information maximum likelihood (FIML) estimation later.

One of the strength of SEM is its ease to specify equality constraints on the parameters. For example, we may impose an equality on both regression coefficients by specifying `f ON OC_alpha (1);` and `f ON JP_alpha (1);`.

```
TITLE:    Mixed-effects model: OC_alpha and JP_alpha
          ! with equal coefficients
DATA:     FILE IS Jaramillo05.dat;
VARIABLE: NAMES id r r_v IDV OC_alpha JP_alpha;
          USEVARIABLES ARE r OC_alpha JP_alpha w2;
          MISSING ARE *;

DEFINE: w2 = SQRT(r_v**(-1));          ! Weight for transformation
          r = w2*r;                    ! Transformed r

ANALYSIS: TYPE=RANDOM;                 ! Use random slope analysis
          ESTIMATOR=ML;                ! Use ML estimation

MODEL:
          [r@0.0];                     ! Intercept fixed at 0
          r@1.0;                       ! Error variance fixed at 1
          f | r ON w2;                 ! f: Study specific random effects
          f*;                          ! var(f): tau^2
          [f*];                        ! beta_0
          f ON OC_alpha (1);           ! beta_1=beta_2
          f ON JP_alpha (1);

OUTPUT: SAMPSTAT;
          CINTERVAL(symmetric);        ! Wald CI
```

```
----------------------- Selected output ----------------------
MODEL FIT INFORMATION

Number of Free Parameters                          3

Loglikelihood

         H0 Value                            -80.579

Information Criteria

         Akaike (AIC)                        167.158
         Bayesian (BIC)                      171.824
         Sample-Size Adjusted BIC            162.456
           (n* = (n + 2) / 24)

MODEL RESULTS
                                                          Two-Tailed
                      Estimate        S.E.  Est./S.E.    P-Value

 F          ON
    OC_ALPHA             0.486       0.295      1.647      0.100
    JP_ALPHA             0.486       0.295      1.647      0.100

 Intercepts
    R                    0.000       0.000    999.000    999.000
    F                   -0.604       0.507     -1.190      0.234

 Residual Variances
    R                    1.000       0.000    999.000    999.000
    F                    0.019       0.006      3.327      0.001
----------------------- Selected output ----------------------
```

The estimated regression coefficient under the equality constraint is 0.486 (SE= 0.295), $p = 0.100$, which is not statistically significant. We may calculate a likelihood ratio (LR) statistic to test the equality constraint by calculating the difference on the $-2*$log-likelihoods (labeled `H0 value` in the output) between the models with and without the constraint: $-2(-80.579 + 80.082) = 0.994$. The degrees of freedom (dfs) for the test statistic is the difference of the numbers of free parameters (labeled `Number of Free Parameters` in the output). The LR statistic is not significant with $\Delta\chi^2(\text{df} = 1) = 0.994, p = 0.319$.

9.2.4 Handling missing values in moderators

When a mixed-effects meta-analysis is conducted, studies with missing values in the moderators are deleted before the analysis. FIML may be used to handle the missing moderators in the SEM framework (see Section 8.2.3 for details). Specifically, we need to include the moderators as part of the model by estimating their means and covariance matrices. As the moderators are now considered as dependent variables in the analysis, the moderators are assumed to be distributed with a multivariate normal distribution. The following M*plus* code is used to handle the missing moderators.

```
TITLE:   Mixed-effects model: OC_alpha and JP_alpha as the predictors
DATA:    FILE IS Jaramillo05.dat;
VARIABLE: NAMES r r_v IDV OC_alpha JP_alpha;
         USEVARIABLES ARE r OC_alpha JP_alpha w2;
         MISSING ARE *;

DEFINE: w2 = SQRT(r_v**(-1));     ! Weight for transformation
         r = w2*r;                 ! Transformed r

ANALYSIS: TYPE=RANDOM;             ! Use random slope analysis
         ESTIMATOR=ML;             ! Use ML estimation

MODEL:
         [r@0.0];                  ! Intercept fixed at 0
         r@1.0;                    ! Error variance fixed at 1
         f | r ON w2;              ! f: Study specific random effects
         f*;                       ! var(f): tau^2
         [f*];                     ! beta_0
         f ON OC_alpha JP_alpha; ! beta1 and beta2

         ! Treat OC_alpha and JP_alpha as observed variables
         ! by estimating their means and variances
         [OC_alpha*];
         [JP_alpha*];
         OC_alpha*;
         JP_alpha*;

OUTPUT: SAMPSTAT;
         CINTERVAL(symmetric);     ! Wald CI
```

```
----------------------- Selected output ----------------------
SUMMARY OF ANALYSIS

Number of groups                                               1
Number of observations                                        61

Number of dependent variables                                  1
Number of independent variables                                3
Number of continuous latent variables                          1

MODEL FIT INFORMATION

Number of Free Parameters                         9

Loglikelihood

          H0 Value                           -5.369
```

```
Information Criteria

         Akaike (AIC)                          28.739
         Bayesian (BIC)                        47.737
         Sample-Size Adjusted BIC              19.424
           (n* = (n + 2) / 24)

MODEL RESULTS
                                                        Two-Tailed
                    Estimate       S.E.  Est./S.E.       P-Value

 F          ON
    OC_ALPHA           0.243      0.361      0.672         0.501
    JP_ALPHA           0.579      0.348      1.664         0.096

 JP_ALPHA WITH
    OC_ALPHA           0.001      0.001      0.876         0.381

 Means
    OC_ALPHA           0.849      0.008    108.032         0.000
    JP_ALPHA           0.864      0.011     81.220         0.000

 Intercepts
    R                  0.000      0.000    999.000       999.000
    F                 -0.519      0.392     -1.326         0.185

 Variances
    OC_ALPHA           0.003      0.001      5.195         0.000
    JP_ALPHA           0.004      0.001      4.277         0.000

 Residual Variances
    R                  1.000      0.000    999.000       999.000
    F                  0.015      0.004      3.976         0.000
----------------------- Selected output -----------------------
```

From the above results, all 61 studies are used in the analysis. The means and covariance matrix between `OC_Alpha` and `JP_Alpha` are also estimated. Although the regression coefficient on `JP_Alpha` is still nonsignificant, its SE is slightly smaller than that based on only 35 studies.

9.3 Multivariate meta-analysis

When there are more than one effect sizes per study, researchers may want to conduct a multivariate meta-analysis to account for the dependence among the effect sizes. The SEM approach discussed in Chapter 5 uses definition variables to fix the known sampling covariance matrix. As M*plus* does not have definition variables, Cheung (2013) extended the transformed variables approach to conduct multivariate meta-analyses in M*plus*.

9.3.1 Fixed-effects model

Suppose that there are p effect sizes per study involved in a multivariate meta-analysis; the fixed-effects model for the ith study is

$$\boldsymbol{y}_i = \boldsymbol{X}_i \beta_F + \boldsymbol{e}_i, \tag{9.10}$$

where $\boldsymbol{y}_i$ is the $p_i \times 1$ vector of the observed effect sizes, $\boldsymbol{X}_i$ is a $p_i \times p$ design matrix with 0 and 1 to select the observed effect sizes, β_F is a $p \times 1$ vector of population effect sizes, and $\boldsymbol{e}_i$ is a $p_i \times 1$ vector of the known sampling error. When all effect sizes are complete, $p_i = p$. When the sample sizes are reasonably large, $\boldsymbol{e}_i$ is assumed to be multivariate normally distributed with a mean vector of zero and a known covariance matrix $\boldsymbol{V}_i$, that is, $\boldsymbol{e}_i \sim \mathcal{N}(0, \boldsymbol{V}_i)$.

Similar to the analysis of a univariate meta-analysis, we transform the effect sizes in such a way that the known sampling errors become i.i.d. First, we calculate a transformation matrix $\boldsymbol{W}_i^{1/2} = \boldsymbol{V}_i^{-1/2}$ by taking the Cholesky decomposition on the inverse of the known sampling covariance matrix $\boldsymbol{V}_i$. Applying the Cholesky decomposition on a covariance matrix is similar to taking a square root on a variance. The fixed-effects model on the transformed effect sizes for the ith study is

$$\begin{aligned} \boldsymbol{W}_i^{1/2}\boldsymbol{y}_i &= \boldsymbol{W}_i^{1/2}\boldsymbol{X}_i\beta_F + \boldsymbol{W}_i^{1/2}\boldsymbol{e}_i, \\ \tilde{\boldsymbol{y}}_i &= \tilde{\boldsymbol{X}}_i\beta_F + \tilde{\boldsymbol{e}}_i, \end{aligned} \tag{9.11}$$

where $\tilde{\boldsymbol{y}}_i = \boldsymbol{W}_i^{1/2}\boldsymbol{y}_i$, $\tilde{\boldsymbol{X}}_i = \boldsymbol{W}_i^{1/2}\boldsymbol{X}_i$, and $\tilde{\boldsymbol{e}}_i = \boldsymbol{W}_i^{1/2}\boldsymbol{e}_i$. It should be noted that the parameter vector β_F are the same after the transformation.

After the transformation, the variance of $\tilde{\boldsymbol{e}}_i$ is

$$\begin{aligned} \mathsf{Var}(\tilde{\boldsymbol{e}}_i) &= \boldsymbol{W}_i^{1/2}\mathsf{Var}(\boldsymbol{e}_i)(\boldsymbol{W}_i^{1/2})^{\mathrm{T}}, \\ &= \boldsymbol{W}_i^{1/2}\boldsymbol{V}_i(\boldsymbol{W}_i^{1/2})^{\mathrm{T}}, \\ &= \boldsymbol{I} \quad \text{as} \quad \boldsymbol{W}_i = \boldsymbol{V}_i^{\mathrm{T}}, \end{aligned} \tag{9.12}$$

where $\boldsymbol{I}$ is an identity matrix with i.i.d. of a known variable of 1. Before the transformation, the multivariate effect sizes are distributed with a known sampling covariance matrix $\boldsymbol{V}_i$. After the transformation, the transformed effect sizes are conditionally independent and with a known variance of 1.

As the transformed effect sizes are conditionally independent with a known variance of 1, Cheung (2013) suggested to stack the transformed effect sizes together for the ease of analysis. Figure 9.4a shows a multivariate meta-analysis with two effect sizes. $\underset{2\times1}{\tilde{x}_1}$ and $\underset{2\times1}{\tilde{x}_2}$ are the transformed design matrix in estimating the intercepts. Unlike the univariate meta-analysis, we cannot use $\sqrt{w_1}$ and $\sqrt{w_2}$ to represent the transformed matrix. It is because the sampling covariance between these two effect sizes is also involved in calculating $\boldsymbol{W}_i^{1/2}$. There are three key features in this model. First, the multivariate effect sizes are stacked together. Therefore, there is only *one* column of $\underset{2\times1}{\tilde{y}_i}$ regardless of how many effect sizes per study there are.

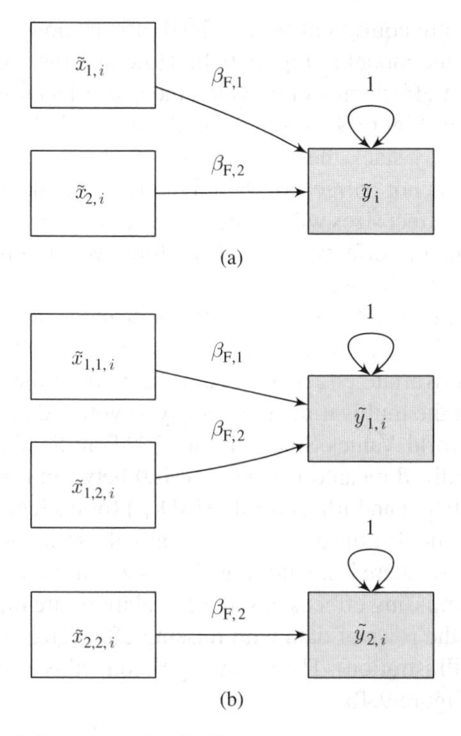

*Figure 9.4 Multivariate fixed-effects meta-analytic models in M*plus.

Second, the number of "subjects" are double of that of the number of studies when there is no missing effect sizes. Third, the sampling error variance is fixed at 1 after applying the transformation.

The above approach, however, is not appropriate for the random-effects model. We introduce a better approach here. This approach uses one row to represent one study. Let us illustrates the model with two effect sizes per study. After the transformation, the fixed-effects model is

$$\begin{bmatrix} \tilde{y}_{1,i} \\ \tilde{y}_{2,i} \end{bmatrix} = \beta_{F,1} \begin{bmatrix} \tilde{x}_{1,1} \\ 0 \end{bmatrix} + \beta_{F,2} \begin{bmatrix} \tilde{x}_{1,2} \\ \tilde{x}_{2,2} \end{bmatrix} + \begin{bmatrix} \tilde{e}_{1,i} \\ \tilde{e}_{2,i} \end{bmatrix}, \tag{9.13}$$

where $\begin{bmatrix} \tilde{x}_{1,1} & \tilde{x}_{1,2} \\ 0 & \tilde{x}_{2,2} \end{bmatrix} = \boldsymbol{W}_i^{1/2}\boldsymbol{X}_i$. It should be noted that there is 0 in $\boldsymbol{W}_i^{1/2}$ because of the Cholesky decomposition. We may treat $\tilde{y}_{1,i}$, $\tilde{y}_{2,i}$, $\tilde{x}_{1,1}$, $\tilde{x}_{1,2}$, and $\tilde{x}_{2,2}$ as *variables*. Figure 9.4b shows the model. As the regression coefficients of both regressing $\tilde{y}_{1,i}$ on $\tilde{x}_{1,2}$ and regressing $\tilde{y}_{2,i}$ on $\tilde{x}_{2,2}$ are $\beta_{F,2}$, an equality constraint is required to ensure that these two paths are the same. When there are p effect sizes per study in the meta-analysis, $p(p+1)/2$ variables on the transformed design matrix are required. The analysis may become complicated and tedious when p is large and with incomplete effect sizes. It should be noted that the number of "subjects" is the same as that of the number of studies under this approach.

These two models are equivalent for the fixed-effects model. It appears that it is easier to implement the model in Figure 9.4a. However, these two models are different for the random-effects model in which the *true* effect sizes may vary across studies. The model in Figure 9.4a treats the transformed effect sizes as independent; therefore, we may stack the transformed effect sizes and treat them as one single variable. This is not correct for the random-effects model. It is because the constraint of the *true* effect sizes will not be imposed when the model in Figure 9.4a is extended to the random-effects model. Therefore, we will only apply the model in Figure 9.4b in the following discussion.

For the univariate meta-analyses, we used the built-in functions to apply the transformation in M*plus*. This does not work for the multivariate meta-analyses. We need to transform the multivariate effect sizes before exporting the data for analysis in M*plus*. We illustrate the multivariate meta-analysis with a data set from the World Values Survey II (World Values Study Group, 1994) in Section 5.7.2. The effect sizes are the standardized mean difference (SMD) between males and females on life satisfaction (SMD_{LS}) and life control (SMD_{LC}) for each country.

The following R code demonstrates how to apply the transformation on the multivariate effect sizes. As there is no missing effect size, the calculations are straightforward. If there are missing effect sizes, the calculations are more tedious because we have to exclude the rows of data with missing effect sizes (see Cheung (2013, Appendix B) for an illustration). The following R code illustrates how to transform the effect sizes for Figure 9.4b.

```
R> library(metaSEM)
R> ## Select the effect sizes
R> y <- wvs94a[, c("lifesat","lifecon")]
R> ## Convert it into a column of effect sizes
R> y <- matrix(t(y), ncol=1)
R> ## Prepare the design matrix
R> X <- matrix(rep(c(1,0,0,1), nrow(wvs94a)), ncol=2, byrow=TRUE)
R> ## Convert the known sampling covariance matrix into
R> ## a block diagonal matrix
R> V <- matrix2bdiag(wvs94a[, c("lifesat_var","inter_cov",
                                 "lifecon_var")])
R> ## Calculate the transformation matrix
R> W0.5 <- chol(solve(V))
R> ## Calculate the transformed effect size
R> y_new <- W0.5 %*% y
R> ## Calculate the transformed design matrix
R> X_new <- W0.5 %*% X
R> ## Center gnp and divide it by 10000 to improve numerical
      stability
R> ## Prepare the gnp
R> gnp <- scale(wvs94a$gnp/10000, scale=FALSE)
R> ## Convert y into one row per study
R> y2 <- matrix(c(t(y_new)), ncol=2, byrow=TRUE)
R> ## Convert X into one row per study
R> x2 <- matrix(c(t(X_new)), ncol=4, byrow=TRUE)
```

```
R> my.wide <- cbind(y2, x2, gnp)
R> ## Add the variable names for ease of reference
R> ## W0.5 = [y1f1, y1f2]
R> ##        [0   , y2f2]
R> colnames(my.wide) <- c("y1", "y2", "y1f1", "y1f2",
                          "y2f1", "y2f2", "gnp")
R> ## Display the first few cases
R> head(my.wide)
```

```
             y1        y2     y1f1      y1f2 y2f1     y2f2        gnp
[1,] -0.8638851 0.8933552 16.74449 -5.667836    0 15.50759 -0.8527838
[2,]  1.5185353 0.1653192 19.66342 -6.345139    0 18.58993 -0.5997838
[3,]  0.2898198 1.1698049 16.74893 -5.578365    0 15.78971 -0.7787838
[4,] -0.7218254 3.2903210 27.23667 -7.289699    0 25.70662  0.4642162
[5,]  1.9122124 3.8052864 22.39287 -7.715437    0 20.89603 -0.8217838
[6,]  0.0181645 0.8482849 21.26301 -9.180409    0 19.08188  0.5202162
```

```
R> ## Write it as a plain text for Mplus
R> ## y2f1 is excluded in my.wide since it contains only 0.
R> ## Missing values are represented by *
R> write.table(my.wide[,-5], "wvs94a.dat", sep=" ", na="*",
               row.names=FALSE,
               col.names=FALSE)
```

The data file was exported as a plain text called `wvs94.dat`. The variables are the two transformed effect sizes (`y1` and `y2`), the three elements of the transformed design matrix $\tilde{x}_{1,1}$ (`y1f1`), $\tilde{x}_{1,2}$ (`y1f2`), and $\tilde{x}_{2,2}$ (`y2f2`), and GNP. As $\tilde{x}_{2,1}$ (`y2f1`) is always 0, we exclude it from the analysis. The following M*plus* code shows how to analyze a fixed-effects multivariate meta-analysis.

```
TITLE:   Fixed-effects model
DATA:    FILE IS wvs94a.dat;
VARIABLE: NAMES y1 y2 y1f1 y1f2 y2f2 GNP;
    USEVARIABLES ARE y1 y2 y1f1 y1f2 y2f2;
    MISSING ARE *;

MODEL:
    y1 ON y1f1;               ! beta_{F,1} in the figure
    y1 ON y1f2 (1);           ! beta_{F,2} in the figure
    y2 ON y2f2 (1);           ! beta_{F,2} in the figure

    [y1@0.0];                 ! Intercept fixed at 0
    [y2@0.0];                 ! Intercept fixed at 0

    y1@1.0                    ! Error variance fixed at 1
    y2@1.0                    ! Error variance fixed at 1

    y1 WITH y2@0;             ! Covariance fixed at 0
```

```
OUTPUT: SAMPSTAT;
    CINTERVAL(symmetric);   ! Wald CI
```

```
------------------------  Selected output  ----------------------
MODEL FIT INFORMATION

Number of Free Parameters                              2

Loglikelihood

          H0 Value                              -202.206
          H1 Value                              -156.495

Information Criteria

          Akaike (AIC)                           408.412
          Bayesian (BIC)                         411.887
          Sample-Size Adjusted BIC               405.624
            (n* = (n + 2) / 24)

MODEL RESULTS
                                                           Two-Tailed
                    Estimate       S.E.  Est./S.E.    P-Value

 Y1        ON
    Y1F1               0.010      0.008      1.156      0.248
    Y1F2               0.071      0.008      8.392      0.000

 Y2        ON
    Y2F2               0.071      0.008      8.392      0.000

 Y1        WITH
    Y2                 0.000      0.000    999.000    999.000

 Intercepts
    Y1                 0.000      0.000    999.000    999.000
    Y2                 0.000      0.000    999.000    999.000

 Residual Variances
    Y1                 1.000      0.000    999.000    999.000
    Y2                 1.000      0.000    999.000    999.000
------------------------  Selected output  ----------------------
```

The estimated common effects with their SEs and p values for SMD_{LS} and SMD_{LC} are 0.010 (SE = 0.008), $p = 0.248$ and 0.071 (SE = 0.008), $p < 0.001$, respectively. Only the estimated common effect on SMD_{LC} is statistically significant. As the estimates are based on the fixed-effects model, readers should be cautious if the fixed-effects model is not appropriate.

9.3.2 Random-effects model

A fixed-effects model can be extended to a random-effects model by using random slopes analysis. The random-effects model on the transformed effect sizes for the ith study is

$$\begin{aligned} \text{Level 1:} \quad & \tilde{\boldsymbol{y}}_i = \tilde{\boldsymbol{X}}_i \boldsymbol{f}_i + \tilde{\boldsymbol{e}}_i, \\ \text{Level 2:} \quad & \boldsymbol{f}_i = \beta_\text{R} + \boldsymbol{u}_i, \end{aligned} \tag{9.14}$$

where $\boldsymbol{f}_i$ is the vector of the *true* effect sizes in the ith study, β_R is the vector of the average population effects under a random-effects model, and $\boldsymbol{u}_i \sim \mathcal{N}(0, \boldsymbol{T}^2)$ is the heterogeneity variance–covariance matrix of the random effects that has to be estimated. Figure 9.5 displays a random-effects model with two effect sizes per study. The regression slopes from the transformed design matrix (three variables) to the transformed effect sizes $\tilde{y}_i$ are $f_{1,i}$ and $f_{2,i}$, respectively. These slopes are considered as random variables with their own means (the average effects $\beta_{\text{R},1}$ and $\beta_{\text{R},2}$ under a random-effects model) and covariance matrix (the heterogeneity covariance matrix).

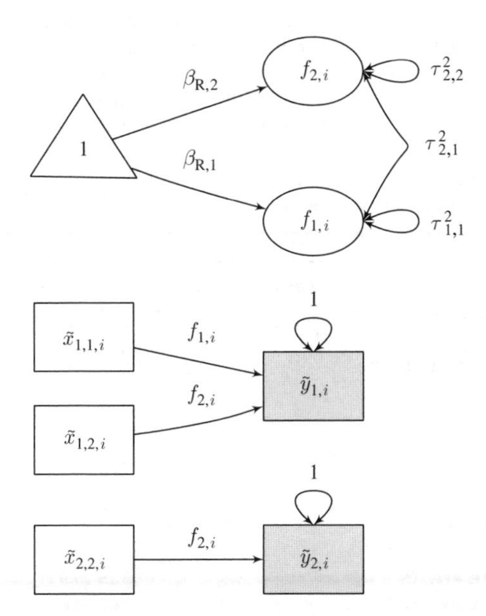

*Figure 9.5 A multivariate random-effects meta-analytic model in M*plus.

The following M*plus* code shows how to conduct a random-effects model with two effect sizes. We specify the covariance between the random effects using `f_LS WITH f_LC*;`.

```
TITLE:   Random-effects model
DATA:    FILE IS wvs94a.dat;
```

```
VARIABLE: NAMES y1 y2 y1f1 y1f2 y2f2 GNP;
    USEVARIABLES ARE y1 y2 y1f1 y1f2 y2f2;
    MISSING ARE *;

ANALYSIS: TYPE=RANDOM;
    ESTIMATOR=ML;           ! Use ML estimation
MODEL:
    f_LS | y1 ON y1f1;
    f_LC | y1 ON y1f2;
    f_LC | y2 ON y2f2;

    [y1@0.0];               ! Intercept fixed at 0
    [y2@0.0];               ! Intercept fixed at 0

    y1@1.0                  ! Error variance fixed at 1
    y2@1.0                  ! Error variance fixed at 1

    y1 WITH y2@0;           ! Covariance fixed at 0

    f_LS*;                  ! tau^2_11 in the figure
    f_LC*;                  ! tau^2_22 in the figure
    f_LS WITH f_LC*;        ! tau^2_21 in the figure

    [f_LS*];                ! beta_{R,1} in the figure
    [f_LC*];                ! beta_{R,2} in the figure

OUTPUT: SAMPSTAT;
    CINTERVAL(symmetric); ! Wald CI
```

```
------------------------ Selected output -----------------------
MODEL FIT INFORMATION

Number of Free Parameters                            5

Loglikelihood

          H0 Value                            -161.898

Information Criteria

          Akaike (AIC)                         333.797
          Bayesian (BIC)                       342.485
          Sample-Size Adjusted BIC             326.827
            (n* = (n + 2) / 24)

MODEL RESULTS
                                                         Two-Tailed
                    Estimate       S.E.  Est./S.E.    P-Value

 F_LS      WITH
    F_LC                0.004       0.002      2.332       0.020
```

```
 Y1         WITH
    Y2                       0.000        0.000      999.000      999.000

 Means
    F_LS                     0.001        0.014        0.097        0.922
    F_LC                     0.069        0.017        4.092        0.000

 Intercepts
    Y1                       0.000        0.000      999.000      999.000
    Y2                       0.000        0.000      999.000      999.000

 Variances
    F_LS                     0.005        0.002        2.684        0.007
    F_LC                     0.008        0.003        3.316        0.001

 Residual Variances
    Y1                       1.000        0.000      999.000      999.000
    Y2                       1.000        0.000      999.000      999.000
------------------------ Selected output -----------------------
```

The estimated average effect sizes with their SEs and p values for SMD_{LS} and SMD_{LC} are 0.001 (SE = 0.014), $p = 0.922$ and 0.069 (SE = 0.017), $p < 0.001$ respectively. The estimated average effect size for SMD_{LC} is significant, while the one for SMD_{LS} is nonsignificant with $\alpha = 0.05$. The SEs of the random-effects model were larger than those of the fixed-effects model. The estimated variance component of the heterogeneity for $\hat{\text{SMD}}_{\text{LS}}$ and $\hat{\text{SMD}}_{\text{LC}}$ is $\begin{bmatrix} 0.005 & \\ 0.004 & 0.008 \end{bmatrix}$. The results are similar to those in Section 5.7.2.

A multivariate approach is usually preferable to a univariate approach in testing research hypotheses. In a multivariate meta-analysis, we may test the null hypothesis that both $SMD_{\text{LS}} = 0$ and $SMD_{\text{LC}} = 0$. This test is preferable to two separate tests because the multivariate test takes the correlation between the effect sizes into account. We can easily conduct this test in M*plus* by using the following code.

```
TITLE:   Random-effects model: Fix population effect at 0
DATA:    FILE IS wvs94a.dat;
VARIABLE: NAMES y1 y2 y1f1 y1f2 y2f2 GNP;
    USEVARIABLES ARE y1 y2 y1f1 y1f2 y2f2;
    MISSING ARE *;

ANALYSIS: TYPE=RANDOM;
    ESTIMATOR=ML;             ! Use ML estimation
MODEL:
    f_LS | y1 ON y1f1;
    f_LC | y1 ON y1f2;
    f_LC | y2 ON y2f2;

    [y1@0.0];                 ! Intercept fixed at 0
```

```
    [y2@0.0];               ! Intercept fixed at 0

    y1@1.0                  ! Error variance fixed at 1
    y2@1.0                  ! Error variance fixed at 1

    y1 WITH y2@0;           ! Covariance fixed at 0

    f_LS*;                  ! tau^2_11 in the figure
    f_LC*;                  ! tau^2_22 in the figure
    f_LS WITH f_LC*;        ! tau^2_21 in the figure

    [f_LS@0];               ! beta_{R,1} fixed at 0
    [f_LC@0];               ! beta_{R,2} fixed at 0

OUTPUT: SAMPSTAT;
    CINTERVAL(symmetric); ! Wald CI
```

```
------------------------ Selected output -----------------------
MODEL FIT INFORMATION

Number of Free Parameters                         3

Loglikelihood

          H0 Value                         -171.123

Information Criteria

          Akaike (AIC)                      348.246
          Bayesian (BIC)                    353.459
          Sample-Size Adjusted BIC          344.064
            (n* = (n + 2) / 24)

MODEL RESULTS
                                                        Two-Tailed
                    Estimate       S.E.  Est./S.E.    P-Value

 F_LS     WITH
    F_LC               0.004      0.002      2.149      0.032

 Y1       WITH
    Y2                 0.000      0.000    999.000    999.000

 Means
    F_LS               0.000      0.000    999.000    999.000
    F_LC               0.000      0.000    999.000    999.000

 Intercepts
    Y1                 0.000      0.000    999.000    999.000
    Y2                 0.000      0.000    999.000    999.000
```

```
Variances
   F_LS                0.005       0.002       2.656       0.008
   F_LC                0.013       0.004       3.681       0.000

Residual Variances
   Y1                  1.000       0.000     999.000     999.000
   Y2                  1.000       0.000     999.000     999.000

----------------------- Selected output ----------------------
```

We compare this model against the model without fixing the average population effect sizes at 0. The LR statistic is $-2(-171.123 + 161.898) = 18.45$, which is significant with $\Delta\chi^2(\text{df} = 2) = 18.45, p < 0.001$. Therefore, at least one effect size is different from zero.

9.3.3 Mixed-effects model

A random-effects model can be extended to a mixed-effects model by including study characteristics as moderators. Suppose that there is a study characteristic z_i

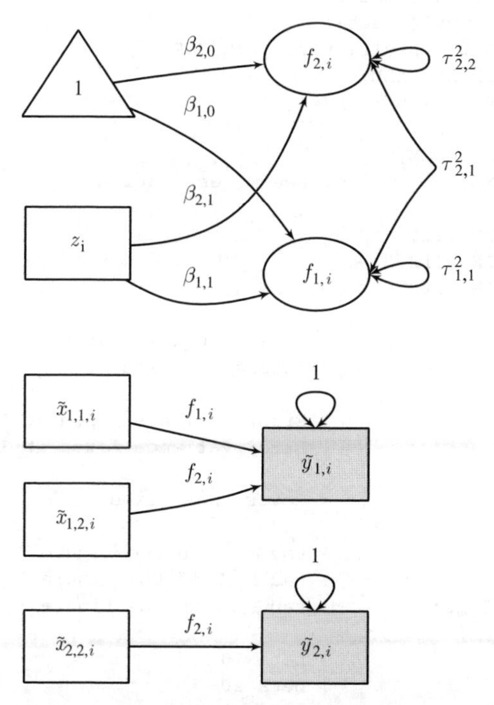

Figure 9.6 A multivariate mixed-effects meta-analytic model in Mplus.

in the ith study; the mixed-effects model on the transformed effect sizes in the ith study is

$$\begin{aligned} \text{Level 1:} \quad & \tilde{\boldsymbol{y}}_i = \tilde{\boldsymbol{X}}_i \boldsymbol{f}_i + \tilde{\boldsymbol{e}}_i, \\ \text{Level 2:} \quad & \boldsymbol{f}_i = \beta_0 + \beta_1 z_i + \boldsymbol{u}_i, \end{aligned} \tag{9.15}$$

where β_0 is the vector of the intercepts when $z_i = 0$, β_1 is the vector of regression coefficients, and $\boldsymbol{u}_i \sim \mathcal{N}(0, \boldsymbol{T}^2)$ is the residual heterogeneity variance–covariance matrix of the random effects after controlling for z_i.

Figure 9.6 shows a mixed-effects model with one moderator z_i. As z_i directly predicts the *true* effect sizes $\boldsymbol{f}_i$, there is no transformation applied to z_i. It should be reminded that all the predictors, z_i and $\tilde{\boldsymbol{x}}$, are correlated in the analysis. The structural model on the average *true* effect sizes for this model is

$$\begin{bmatrix} f_{1,i} \\ f_{2,i} \end{bmatrix} = \begin{bmatrix} \beta_{1,0} \\ \beta_{2,0} \end{bmatrix} + \begin{bmatrix} \beta_{1,1} \\ \beta_{2,1} \end{bmatrix} z_i + \begin{bmatrix} u_{1,i} \\ u_{2,i} \end{bmatrix}. \tag{9.16}$$

The following M*plus* code shows a mixed-effects meta-analysis using *GNP* as a moderator. To improve the numerical stability and interpretations, gross national product (GNP) has been centered and divided by 10,000 before exporting the data set.

```
TITLE:   Mixed-effects model
DATA:    FILE IS wvs94a.dat;
VARIABLE: NAMES y1 y2 y1f1 y1f2 y2f2 GNP;
    USEVARIABLES ARE ALL;
    MISSING ARE *;

ANALYSIS: TYPE=RANDOM;
    ESTIMATOR=ML;               ! Use ML estimation
MODEL:
    f_LS | y1 ON y1f1;
    f_LC | y1 ON y1f2;
    f_LC | y2 ON y2f2;

    [y1@0.0];                   ! Intercept fixed at 0
    [y2@0.0];                   ! Intercept fixed at 0

    y1@1.0                      ! Error variance fixed at 1
    y2@1.0                      ! Error variance fixed at 1

    y1 WITH y2@0;               ! Covariance fixed at 0

    f_LS*;                      ! tau^2_11 in the figure
    f_LC*;                      ! tau^2_22 in the figure
    f_LS WITH f_LC*;            ! tau^2_21 in the figure

    [f_LS*];                    ! beta_10
    [f_LC*];                    ! beta_20
```

```
    f_LS ON GNP;              ! beta_11
    f_LC ON GNP;              ! beta_21

OUTPUT: SAMPSTAT;
    CINTERVAL(symmetric); ! Wald CI
```

```
------------------------ Selected output ------------------------
MODEL FIT INFORMATION

Number of Free Parameters                         7

Loglikelihood

         H0 Value                          -141.277

Information Criteria

         Akaike (AIC)                       296.554
         Bayesian (BIC)                     307.830
         Sample-Size Adjusted BIC           285.952
           (n* = (n + 2) / 24)

MODEL RESULTS
                                                   Two-Tailed
                    Estimate       S.E.  Est./S.E.    P-Value

 F_LS       ON
    GNP               -0.024      0.015     -1.571      0.116

 F_LC       ON
    GNP               -0.037      0.018     -2.073      0.038

 F_LS     WITH
    F_LC               0.004      0.002      2.137      0.033

 Y1       WITH
    Y2                 0.000      0.000    999.000    999.000

 Intercepts
    Y1                 0.000      0.000    999.000    999.000
    Y2                 0.000      0.000    999.000    999.000
    F_LS               0.001      0.015      0.089      0.929
    F_LC               0.071      0.017      4.119      0.000

 Residual Variances
    Y1                 1.000      0.000    999.000    999.000
    Y2                 1.000      0.000    999.000    999.000
    F_LS               0.005      0.002      2.559      0.011
    F_LC               0.007      0.002      3.026      0.002
------------------------ Selected output ------------------------
```

The estimated regression coefficients with their SEs and p values from GNP on SMD_{LS} and SMD_{LC} are -0.024 (SE $= 0.015$), $p = 0.116$ and -0.037 (SE $= 0.018$), $p = 0.038$, respectively. Only the regression coefficient on SMD_{LC} is statistically significant.

9.3.4 Mediation and moderation models on the effect sizes

The mediation model on the *true* effect sizes discussed in Section 5.6.2 can be fitted in M*plus*. Suppose that the *true* effect sizes $f_{1,i}$ and $f_{2,i}$ and the observed variable z_i are the dependent variable, the mediator, and the independent variable, respectively; the mediation model on the transformed variables is

$$\text{Measurement model: } \underset{2\times 1}{\begin{bmatrix}\tilde{y}_{1,i}\\ \tilde{y}_{2,i}\end{bmatrix}} = \underset{2\times 2}{\tilde{\boldsymbol{X}}} \underset{2\times 1}{\begin{bmatrix}f_{1,i}\\ f_{2,i}\end{bmatrix}} + \underset{2\times 1}{\begin{bmatrix}\tilde{e}_{1,i}\\ \tilde{e}_{2,i}\end{bmatrix}} \quad \text{and}$$

$$\begin{aligned}\text{Structural model: } \begin{bmatrix}f_{1,i}\\ f_{2,i}\end{bmatrix} &= \begin{bmatrix}\beta_{1,0}\\ \beta_{2,0}\end{bmatrix} + \begin{bmatrix}0 & \beta_{1,2}\\ 0 & 0\end{bmatrix}\begin{bmatrix}f_{1,i}\\ f_{2,i}\end{bmatrix}\\ &\quad + \begin{bmatrix}\gamma_1\\ \gamma_2\end{bmatrix} z_i + \begin{bmatrix}u_{1,i}\\ u_{2,i}\end{bmatrix},\end{aligned} \tag{9.17}$$

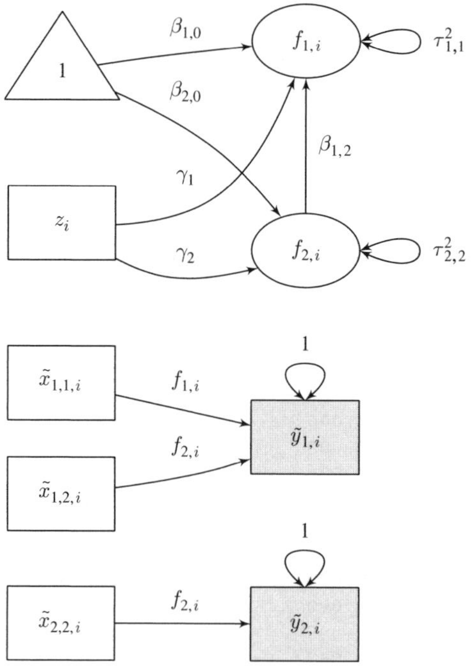

*Figure 9.7 A mediation model with two effect sizes in M*plus*.*

where $\text{Cov}\left(\begin{bmatrix}\tilde{e}_{1,i}\\ \tilde{e}_{2,i}\end{bmatrix}\right) = \begin{bmatrix}1 & 0\\ 0 & 1\end{bmatrix}$ is the known conditional sampling covariance matrix of the transformed errors and $\text{Cov}\left(\begin{bmatrix}u_{1,i}\\ u_{2,i}\end{bmatrix}\right) = \begin{bmatrix}\tau^2_{1,1} & 0\\ 0 & \tau^2_{2,2}\end{bmatrix}$ is the residual heterogeneity covariance matrix.

Figure 9.7 shows the model. The measurement model is used to handle the conditional sampling covariance matrix between the effect sizes, while the structural model is used to model the direct and indirect effects. Under the above model specification, the direct effect from z_i to the *true* effect size $f_{1,i}$ is γ_1, while the indirect effect via the *true* effect size $f_{2,i}$ is $\gamma_2\beta_{1,2}$. The total effect is $\gamma_2\beta_{1,2} + \gamma_1$. We may label the parameters and define new functions of the parameters (the direct, indirect, and total effects) in M*plus*. Approximate SEs and CIs on these effects are automatically calculated based on the delta method.

```
TITLE:   Mediation model on the "true" effect sizes
DATA:    FILE IS wvs94a.dat;
VARIABLE: NAMES y1 y2 y1f1 y1f2 y2f2 GNP;
    USEVARIABLES ARE ALL;
    MISSING ARE *;

ANALYSIS: TYPE=RANDOM;
    ESTIMATOR=ML;              ! Use ML estimation
MODEL:
    f_LS | y1 ON y1f1;
    f_LC | y1 ON y1f2;
    f_LC | y2 ON y2f2;

    [y1@0.0];                  ! Intercept fixed at 0
    [y2@0.0];                  ! Intercept fixed at 0

    y1@1.0                     ! Error variance fixed at 1
    y2@1.0                     ! Error variance fixed at 1

    y1 WITH y2@0;              ! Covariance fixed at 0

    f_LS*;                     ! tau^2_11
    f_LC*;                     ! tau^2_22

    [f_LS*];                   ! beta_10
    [f_LC*];                   ! beta_20

    f_LS ON GNP (gamma_1);      ! gamma_1
    f_LC ON GNP (gamma_2);      ! gamma_2
    f_LS ON f_LC (beta_12);     ! beta_12

MODEL CONSTRAINT:              ! Define new functions of parameters
    NEW (ind, dir, total);
    ind=gamma_2*beta_12;       ! Indirect effect
    dir=gamma_1;               ! Direct effect
    total=ind+dir;             ! Total effect
```

```
OUTPUT: SAMPSTAT;
    CINTERVAL(symmetric); ! Wald CI
```

```
------------------------ Selected output ----------------------

MODEL FIT INFORMATION

Number of Free Parameters                        7

Loglikelihood

          H0 Value                        -141.277

Information Criteria

          Akaike (AIC)                     296.554
          Bayesian (BIC)                   307.830
          Sample-Size Adjusted BIC         285.952
            (n* = (n + 2) / 24)

MODEL RESULTS
                                                    Two-Tailed
                    Estimate       S.E.  Est./S.E.    P-Value

 F_LS       ON
    F_LC               0.480      0.168      2.859      0.004

 F_LS       ON
    GNP               -0.006      0.015     -0.421      0.674

 F_LC       ON
    GNP               -0.037      0.018     -2.073      0.038

 Y1       WITH
    Y2                 0.000      0.000    999.000    999.000

 Intercepts
    Y1                 0.000      0.000    999.000    999.000
    Y2                 0.000      0.000    999.000    999.000
    F_LS              -0.033      0.017     -1.900      0.057
    F_LC               0.071      0.017      4.119      0.000

 Residual Variances
    Y1                 1.000      0.000    999.000    999.000
    Y2                 1.000      0.000    999.000    999.000
    F_LS               0.003      0.001      2.171      0.030
    F_LC               0.007      0.002      3.026      0.002

 New/Additional Parameters
    IND               -0.018      0.011     -1.657      0.097
```

```
    DIR                       -0.006       0.015       -0.421       0.674
    TOTAL                     -0.024       0.015       -1.571       0.116
------------------------ Selected output ------------------------
```

The estimated indirect, direct, and total effects with their SEs are −0.018(SE = 0.011), −0.006(SE = 0.015), and −0.024(SE = 0.015), respectively. None of them is statistically significant at $\alpha = 0.05$ using the Wald test.

We can fit the moderating model discussed in Section 5.6.3 in M*plus*. The models are

$$
\begin{aligned}
\text{Measurement model: } & \begin{bmatrix} \tilde{y}_{1,i} \\ \tilde{y}_{2,i} \end{bmatrix} = \tilde{X} \begin{bmatrix} f_{1,i} \\ f_{2,i} \end{bmatrix} + \begin{bmatrix} \tilde{e}_{1,i} \\ \tilde{e}_{2,i} \end{bmatrix}, \\
\text{Structural model: } & \begin{bmatrix} f_{1,i} \\ f_{2,i} \end{bmatrix} = \begin{bmatrix} \beta_{1,0} \\ \beta_{2,0} \end{bmatrix} + \begin{bmatrix} 0 & \beta_{1,2} \\ 0 & 0 \end{bmatrix} \begin{bmatrix} f_{1,i} \\ f_{2,i} \end{bmatrix} + \begin{bmatrix} \gamma_1 \\ 0 \end{bmatrix} z_i \\
& + z_i \begin{bmatrix} 0 & \omega_{1,2} \\ 0 & 0 \end{bmatrix} \begin{bmatrix} f_{1,i} \\ f_{2,i} \end{bmatrix} + \begin{bmatrix} u_{1,i} \\ u_{2,i} \end{bmatrix}.
\end{aligned}
\tag{9.18}
$$

Figure 9.8 displays the graphical model. Using the notation in M*plus*, the small filled circle represents the interaction between the observed variable z_i and the latent variable $f_{2,i}$. Therefore, $\omega_{1,2}$ is the regression coefficient of the moderating effect.

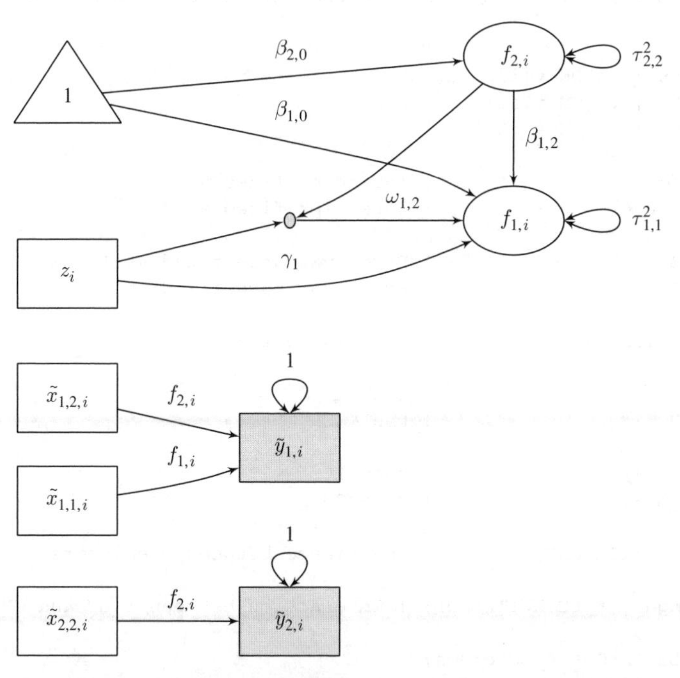

*Figure 9.8 A moderation model with two effect sizes in M*plus.

From the above equations, the equation for $f_{1,i}$ is

$$f_{1,i} = \beta_{1,0} + \gamma_1 z_i + (\beta_{1,2} + \omega_{1,2} z_i) f_{2,i} + u_{1,i}. \tag{9.19}$$

The following M*plus* code shows how to analyze the moderating model. We define the interaction effect between f_{LC} and GNP using `omega | f_LC XWITH GNP;`. Numerical integration is required to analyze this model. We specify 100 points for the integration using `INTEGRATION=100;`. As there are four processors in my server, we may speed up the analysis by using all processors with `PROCESSORS=4;`. This process is still computational intensive. It may take some time to complete.

```
TITLE:  Mediation model on the "true" effect sizes
DATA:   FILE IS wvs94a.dat;
VARIABLE: NAMES y1 y2 y1f1 y1f2 y2f2 GNP;
    USEVARIABLES ARE ALL;
    MISSING ARE *;

ANALYSIS: TYPE=RANDOM;
    ESTIMATOR=ML;            ! Use ML estimation
    ALGORITHM=INTEGRATION;
    PROCESSORS=4;           ! Use 4 processors to speed up the analysis
    INTEGRATION=100;         ! Number of points for integration
    CHOLESKY=ON;            ! Ensure the variances are positive definite

MODEL:
    f_LS | y1 ON y1f1;
    f_LC | y1 ON y1f2;
    f_LC | y2 ON y2f2;

    [y1@0.0];                ! Intercept fixed at 0
    [y2@0.0];                ! Intercept fixed at 0

    y1@1.0                   ! Error variance fixed at 1
    y2@1.0                   ! Error variance fixed at 1

    y1 WITH y2@0;            ! Covariance fixed at 0

    f_LS*;                   ! tau^2_11
    f_LC*;                   ! tau^2_22

    [f_LS*];                 ! beta_10
    [f_LC*];                 ! beta_20

    f_LC WITH GNP;           ! Covariance between predictors

    omega | f_LC XWITH GNP; ! Interaction between f_LC and GNP

    f_LS ON GNP f_LC omega;
```

```
OUTPUT: SAMPSTAT;
    CINTERVAL(symmetric); ! Wald CI
```

```
------------------------ Selected output ----------------------
*** WARNING
  Data set contains cases with missing on variables used to define
  interactions.  These cases were not included in the analysis.
  Number of such cases:  5
   1 WARNING(S) FOUND IN THE INPUT INSTRUCTIONS

MODEL FIT INFORMATION

Number of Free Parameters                        10

Loglikelihood

          H0 Value                         -192.133

Information Criteria

          Akaike (AIC)                      404.267
          Bayesian (BIC)                    420.376
          Sample-Size Adjusted BIC          389.122
            (n* = (n + 2) / 24)

MODEL RESULTS
                                                      Two-Tailed
                    Estimate       S.E.  Est./S.E.     P-Value

 F_LS       ON
    F_LC               0.471      0.170      2.776       0.006
    OMEGA              0.072      0.171      0.421       0.674

 F_LS       ON
    GNP               -0.010      0.018     -0.580       0.562

 F_LC     WITH
    GNP               -0.035      0.018     -1.888       0.059

 Y1       WITH
    Y2                 0.000      0.000    999.000     999.000

 Means
    GNP                0.000      0.158      0.000       1.000
    F_LC               0.070      0.018      3.895       0.000

 Intercepts
    Y1                 0.000      0.000    999.000     999.000
    Y2                 0.000      0.000    999.000     999.000
    F_LS              -0.029      0.019     -1.563       0.118
```

```
Varianc es
    GNP                  0.919      0.214      4.301      0.000
    F_LC                 0.009      0.003      3.119      0.002

Residual Variances
    Y1                   1.000      0.000    999.000    999.000
    Y2                   1.000      0.000    999.000    999.000
    F_LS                 0.003      0.001      2.047      0.041
------------------------ Selected output ----------------------
```

The estimated regression equation for the average $f_{1,i}$ is

$$\hat{f}_{1,i} = -0.029 - 0.010\text{GNP} + (0.471 + 0.072\text{GNP})f_{2,i}.$$

The estimated moderating effect with its SE is 0.072 (SE= 0.171), $p = 0.674$. There is no evidence in justifying the moderating effect.

9.4 Three-level meta-analysis

When multiple effect sizes are reported in a study, we cannot assume that the effect sizes are independent. If the conditional sampling covariance matrix can be estimated, we may use the multivariate meta-analysis to model the dependence. If the degree of dependence is unknown, we can apply a three-level meta-analysis to conduct the analysis by taking the unknown degree into account (see Chapter 6 and Cheung (2014) for details). The transformed variables approach can be applied to the three-level meta-analysis in M*plus*. We illustrate how to conduct a three-level random-effects and mixed-effects models in this section.

9.4.1 Random-effects model

A three-level meta-analysis of the transformed effect size of the ith effect size in the jth cluster is

$$\begin{aligned}
&\text{Level 1:} && \tilde{y}_{ij} = \lambda_{ij}\sqrt{w_{ij}} + \tilde{e}_{ij},\\
&\text{Level 2:} && \lambda_{ij} = f_j + u_{(2)ij},\\
&\text{Level 3:} && f_j = \beta_0 + u_{(3)j},
\end{aligned} \tag{9.20}$$

where $\tilde{y}_{ij} = \sqrt{w_{ij}}y_{ij}$ is the transformed effect size, $\sqrt{w_{ij}} = 1/\sqrt{v_{ij}}$ is the transformation variable, $\tilde{e}_{ij} = \sqrt{w_{ij}}e_{ij}$ is the transformed residual, λ_{ij} is the *true* effect size, $\text{Var}(e_{ij}) = v_{ij}$ is the known sampling variance for the ith effect size in the jth cluster, f_j is the *true* effect size in the jth cluster, β_0 is the average population effect, and $\text{Var}(u_{(2)ij}) = \tau^2_{(2)}$ and $\text{Var}(u_{(3)j}) = \tau^2_{(3)}$ are the level-2 and level-3 heterogeneity variances, respectively.

The key feature of this approach is that the known sampling variance v_{ij} is transformed into a known variance of 1. After the transformation, λ_{ij} is considered as a random slope at level 2; f_j is a random slope at level 3. Both the level-2 and level-3 variances of the random slopes are estimated. There is only one fixed effect, which is β_0.

We illustrate the three-level meta-analysis using the data set from Bornmann et al. (2007), which was also used in Section 6.5. The effect size was a log odds ratio that measured the odds of being approved among the female applicants divided by the odds of being approved among the male applicants. The data file was exported as a plain text `Bornmann07.dat`. The following R code was used to export the data.

```
R> library(metaSEM)
R> ## Select the relevant variables to export
R> my.df <- Bornmann07[, c(3,1,4:6)]
R> ## Standardize "year"
R> my.df$Year <- scale(my.df$Year)
R> Fellow <- ifelse(Bornmann07$Type=="Fellowship", yes=1, no=0)
R> D_Phy <- ifelse(Bornmann07$Discipline=="Physical sciences",
                   yes=1, no=0)
R> D_Life <- ifelse(Bornmann07$Discipline=="Life sciences/biology",
                    yes=1, no=0)
R> D_Soc<- ifelse(Bornmann07$Discipline==
                             "Social sciences/humanities",
                  yes=1, no=0)
R> D_Mul <- ifelse(Bornmann07$Discipline=="Multidisciplinary",
                   yes=1, no=0)
R> C_USA <- ifelse(Bornmann07$Country=="United States", yes=1, no=0)
R> C_Aus <- ifelse(Bornmann07$Country=="Australia", yes=1, no=0)
R> C_Can <- ifelse(Bornmann07$Country=="Canada", yes=1, no=0)
R> C_Eur <- ifelse(Bornmann07$Country=="Europe", yes=1, no=0)
R> C_UK <- ifelse(Bornmann07$Country=="United Kingdom", yes=1, no=0)
R> my.df <- cbind(my.df, Fellow, D_Phy, D_Life, D_Soc, D_Mul,
                  C_USA, C_Aus, C_Can, C_Eur, C_UK)
R> ## Show a few cases
R> head(my.df)
```

```
  Cluster Id    logOR          v        Year Fellow D_Phy D_Life
1       1  1 -0.40108 0.01391692 -0.08020158      1     1      0
2       1  2 -0.05727 0.03428793 -0.08020158      1     1      0
3       1  3 -0.29852 0.03391122 -0.08020158      1     1      0
4       1  4  0.36094 0.03404025 -0.08020158      1     1      0
5       1  5 -0.33336 0.01282103 -0.08020158      1     0      0
6       1  6 -0.07173 0.01361189 -0.08020158      1     1      0
  D_Soc D_Mul C_USA C_Aus C_Can C_Eur C_UK
1     0     0     0     0     0     1    0
2     0     0     0     0     0     1    0
3     0     0     0     0     0     1    0
4     0     0     0     0     0     1    0
5     1     0     0     0     0     1    0
6     0     0     0     0     0     1    0
```

```
R> ## Write to an external file
R> write.table(my.df, "Bornmann07.dat", row.names=FALSE,
               col.names=FALSE, na="NA", sep="\t")
```

To conduct a three-level meta-analysis in M*plus*, we need to specify the level 2 (`ID`) and the level 3 (`Cluster`) in the `cluster` command. The `ID` is just a unique label for each effect size, while the `Cluster` is a label on how the effect sizes are nested. Moreover, we specify the three-level modeling with random slopes using `ANALYSIS: TYPE=THREELEVEL RANDOM;`.

```
TITLE: Random-effects model
DATA:   FILE IS Bornmann07.dat;
VARIABLE: NAMES Cluster ID y v Year Fellow D_Phy D_Life D_Soc D_Mul
          C_USA C_Aus C_Can C_Eur C_UK;

USEVARIABLES y w2;               ! Use both y and w2 in the analysis
        cluster = Cluster ID;      ! Level 2: ID; Level 3: Cluster
        within = y w2;             ! Define within level variables

DEFINE: w2 = SQRT(v**(-1));        ! Define the transformation weight
        y = w2*y;                  ! Transform the effect size
                                   ! Use three-level modeling
ANALYSIS: TYPE=THREELEVEL RANDOM; ! Activate random slope analysis
        ESTIMATOR = ML;

MODEL:  %WITHIN%
        [y@0.0];
        y@1.0;
        f | y ON w2;               ! Define random slope

        %BETWEEN ID%               ! Level 2 variance
        f*;                        ! Optional; default model

        %BETWEEN Cluster%          ! Level 3 variance
        f*;                        ! Optional; default model

OUTPUT: SAMPSTAT;
        TECH1 TECH8;
```

```
------------------------ Selected output ----------------------
MODEL FIT INFORMATION

Number of Free Parameters                              3

Loglikelihood

         H0 Value                               -119.470
```

```
Information Criteria

        Akaike (AIC)                          244.939
        Bayesian (BIC)                        251.508
        Sample-Size Adjusted BIC              242.064
          (n* = (n + 2) / 24)

MODEL RESULTS
                                                        Two-Tailed
                    Estimate       S.E.  Est./S.E.    P-Value

Within Level

 Intercepts
    Y                  0.000      0.000    999.000    999.000

 Residual Variances
    Y                  1.000      0.000    999.000    999.000

Between ID Level

 Variances
    F                  0.004      0.003      1.395      0.163

Between CLUSTER Level

 Means
    F                 -0.101      0.040     -2.511      0.012

 Variances
    F                  0.014      0.009      1.546      0.122
 ----------------------- Selected output ----------------------
```

The estimated average effect with its SE based on the three-level model is $-0.101(\text{SE} = 0.040)$, $p = 0.012$. The result shows that there is a slight favor to male applicants. The estimated level-2 and level-3 heterogeneity variances, $\tau^2_{(2)}$ and $\tau^2_{(3)}$, are 0.004 and 0.014, respectively.

As there are two variance components, it is of interest to test the null hypothesis $H_0 : \tau^2_{(2)} = \tau^2_{(3)}$. This hypothesis is particularly relevant in cross-cultural meta-analysis where $\tau^2_{(2)}$ and $\tau^2_{(3)}$ represent the intracultural and between-cultural variations, respectively. We use the following M*plus* code to test the equality constraint.

```
TITLE:    Random-effects model: tau^2_3=tau^2_2
DATA:     FILE IS Bornmann07.dat;
VARIABLE: NAMES Cluster ID y v Year Fellow D_Phy D_Life D_Soc D_Mul
          C_USA C_Aus C_Can C_Eur C_UK;

USEVARIABLES y w2;
```

```
       cluster = Cluster ID;         ! Level 2: ID; Level 3: Cluster
       within = y w2;                ! Define within level variables

DEFINE: w2 = SQRT(v**(-1));
       y = w2*y;
                                     ! Use three-level modeling
ANALYSIS: TYPE=THREELEVEL RANDOM;    ! Activate random slope function
       ESTIMATOR = ML;

MODEL:  %WITHIN%
       [y@0.0];
       y@1.0;
       f | y ON w2;                  ! Define random slope

       %BETWEEN ID%                  ! Level 2 variance
       f* (1);                       ! tau^2_3=tau^2_2

       %BETWEEN Cluster%             ! Level 3 variance
       f* (1);                       ! tau^2_3=tau^2_2

OUTPUT: SAMPSTAT;
       TECH1 TECH8;
```

```
------------------------ Selected output ----------------------
MODEL FIT INFORMATION

Number of Free Parameters                          2

Loglikelihood

         H0 Value                           -120.149

Information Criteria

         Akaike (AIC)                        244.298
         Bayesian (BIC)                      248.678
         Sample-Size Adjusted BIC            242.381
           (n* = (n + 2) / 24)

MODEL RESULTS
                                                        Two-Tailed
                    Estimate       S.E.  Est./S.E.    P-Value

Within Level

 Intercepts
    Y                  0.000      0.000    999.000    999.000

 Residual Variances
    Y                  1.000      0.000    999.000    999.000
```

```
Between ID Level

 Variances
    F                    0.007         0.003         2.455         0.014

Between CLUSTER Level

 Means
    F                   -0.096         0.035        -2.775         0.006

 Variances
    F                    0.007         0.003         2.455         0.014
------------------------ Selected output -----------------------
```

The pooled estimate on the level-2 and level-3 heterogeneity variances is 0.007. The LR statistic on testing the equality constraint is $-2(-120.149 + 119.470) = 1.358$ which is not significant with $\Delta\chi^2(\text{df} = 1) = 1.358, p = 0.244$.

9.4.2 Mixed-effects model

A three-level random-effects meta-analysis can be extended to a mixed-effects model by using study characteristics as moderators. The mixed-effects model of the transformed effect sizes of the ith effect size in the jth cluster is

$$\begin{aligned} \text{Level 1:} \quad & \tilde{y}_{ij} = \lambda_{ij}\sqrt{w_{ij}} + \tilde{e}_{ij}, \\ \text{Level 2:} \quad & \lambda_{ij} = f_j + u_{(2)ij}, \\ \text{Level 3:} \quad & f_j = \beta_0 + \beta_1 z_j + u_{(3)j}, \end{aligned} \tag{9.21}$$

where z_j is a moderator in the jth cluster. Now, $\tau^2_{(2)}$ and $\tau^2_{(3)}$ represent the level-2 and level-3 residual heterogeneity after controlling for z_j, respectively. If the moderators are level-2 variables, we include them in the level-2 model; otherwise, we include them in the level-3 model.

The following M*plus* code demonstrates how to conduct a mixed-effects model with a predictor `Fellow`, which is a level-2 moderator. `Fellow` is an indicator for the type of applications. If the type is fellowship, it is 1; if it is grant, it is 0. As `Fellow` is a level-2 variable, it is defined in `BETWEEN = (ID) Fellow;`.

```
TITLE:   Mixed-effects model: Fellow as the predictor
DATA:    FILE IS Bornmann07.dat;
VARIABLE: NAMES Cluster ID y v Year Fellow D_Phy D_Life D_Soc D_Mul
          C_USA C_Aus C_Can C_Eur C_UK;

USEVARIABLES y Fellow w2;
        cluster = Cluster ID;           ! Level 2: ID; Level 3: Cluster
        within = y w2;                  ! Define within level variables
        BETWEEN = (ID) Fellow;          ! Fellow is a level-2 variable
```

```
DEFINE: w2 = SQRT(v**(-1));
        y = w2*y;
                                     ! Use three-level modeling
ANALYSIS: TYPE=THREELEVEL RANDOM;    ! Activate random slope function
        ESTIMATOR = ML;

MODEL:  %WITHIN%
        [y@0.0];
        y@1.0;
        f | y ON w2;                 ! Define random slope

        %BETWEEN ID%                 ! Level 2 variable
        f ON Fellow;

        %BETWEEN Cluster%            ! Level 3 variance
        f*;                          ! Optional, default model

OUTPUT: SAMPSTAT;
        TECH1 TECH8;
```

```
------------------------ Selected output ----------------------
MODEL FIT INFORMATION

Number of Free Parameters                          4

Loglikelihood

          H0 Value                          -115.381

Information Criteria

          Akaike (AIC)                       238.763
          Bayesian (BIC)                     247.521
          Sample-Size Adjusted BIC           234.928
            (n* = (n + 2) / 24)

MODEL RESULTS
                                                         Two-Tailed
                    Estimate       S.E.  Est./S.E.      P-Value

Within Level

 Intercepts
    Y                  0.000      0.000    999.000      999.000

 Residual Variances
    Y                  1.000      0.000    999.000      999.000
```

```
Between ID Level

 F        ON
    FELLOW              -0.196        0.054       -3.611        0.000

 Residual Variances
    F                    0.004        0.002        1.454        0.146

Between CLUSTER Level

 Means
    F                   -0.007        0.037       -0.178        0.859

 Variances
    F                    0.003        0.003        0.933        0.351
-----------------------  Selected output  ----------------------
```

The estimated intercept with its SE is $-0.007(SE = 0.037)$, $p = 0.859$. This indicates that there is no evidence on the gender difference on *grant*. The estimated regression slope with its SE is $-0.196(SE = 0.054)$, $p < 0.001$. Thus, there is a significant difference between the log odds ratios on *fellowship* and *grant*.

9.5 Concluding remarks and further readings

This chapter introduced the transformed variables approach to conduct the SEM-based meta-analyses in M*plus*. After applying the transformation, M*plus* can be used to analyze univariate, multivariate, and three-level meta-analyses. Both fixed- and random-effects models can be used in the analyses. Other SEM packages, such as LISREL (Jöreskog and Sörbom, 1996) and EQS (Bentler, 2006), may be used to conduct a fixed-effects meta-analysis using the transformed variables approach or a multiple-group approach using the WLS estimation (Cheung, 2010). Recently, Palmer and Sterne (2014) showed how the SEM-based meta-analyses could be implemented in Stata. Stata users may refer to their work for more details.

There are many potential applications and extensions of using M*plus* to conduct meta-analysis. For example, FIML and robust statistics can be used to handle missing predictors and nonnormal data. A more recent development is the integration of Bayesian inferences in M*plus* (Muthén and Asparouhov, 2012). Bayesian statistics is an alternative choice to conducting meta-analysis (e.g., Mak et al., 2009; Bujkiewicz et al., 2013; Sutton and Abrams, 2001). By applying the transformed variables approach, researchers can conduct a Bayesian meta-analysis in M*plus*. As M*plus* has only recently been introduced as a software to conduct meta-analyses (Cheung, 2008, 2013), more studies should explore how to apply M*plus* to conduct meta-analysis.

References

Bentler PM 2006. *EQS 6 structural equations program manual*. Multivariate Software, Encino, CA.

Boker S, Neale M, Maes H, Wilde M, Spiegel M, Brick T, Spies J, Estabrook R, Kenny S, Bates T, Mehta P and Fox J 2011. OpenMx: an open source extended structural equation modeling framework. *Psychometrika* **76**(2), 306–317.

Bornmann L, Mutz R and Daniel HD 2007. Gender differences in grant peer review: a meta-analysis. *Journal of Informetrics* **1**(3), 226–238.

Bujkiewicz S, Thompson JR, Sutton AJ, Cooper NJ, Harrison MJ, Symmons DP and Abrams KR 2013. Multivariate meta-analysis of mixed outcomes: a Bayesian approach. *Statistics in Medicine* **32**(22), 3926–3943.

Cheung MWL 2008. A model for integrating fixed-, random-, and mixed-effects meta-analyses into structural equation modeling. *Psychological Methods* **13**(3), 182–202.

Cheung MWL 2010. Fixed-effects meta-analyses as multiple-group structural equation models. *Structural Equation Modeling: A Multidisciplinary Journal* **17**(3), 481–509.

Cheung MWL 2013. Multivariate meta-analysis as structural equation models. *Structural Equation Modeling: A Multidisciplinary Journal* **20**(3), 429–454.

Cheung MWL 2014. Modeling dependent effect sizes with three-level meta-analyses: a structural equation modeling approach. *Psychological Methods* **19**(2), 211–229.

Hedges LV and Olkin I 1985. *Statistical methods for meta-analysis*. Academic Press, Orlando, FL.

Higgins JPT and Thompson SG 2002. Quantifying heterogeneity in a meta-analysis. *Statistics in Medicine* **21**(11), 1539–1558.

Jaramillo F, Mulki JP and Marshall GW 2005. A meta-analysis of the relationship between organizational commitment and salesperson job performance: 25 years of research. *Journal of Business Research* **58**(6), 705–714.

Jöreskog KG and Sörbom D 1996. *LISREL 8: a user's reference guide*. Scientific Software International, Inc., Chicago, IL.

Kalaian HA and Raudenbush SW 1996. A multivariate mixed linear model for meta-analysis. *Psychological Methods* **1**(3), 227–235.

Mak A, Cheung MWL, Ho RCM, Cheak AAC and Lau CS 2009. Bisphosphonates and atrial fibrillation: Bayesian meta-analyses of randomized controlled trials and observational studies. *BMC Musculoskeletal Disorders* **10**(1), 1–12.

Muthén B and Asparouhov T 2012. Bayesian structural equation modeling: a more flexible representation of substantive theory. *Psychological Methods* **17**(3), 313–335.

Muthén BO and Muthén LK 2012. *Mplus user's guide*, 7th edn. Muthén & Muthén, Los Angeles, CA.

Palmer TM and Sterne JAC 2014. Fitting fixed and random effects meta-analysis models using structural equation modeling with the SEM and GSEM commands. Manuscript submitted for publication.

Raudenbush SW, Becker BJ and Kalaian H 1988. Modeling multivariate effect sizes. *Psychological Bulletin* **103**(1), 111–120.

Sutton AJ and Abrams KR 2001. Bayesian methods in meta-analysis and evidence synthesis. *Statistical Methods in Medical Research* **10**(4), 277–303.

World Values Study Group 1994. *World Values Survey, 19811984 and 19901993 [Computer file]*. Inter-university Consortium for Political and Social Research, Ann Arbor, MI.

Appendix A

A brief introduction to R, OpenMx, and metaSEM packages

This appendix provides a short introduction to the R statistical environment (R Development Core Team, 2014), the OpenMx package (Boker et al., 2011), and the metaSEM package (Cheung, 2014). R is a statistical environment for data manipulation, calculations, and graphical display. It provides a complete system for statistical analyses and statistical programming. Many state-of-the-art statistical techniques have been implemented in R via packages. Readers can easily download and apply these techniques. If you are not familiar with R, you may get it started by reading some books in R, for example, Adler (2010), Dalgaard (2008), and Zuur et al. (2009). There is an updated list of books in R available at http://www.r-project.org/doc/bib/R-books.html. Moreover, there are also lots of free resources available to learning R in the R Web site. Readers are strongly recommended to go through at least *An Introduction to R* which is available at http://cran.r-project.org/doc/manuals/r-release/R-intro.html. This appendix is based on R (3.1.1), metafor (1.9-3), OpenMx (2.0.0-3654), and metaSEM (0.9-0). The output format may be slightly different from the version that you are using. The Web sites for the software are

(i) R: http://www.r-project.org/;

(ii) OpenMx: http://openmx.psyc.virginia.edu/; and

(iii) metaSEM: http://courses.nus.edu.sg/course/psycwlm/Internet/metaSEM/.

Meta-Analysis: A Structural Equation Modeling Approach, First Edition. Mike W. -L. Cheung.

Companion Website: www.wiley.com/go/cheung/meta_analysis

A.1 R

R can be used interactively or noninteractively in batch mode. The batch mode is usually for routine analyses and programming. We only focus the interactive mode here. R comes with a standard command line interface for inputs and outputs. Many new users may find it more convenient to use a graphical user interface, such as RStudio. Users may call R by typing R in the terminal in Unix-like systems or clicking the R icon in Windows-like systems. The following are a few features in R:

1. Everything after # is considered as comments and ignored by R. It is always a good practice to document the analyses using comments;
2. Everything is an object in R. We may store the results via the <- assign operator for further processing; and
3. Everything is case sensitive in R; x and X are different objects.

We illustrate R with a few examples:

```
R> ## Assign 10 to x
R> x <- 10
R> ## Display the content of x
R> x
```

```
[1] 10
```

```
R> ## Multiply 5 to x, store it as X,
R> ## and display the content of X using ()
R> ( X <- 5*x )
```

```
[1] 50
```

```
R> ## x and X are different in R
R> x
```

```
[1] 10
```

Most of the operators in R were built for vectorization. R usually runs faster for vectorized objects. Moreover, the syntax is also more compact by using vectorization.

```
R> ## x is a vector of data
R> ## c() is used to combine the data
R> ( x <- c(5, 6, 9, 10) )
```

```
[1]  5  6  9 10
```

```
R> ## Square the data in x
R> (x^2)
```

```
[1]  25  36  81 100
```

```
R> ## Create another vector of data
R> ( y <- c(5, 9, 8, 11) )
```

```
[1]  5  9  8 11
```

```
R> ## Multiply the elements one by one
R> ( x*y )
```

```
[1]  25  54  72 110
```

Data frames are usually used to represent data in statistical analyses. The columns and rows represent the *variables* and *subjects*, respectively. Many functions in R read data frames as inputs. For example, we may create a data frame and use the data for further analyses. `y=y` in the following example means that we create a variable named `y` in the data frame copied from the data `y`. If we use `k=y`, the content of `y` will be copied to a variable named `k` in the data frame.

```
R> ## Create a data frame called my.df1
R> ( my.df1 <- data.frame(y=y, x=x) )
```

```
   y  x
1  5  5
2  9  6
3  8  9
4 11 10
```

```
R> ## Get the summary of the data
R> summary(my.df1)
```

```
       y                x
 Min.   : 5.00   Min.   : 5.00
 1st Qu.: 7.25   1st Qu.: 5.75
 Median : 8.50   Median : 7.50
```

```
 Mean    : 8.25  Mean    : 7.50
 3rd Qu.: 9.50  3rd Qu.: 9.25
 Max.   :11.00  Max.   :10.00
```

```
R> ## Calculate the covariance matrix of the variables
R> cov(my.df1)
```

```
     y     x
y 6.25 4.500
x 4.50 5.667
```

```
R> ## Calculate the means of the variables
R> ## apply(): apply the function either by rows or by columns
R> ## MARGIN: "1" for by rows and "2" for by columns
R> ## FUN: name of the function
R> apply(my.df1, MARGIN=2, FUN=mean)
```

```
   y    x
8.25 7.50
```

We conduct a regression analysis with the `lm()` (linear model) function by regression y on x. We provide two arguments to the `lm()` function. The first argument `y~x` means that y is modeled by x that specifies a linear relationship between x and y, while the second argument `data=my.df1` supplies the data. After the analysis, we plot the data with the `plot()` function and draw the best fitted line with the `abline()` function.

```
R> ## Conduct a simple regression by regressing y on x
R> my.lm <- lm(y~x, data=my.df1)
R> ## Get the summary
R> summary(my.lm)
```

```
Call:
lm(formula = y ~ x, data = my.df1)

Residuals:
     1      2      3      4
-1.265  1.941 -1.441  0.765

Coefficients:
            Estimate Std. Error t value Pr(>|t|)
(Intercept)    2.294      3.780    0.61     0.61
x              0.794      0.486    1.63     0.24

Residual standard error: 2 on 2 degrees of freedom
```

```
Multiple R-squared:  0.572, Adjusted R-squared:  0.358
F-statistic: 2.67 on 1 and 2 DF,  p-value: 0.244
```

```
R> ## Get the regression coefficient
R> coef(my.lm)
```

```
(Intercept)           x
     2.2941      0.7941
```

```
R> ## Figure A.1
R> ## Plot the data on y against x
R> plot(y~x, data=my.df1)
R> ## Draw the best fitted line using the my.lm object
R> abline(my.lm)
```

When the basic functions in R are not sufficient for the analyses, we may extend the power of R by installing packages. Packages store functions and data sets for similar analyses. The Comprehensive R Archive Network (CRAN) repository features more than 6100 contributed packages at the time of writing. They can be installed using the `install.packages()` function. We can also update the installed packages using the `update.packages()` function. As the `OpenMx` and `metaSEM` packages are not available in the CRAN repository yet, readers many refer to the Web sites of `OpenMx` and `metaSEM` on how to install them. As an example, we install the `metafor` package (Viechtbauer, 2010) and use it to calculate the standardized mean difference (SMD) for two studies. Each row

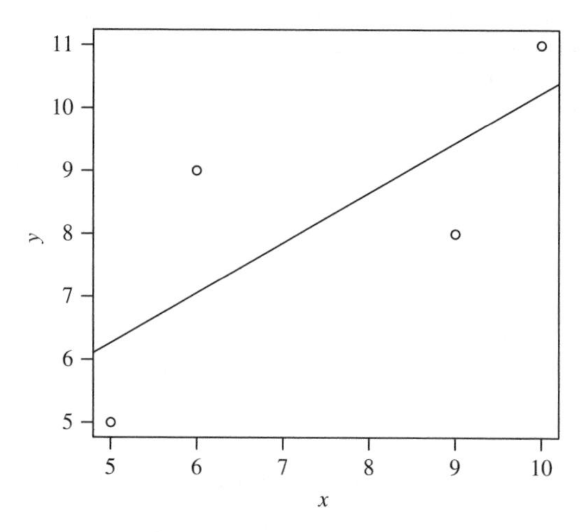

Figure A.1 A plot of y against x with the best fitted regression line.

represents one study. The calculated effect size and its sampling variance are called `yi` and `vi`, respectively.

```
R> ## We only need to install it once.
R> install.packages("metafor")
```

```
R> ## Load the library before use
R> library("metafor")
R> ## Prepare a dataset
R> ## Study ID in the meta-analysis
R> Study <- c("Author (2012)", "Author (2014)")
R> ## Means of the treatment group
R> m1i <- c(12, 14)
R> ## SDs of the treatment group
R> sd1i <- c(4, 5)
R> ## Sample sizes of the treatment group
R> n1i <- c(40, 50)
R> ## Means of the control group
R> m2i <- c(10, 11)
R> ## SDs of the control group
R> sd2i <- c(4, 5)
R> ## Sample sizes of the control groups
R> n2i <- c(40, 50)
R> ## Create a data frame
R> ( my.df2 <- data.frame(Study=Study, m1i=m1i, sd1i=sd1i, n1i=n1i,
                         m2i=m2i, sd2i=sd2i, n2i=n2i) )
```

```
          Study m1i sd1i n1i m2i sd2i n2i
1 Author (2012)  12    4  40  10    4  40
2 Author (2014)  14    5  50  11    5  50
```

```
R> ## Calculate the SMD
R> escalc(measure="SMD", m1i=m1i, sd1i=sd1i, n1i=n1i,
          m2i=m2i, sd2i=sd2i, n2i=n2i, data=my.df2)
```

```
          Study m1i sd1i n1i m2i sd2i n2i     yi     vi
1 Author (2012)  12    4  40  10    4  40 0.4952 0.0515
2 Author (2014)  14    5  50  11    5  50 0.5954 0.0418
```

R provides a comprehensive help manual. There are several functions helping users to search the appropriate manuals.

```
R> ## Read the help manual on the log() function
R> help(log)
R> ?log
R> ## Run the examples in the manual
R> example(log)
```

```
R> ## Search the relevant functions with log()
R> help.search("log")
R> ## Start the html browser
R> help.start()
```

A.2 OpenMx

As the metaSEM package is based on the OpenMx package, it is helpful to learn a bit the syntax of the OpenMx package. The OpenMx package is a matrix optimizer that can be used to fit general structural equation models. We illustrate the basic idea with a simple regression model. We use the RAM formulation for model specifications (see Section 2.2.3). The variables are arranged as x and y in the model specification.

```
R> ## Load the library
R> library("OpenMx")
R> ## Sample covariance matrix
R> ( my.cov <- matrix(c(4.5, 2.0, 2.0, 3.6), nrow=2, ncol=2,
                      dimnames=list(c("x","y"), c("x","y"))) )
```

```
    x   y
x 4.5 2.0
y 2.0 3.6
```

```
R> ## Sample means of the variables
R> my.means <- c(5, 7)
R> ## Add the names for the means
R> names(my.means) <- c("x", "y")
R> my.means
```

```
x y
5 7
```

```
R> ## Prepare the matrices in the RAM formulation
R> ## A matrix: asymmetric matrix representing regression
R> ## coefficients and factor loadings
R> ## type: type of the matrix; "Full": full matrix
R> ## free: whether the parameters are free or fixed
R> ## values: starting values for free parameters and fixed values
R> ##  for fixed parameters
R> ## labels: labels of the parameters; they are constrained equally
R> ##  if the labels are the same
R> ## byrow: whether the data are arranged by row (or by column)
R> ## name: name of the matrix
```

```
R> A1 <- mxMatrix(type="Full", nrow=2, ncol=2,
                  free=c(FALSE, FALSE,
                         TRUE, FALSE),
                  values=0,
                  labels=c(NA, NA,
                           "beta1", NA),
                  byrow=TRUE,
                  name="A1")
R> ## S matrix: symmetric matrix representing variance covariance of
R> ##  the variables
R> ## type: "Symm" means symmetric matrix
R> S1 <- mxMatrix(type="Symm", nrow=2, ncol=2,
                  values=c(1, 0,
                           0, 1),
                  free=c(TRUE, FALSE,
                         FALSE, TRUE),
                  labels=c("VarX", NA,
                           NA, "ErrorVarY"),
                  byrow=TRUE,
                  name="S1")
R> ## M matrix: mean vector of the variables
R> M1 <- mxMatrix(type="Full", nrow=1, ncol=2,
                  free=c(TRUE, TRUE),
                  values=c(0, 0),
                  labels=c("meanx", "beta0"),
                  name="M1")
R> ## F matrix: a selection matrix to select the observed variables
R> ## type: "Iden" means identity matrix
R> F1 <- mxMatrix(type="Iden", nrow=2, ncol=2, name="F1")
R> ## Create a model
R> ## The sample size is 100.
R> reg.model <- mxModel("Simple Regression",
                        mxData(observed=my.cov, type="cov",
                               numObs=100, means=my.means),
                        A1, S1, M1, F1,
                        mxExpectationRAM(A="A1", S="S1",
                                         F="F1", M="M1",
                        dimnames = c("x","y")),
                        mxFitFunctionML())
R> ## Run the analysis
R> reg.fit <- mxRun(reg.model, silent=TRUE)
```

```
R> ## Display the results
R> summary(reg.fit)
```

```
----------------------- Selected output -----------------------

free parameters:
      name matrix row col Estimate Std.Error lbound ubound
1     beta1    A1   2   1   0.4444   0.07801
```

```
2       VarX    S1   1   1    4.5000    0.63960
3 ErrorVarY     S1   2   2    2.7111    0.38534
4      meanx    M1   1   x    5.0000    0.21213
5      beta0    M1   1   y    4.7778    0.42338

observed statistics:  5
estimated parameters:  5
degrees of freedom:  0
-2 log likelihood:  445.6
saturated -2 log likelihood:  445.6
number of observations:  100
chi-square:  1.745e-09
chi-square degrees of freedom:  0
chi-square p-value:  0

-----------------------  Selected output  ----------------------
```

A.3 metaSEM

The metaSEM conducts meta-analyses by formulating meta-analytic models as structural equation models. There are two main types of analyses in the package—SEM-based meta-analyses, for example, univariate, multivariate, and three-level meta-analyses, and meta-analytic structural equation modeling with the two-stage structural equation modeling (TSSEM) approach. This section shows how to read external data for a meta-analysis.

When we are conducting a meta-analysis, the data are likely externally stored in some formats such as Excel. Users can save the data in comma-separated values (CSV) format. For example, the metaSEM package includes a sample data set Becker83. The data set includes studies on sex differences in conformity using the fictitious norm group paradigm reported by Becker (1983). As an illustration, we exported the data to an external file named Becker83.csv. The content of the file is displayed as follows.

```
"study","di","vi","percentage","items"
1,-0.33,0.03,25,2
2,0.07,0.03,25,2
3,-0.3,0.02,50,2
4,0.35,0.02,100,38
5,0.69,0.07,100,30
6,0.81,0.22,100,45
7,0.4,0.05,100,45
8,0.47,0.07,100,45
9,0.37,0.05,100,5
10,-0.06,0.03,100,5
```

Suppose that the data file is stored in the same directory as the R session; we can read the file with the following syntax. If the file is stored in a different directory, we need to specify the directory. Please note that the path is specified differently for Windows-like systems. Instead of using "d:\abc\def" in Windows-like systems, we have to use either "d:\\abc\\def" or "/d/abc/def".

```
R> ## Load the library
R> library("metaSEM")
R> ## Display the current directory
R> getwd()
R> ## Set the working directory to /mydirectory/R/data/
R> ##  in Unix-like systems
R> # setwd("/mydirectory/R/data/")
R> ## Set the working directory to D:\mydirectory\R\data
R> ##  in Windows-like systems
R> # setwd("D:\\mydirectory\\R\\data")
R> # setwd("/D/mydirectory/R/data")
R>
R> ## Read the data in CSV format
R> ( my.df3 <- read.csv("Becker83.csv") )
```

```
   study    di   vi percentage items
1      1 -0.33 0.03         25     2
2      2  0.07 0.03         25     2
3      3 -0.30 0.02         50     2
4      4  0.35 0.02        100    38
5      5  0.69 0.07        100    30
6      6  0.81 0.22        100    45
7      7  0.40 0.05        100    45
8      8  0.47 0.07        100    45
9      9  0.37 0.05        100     5
10    10 -0.06 0.03        100     5
```

```
R> ## Mixed-effects meta-analysis with log(items) as the predictor
R> summary( meta(y=di, v=vi, x=log(items), data=my.df3) )
```

```
-----------------------  Selected output  ----------------------

95% confidence intervals: z statistic approximation
Coefficients:
            Estimate Std.Error     lbound     ubound z value Pr(>|z|)
Intercept1 -3.20e-01  1.10e-01 -5.35e-01 -1.05e-01   -2.92   0.0036
Slope1_1    2.11e-01  4.51e-02  1.23e-01  2.99e-01    4.68  2.9e-06
Tau2_1_1    1.00e-10  2.01e-02 -3.94e-02  3.94e-02    0.00   1.0000

Intercept1 **
Slope1_1   ***
Tau2_1_1
---
Signif. codes:  0 '***' 0.001 '**' 0.01 '*' 0.05 '.' 0.1 ' ' 1
```

```
Q statistic on the homogeneity of effect sizes: 30.65
Degrees of freedom of the Q statistic: 9
P value of the Q statistic: 0.0003399

Explained variances (R2):
                          y1
Tau2 (no predictor)     0.08
Tau2 (with predictors) 0.00
R2                      1.00

Number of studies (or clusters): 10
Number of observed statistics: 10
Number of estimated parameters: 3
Degrees of freedom: 7
-2 log likelihood: -4.208
OpenMx status1: 0 ("0" or "1": The optimization is considered fine.
Other values indicate problems.)

------------------------ Selected output ----------------------
```

When we are conducting TSSEM, the data are usually in the format of correlation (or covariance) matrices. The data are more complicated than conventional meta-analyses. The `metaSEM` package provides three functions (`readFullMat()`, `readStackVec()`, and `readLowTriMat()`) to read correlation matrices stored externally. For example, the `metaSEM` package includes a sample data set `Hunter83` that has been used in Section 7.6.3. The correlation matrices of this data set are stored as lower triangle matrices stacked together in a file `Hunter83.txt`. When there are missing values, they are represented by `NA` in the file. Part of the content of `Hunter83` is displayed as follows.

```
1
.47 1
.35 .49 1
.15 .27 .19 1
1
.39 1
NA NA NA
.28 .39 NA 1
1
.33 1
NA NA NA
.29 .25 NA 1
1
NA NA
.51 NA 1
.29 NA .27 1
1
NA NA
.69 NA 1
```

```
.40 NA .38 1
NA
NA 1
NA .72 1
NA .05 .32 1
1
.62 1
.50 .62 1
NA NA NA NA
```

We read this file using the `readLowTriMat()` function in the `metaSEM` package. As the analyses have been illustrated in Section 7.6.3, the output are not listed here.

```
R> library("metaSEM")
R> ## Read lower triangle matrices
R> ## no.var: no. of variables for the lower triangle matrices
R> my.df4 <- readLowTriMat("Hunter83.txt", no.var=4)
R> ## Add the variable names to improve readability
R> my.df4 <- lapply(my.df4, function(x)
                    {dimnames(x) <- list(c("A","K","W","S"),
                                         c("A","K","W","S"))
                     x})
R> ## Sample sizes for the studies
R> my.n4 <- c(443,186,292,233,368,360,380,366,456,78,384,59,160,210)
R> #### Random-effects model with diagonal elements only
R> ## First stage analysis
R> random1 <- tssem1(my.df4, my.n4, method="REM", RE.type="Diag")
R> ## rerun to remove error code
R> random1 <- rerun(random1, silent=TRUE)
R> summary(random1)
R> A2 <- create.mxMatrix(c(0, 0, 0, 0,
                          "0.1*A2J", 0, 0, 0,
                          "0.1*A2W", "0.1*J2W", 0, 0,
                          0, "0.1*J2S", "0.1*W2S", 0),
                          type="Full", nrow=4, ncol=4, byrow=TRUE)
R> S2 <- create.mxMatrix(c(1,"0.1*Var_J",
                          "0.1*Var_W", "0.1*Var_S"),
                          type="Diag")
R> ## Second stage analysis
R> ## Model without direct effect from Ability to Supervisor
R> ## diag.constraints=TRUE is required as there are mediators
R> summary( tssem2(random1, Amatrix=A2, Smatrix=S2,
                  intervals.type="LB", diag.constraints=TRUE) )
```

References

Adler J 2010. *R in a nutshell.* O'Reilly, Sebastopol, CA.

Becker BJ 1983. Influence again: a comparison of methods for meta-analysis. Paper presented at the annual meeting of the American Educational Research Association, Montreal.

Boker S, Neale M, Maes H, Wilde M, Spiegel M, Brick T, Spies J, Estabrook R, Kenny S, Bates T, Mehta P and Fox J 2011. OpenMx: an open source extended structural equation modeling framework. *Psychometrika* **76**(2), 306–317.

Cheung MWL 2014. metaSEM: An R package for meta-analysis using structural equation modeling. Frontiers in Psychology 5(1521).

Dalgaard P 2008. *Introductory statistics with R*, 2nd edn. Springer-Verlag, New York.

R Development Core Team 2014. *R: a language and environment for statistical computing*, R Foundation for Statistical Computing, Vienna, Austria. ISBN: 3-900051-07-0.

Viechtbauer W 2010. Conducting meta-analyses in R with the metafor package. *Journal of Statistical Software* **36**(3), 1–48.

Zuur AF, Ieno EN and Meesters EHWG 2009. *A beginner's guide to R*. Springer, Dordrecht, London.

Index

abbreviations, xvii
advanced topics in SEM-based meta-analysis, 279–310
  illustrations using R, 294–309
  moderators, missing values, 289–94, 300–9
  restricted maximum likelihood estimation (REML), 279–89
Akaike information criterion (AIC), 38
analysis of covariance (ANCOVA), 5
analysis of variance (ANOVA), 4–5
average population effect, random-effects model, 90
averaged effect sizes models, 2

Bayesian information criterion (BIC), 38
BCG vaccine for preventing tuberculosis
  effect sizes, 149–52
  multivariate meta-analysis, 146–56
  plotting the figures, 153–6
  random-effects model, 147–8
BIC *see* Bayesian information criterion
Big Five model
  confirmatory factor analysis (CFA), 244–58
  meta-analytic structural equation modeling (MASEM), 244–58
binary variables
  effect sizes, 56–7
  odds ratio (OR), 56–7
  proportion, 56

CFA *see* confirmatory factor analysis
CFI *see* comparative fit index
CIs *see* confidence intervals
common effects model, fixed-effects model, 94, 124
comparative fit index (CFI), 22–3
confidence intervals (CIs), 29–34
  likelihood-based confidence intervals (LBCIs), 30–4
  Wald confidence intervals (Wald CIs), 30–4
confirmatory factor analysis (CFA), 19–20
  Big Five model, 244–58
  illustrations using R, 244–58
  meta-analytic structural equation modeling (MASEM), 244–58

Meta-Analysis: A Structural Equation Modeling Approach, First Edition. Mike W. -L. Cheung.

Companion Website: www.wiley.com/go/cheung/meta_analysis

conflicting research findings, meta-analytic structural equation modeling (MASEM), for solution of, 215–17
correlation coefficient
  effect sizes, 55
  vs. Fisher's z score, 242–3
  Fisher's z transformation, 55
correlation matrix
  vs. covariance matrix, meta-analytic structural equation modeling (MASEM), 221
  effect sizes, 59–60, 68, 69, 77–8
  Fisher's z transformation, 59–60
  illustrations using R, 77–8
  structural equation modeling (SEM), 68, 69
correlation structure analysis *see* structural equation modeling (SEM)
covariance matrix, vs. correlation matrix, meta-analytic structural equation modeling (MASEM), 221
covariance structure analysis *see* structural equation modeling (SEM)

datasets, 7, 8
definition variables, extension in SEM, 39–40
delta method, estimating sampling variances/covariances, 61–4
dependence, effect sizes, 121–2, 180–3
  averaging the dependent effect sizes within studies, 181–2
  elimination, 182
  ignoring the dependence, 181
  selecting one effect size per study, 182
  shifting the unit of analysis, 182–3
  types, 121–2
discrepancy functions, 25–7

effect sizes, 48–78
  binary variables, 56–7
  correlation coefficient, 55
  correlation matrix, 59–60, 68, 69, 77–8
  defining, 48–9
  dependence, 121–2, 180–3
  Fisher's z transformation, 55, 59–60, 61–4
  fixed-effects model, 85, 126
  Graduate Record Examinations (GRE), 51
  heterogeneity, random-effects model, 92–3, 132–3
  heterogeneity, three-level meta-analysis, 186–7
  homogeneity, fixed-effects model, 85, 125–6
  homogeneity testing, 85, 125–6, 184–5
  illustrations using R, 68–78
  mean differences, 50–5, 57–9
  mediation models, 140–5, 340–6
  moderation models, 140–5, 146, 340–6
  M*plus*, 340–6
  multiple-endpoint (ME) studies, 58–9, 67–8, 73–7
  multiple treatment (MT) studies, 58, 66–7, 71–7
  multivariate meta-analysis, 57–60, 149–52
  odds ratio (OR), 60
  properties, 49
  random-effects model, 92–3, 132–3
  raw data distribution, 50
  raw mean difference (RMD), 50–2
  regression model,, 141–3, 162–6

repeated measures (RM), 53–5, 65–6, 69–71
sample size, 49–50
sampling variances/covariances, estimating, 60–8
standardized mean difference (SMD), 50, 52–3, 61
structural equation modeling (SEM), 61, 64–8
types, 50
univariate meta-analysis, 50–7
equations, structural equation modeling (SEM), 14
explained variance
mixed-effects model, 98–9, 188
multivariate meta-analysis, 135–6
three-level meta-analysis, 188
extensions in SEM, 38–42
definition variables, 39–40
full information maximum likelihood (FIML), 41–2
phantom variables, 38–9

FIML *see* full information maximum likelihood
Fisher's z score, vs. correlation coefficient, 242–3
Fisher's z transformation
correlation coefficient, 55
correlation matrix, 59–60
effect sizes, 55, 59–60, 61–4
fixed-effects model
common effects model, 94, 124
effect sizes homogeneity, 85, 125–6
estimation, 83–5, 126–7
generalized least squares (GLS), vs. two-stage structural equation modeling (TSSEM), 237–9
hypotheses testing, 83–5, 126–7
meta-analytic structural equation modeling (MASEM), 223–33, 244–51, 258–9, 266–8
mixed-effects model, 96–9
M*plus*, 314–17, 328–32
multivariate meta-analysis, 124–7, 136–7, 328–32
vs. random-effects model, 2, 93–6
sampling variance, known vs. estimated, 85–7
structural equation modeling approach, multivariate meta-analysis, 136–7
structural equation modeling approach, univariate meta-analysis, 100–1
two-stage structural equation modeling (TSSEM), 223–33, 244–51, 258–9, 266–8
two-stage structural equation modeling (TSSEM), vs. generalized least squares (GLS), 237–9
univariate meta-analysis, 83–7, 100–1, 107–8, 314–17
fixed-effects model with clusters
meta-analytic structural equation modeling (MASEM), 251–3, 260–1
two-stage structural equation modeling (TSSEM), 251–3, 260–1
full information maximum likelihood (FIML), extension in SEM, 41–2

generalized least squares (GLS), 25, 126
meta-analytic structural equation modeling (MASEM), 221–3
vs. two-stage structural equation modeling (TSSEM), 237–9

goodness-of-fit indices, 35–8
  comparative fit index (CFI), 22–3
  Tucker–Lewis index (TLI), 22–3
Graduate Record Examinations (GRE), effect sizes, 51

illustrations using R
  advanced topics in SEM-based meta-analysis, 294–309
  confirmatory factor analysis (CFA), 244–58
  correlation matrix, 77–8
  effect sizes, 68–78
  likelihood-based confidence intervals (LBCIs), 203–4
  meta-analytic structural equation modeling (MASEM), 244–73
  moderators, missing values, 300–9
  multiple-endpoint (ME) studies, 73–7
  multiple treatment (MT) studies, 71–7
  multivariate meta-analysis, 145–74, 297–9
  odds ratio (OR) of atrial fibrillation between bisphosphonate and non-bisphosphonate users, 105–8
  random-effects model, 202–3
  repeated measures (RM), 69–71
  restricted maximum likelihood estimation (REML), 295–300
  salesperson job performance, correlation with organizational commitment, 105–8
  testing, 204–10
  three-level meta-analysis, 200–10, 299–300
  univariate meta-analysis, 105–16, 295–7
independent vs. nonindependent effect sizes models, 2
input correlation matrix, meta-analytic structural equation modeling (MASEM), 220

Lagrange multiplier test, 28
latent growth model, 22–3, 24
likelihood-based confidence intervals (LBCIs), 30–4
  three-level meta-analysis, 203–4
  univariate meta-analysis, 110
likelihood ratio (LR) test, 28–34
  LR test statistic, 34–5

MANOVA *see* multivariate analysis of variance
MAR *see* missing at random
MASEM *see* meta-analytic structural equation modeling
matrix representations
  reticular action model (RAM), 16–18
  structural equation modeling (SEM), 15–18
maximum likelihood estimation (MLE), 25–6
  moderators, 291–4
  random-effects model, 89
  vs. restricted maximum likelihood estimation (REML), 242
MCAR *see* missing completely at random
ME studies *see* multiple-endpoint studies
mean differences
  effect sizes, 50–5, 57–9
  multivariate meta-analysis, 57–9
mediation models
  effect sizes, 140–5, 340–6
  M*plus*, 340–6
  multivariate meta-analysis, 140–5, 161–74
meta-analysis
  applications, 1–6

benefits to users, 6
history, 1–6
meta-analytic structural equation modeling (MASEM), 4, 214–73
Big Five model, 244–58
confirmatory factor analysis (CFA), 244–58
conflicting research findings, for solution of, 215–17
conventional approaches, 218–23
correlation matrix instead of a covariance matrix, 221
fixed-effects model, 223–33, 237–9, 244–51, 258–9, 266–8
fixed-effects model with clusters, 251–3, 260–1
generalized least squares (GLS), 221–3
illustrations using R, 244–73
input correlation matrix, 220
vs. multiple-group structural equation modeling, 236–7
path model for cognitive ability to supervisor rating, 265–73
random-effects model, 233–5, 239–42, 253–8, 261–5, 268–73
sample size, 220
sampling variation, 220
steps, 217
two-stage structural equation modeling (TSSEM), 223–35
univariate approaches, 218–21
unreliability correction, 243–4
metaSEM, 364–7
missing at random (MAR), 123, 290
missing completely at random (MCAR), 123
moderators, 289–91, 307–9
missing values, moderators, 289–94, 300–9, 325–7
mixed-effects model
estimation, 97–8
explained variance, 98–9, 188
hypotheses testing, 97–8
M*plus*, 322–5, 337–40, 351–3
multivariate meta-analysis, 134–6, 138–40, 160–1, 337–40
structural equation modeling approach, 193–5
structural equation modeling approach, multivariate meta-analysis, 138–40
structural equation modeling approach, univariate meta-analysis, 102–4
three-level meta-analysis, 187–8, 193–5, 351–3
univariate meta-analysis, 96–9, 102–4, 110, 322–5
MLE *see* maximum likelihood estimation
moderation models
effect sizes, 140–5, 146, 340–6
M*plus*, 340–6
multivariate meta-analysis, 140–5, 146, 161–74
moderators
illustrations using R, 300–9
maximum likelihood estimation (MLE), 291–4
missing completely at random (MCAR), 289–91, 307–9
missing values, 289–94, 300–9, 325–7
M*plus*, 325–7
multivariate meta-analysis, 292–3, 304–7
three-level meta-analysis, 307–9
univariate meta-analysis, 111–13, 292–3, 301–4
M*plus*, 313–53
effect sizes, 340–6
fixed-effects model, 314–17, 328–32
mediation models, 340–6

M*plus*, (*continued*)
mixed-effects model, 322–5, 337–40, 351–3
moderation models, 340–6
moderators, 325–7
multivariate meta-analysis, 327–46
random-effects model, 317–21, 333–7, 346–51
three-level meta-analysis, 346–53
univariate meta-analysis, 314–27
MT studies *see* multiple treatment studies
multi-sample analysis *see* multiple-group analysis
multiple-endpoint (ME) studies
effect sizes, 58–9, 67–8, 73–7
illustrations using R, 73–7
structural equation modeling (SEM), 67–8
multiple-group analysis, 23–5, 28
multiple-group structural equation modeling, vs. meta-analytic structural equation modeling (MASEM), 236–7
multiple treatment (MT) studies
effect sizes, 58, 66–7, 71–7
illustrations using R, 71–7
structural equation modeling (SEM), 66–7
multivariate analysis of variance (MANOVA), multivariate meta-analysis, 57
multivariate meta-analysis, 121–74
BCG vaccine for preventing tuberculosis, 146–56
effect sizes, 57–60, 149–52
explained variance, 135–6
extensions, 140–5
fixed-effects model, 124–7, 136–7, 328–32
illustrations using R, 145–74, 297–9
mean differences, 57–9
mediation models, 140–5, 161–74
mixed-effects model, 134–6, 138–40, 160–1, 337–40
moderation models, 140–5, 146, 161–74
moderators, 292–3, 304–7
M*plus*, 327–46
multivariate analysis of variance (MANOVA), 57
random-effects model, 127–34, 137–8, 147–8, 157–60, 333–7
restricted maximum likelihood estimation (REML), 287
standardized mean difference (SMD) between males and females on life satisfaction and life control, 156–61
structural equation modeling approach, 136–40
vs. three-level meta-analysis, 195–200
vs. univariate meta-analysis, 122–4
variance component testing, 152–3

not missing at random (NMAR), 290

odds ratio (OR)
atrial fibrillation among bisphosphonate and non-bisphosphonate users, 105–8
binary variables, 56–7
effect sizes, 60
illustrations using R, 105–8
univariate meta-analysis, 105–8
OpenMx, 362–4
OR *see* odds ratio

path analysis, 18–19
path diagrams, structural equation modeling (SEM), 15, 16
path model for cognitive ability to supervisor rating,

meta-analytic structural equation modeling (MASEM), 265–73
phantom variables, extension in SEM, 38–9
pooling correlation matrices, two-stage structural equation modeling (TSSEM), 224–7, 234–5

R statistical environment, 357–62
  see also illustrations using R
RAM *see* reticular action model
random-effects model
  average population effect, 90
  BCG vaccine for preventing tuberculosis, 147–8
  effect sizes heterogeneity, 92–3, 132–3
  estimation, 88–90, 131–2
  vs. fixed-effects model, 2, 93–6
  heterogeneity variance, 87–8
  hypotheses testing, 88–90, 131–2
  maximum likelihood (ML) estimation, 89
  meta-analytic structural equation modeling (MASEM), 233–5, 239–42, 253–8, 261–5, 268–73
  mixed-effects model, 96–9
  M*plus*, 317–21, 333–7, 346–51
  multivariate meta-analysis, 127–34, 137–8, 147–8, 157–60, 333–7
  restricted maximum likelihood estimation (REML), 89
  sampling covariances, 133–4
  structural equation modeling approach, multivariate meta-analysis, 137–8
  structural equation modeling approach, three-level meta-analysis, 189–93
  structural equation modeling approach, univariate meta-analysis, 101–2
  three-level meta-analysis, 183–7, 189–93, 202–3, 346–51
  two-stage structural equation modeling (TSSEM), 233–5, 253–8, 261–5, 268–73
  univariate meta-analysis, 87–93, 101–2, 106–7, 109–10, 317–21
  unweighted method of moments (UMM), 87–8
  variance component, nonnegative definite, 129–31
  variance component structure, 128–9
  variance component testing, 90–2, 132, 152–3
  weighted method of moments (WMM), 87–9
raw mean difference (RMD), effect sizes, 50–2
regression model
  effect sizes, 141–3, 162–6
  SAT (Math), two-stage structural equation modeling (TSSEM), 258–65
REML *see* restricted maximum likelihood estimation
repeated measures (RM)
  effect sizes, 53–5, 65–6, 69–71
  illustrations using R, 69–71
  structural equation modeling (SEM), 65–6
residual maximum likelihood estimation *see* restricted maximum likelihood estimation
restricted maximum likelihood estimation (REML), 279–89
  applying, 281–3

restricted maximum likelihood estimation (REML), (*continued*)
illustrations using R, 295–300
implementation in structural equation modeling, 283–9
implementation issues, 288–9
vs. maximum likelihood estimation (MLE), 242
multivariate meta-analysis, 287
random-effects model, 89
reasons for/against, 280–1
three-level meta-analysis, 287–8
univariate meta-analysis, 284–6
reticular action model (RAM), 16–18
latent growth model, 23
RM *see* repeated measures
RMD *see* raw mean difference
RMR *see* root mean square residual
RMSEA *see* root mean square error of approximation
root mean square error of approximation (RMSEA), 36
root mean square residual (RMR), 36–7

salesperson job performance, correlation with organizational commitment, 108–16
sample size
effect sizes, 49–50
meta-analytic structural equation modeling (MASEM), 220
sampling variances/covariances
delta method, 61–4
estimating, 60–8
fixed-effects model, known vs. estimated, 85–7
random-effects model, known vs. estimated, 133–4
structural equation modeling (SEM), 64–8
sampling variation, meta-analytic structural equation modeling (MASEM), 220
SAT (Math)
regression model, 258–65
two-stage structural equation modeling (TSSEM), 258–65
Satorra–Bentler scaled statistic, 35
SEM *see* structural equation modeling
SMD *see* standardized mean difference
SRMR *see* standardized root mean square residual
standardized mean difference (SMD)
effect sizes, 50, 52–3, 61
between males and females on life satisfaction and life control, multivariate meta-analysis, 156–61
standardized root mean square residual (SRMR), 37–8
statistical models
averaged effect sizes models, 2
dimensions, 2
fixed-effects model vs. random-effects model, 2
independent vs. nonindependent effect sizes models, 2
structural equation model, 21–2
structural equation modeling approach, multivariate meta-analysis, 136–40
fixed-effects model, 136–7
mixed-effects model, 138–40
random-effects model, 137–8
structural equation modeling approach, three-level meta-analysis, 188–95
mixed-effects model, 193–5
random-effects model, 189–93

structural equation modeling approach, univariate meta-analysis, 100–4
  fixed-effects model, 100–1
  mixed-effects model, 102–4
  random-effects model, 101–2
structural equation modeling (SEM)
  applications, 2–6, 14
  benefits to users, 6
  correlation matrix, 68, 69
  effect sizes, 61, 64–8
  history, 2–6
  model specification, 14–18
  multiple-endpoint (ME) studies, 67–8
  multiple treatment (MT) studies, 66–7
  repeated measures (RM), 65–6
  sampling variances/covariances, estimating, 64–8

three-level meta-analysis, 179–211
  data inspection, 201–2
  effect sizes, dependence, 180–3
  explained variance, 188
  illustrations using R, 200–10, 299–300
  likelihood-based confidence intervals (LBCIs), 203–4
  mixed-effects model, 187–8, 193–5, 351–3
  moderators, 307–9
  M*plus*, 346–53
  vs. multivariate meta-analysis, 195–200
  random-effects model,, 183–7, 189–93, 202–3, 346–51
  restricted maximum likelihood estimation (REML), 287–8
  structural equation modeling approach, 188–95
  testing, 204–10
  three-level model, 183–8
TLI *see* Tucker–Lewis index
TSSEM *see* two-stage structural equation modeling
Tucker–Lewis index (TLI), 22–3
two-stage structural equation modeling (TSSEM)
  fitting structural models, 227–33, 235
  fixed-effects model, 223–33, 244–51, 258–9, 266–8
  fixed-effects model with clusters, 251–3, 260–1
  vs. generalized least squares (GLS), 237–9
  meta-analytic structural equation modeling (MASEM), 223–35
  pooling correlation matrices, 224–7, 234–5
  random-effects model, 233–5, 253–8, 261–5, 268–73
  regression model, SAT (Math), 258–65
  subgroup analysis, 233

UMM *see* unweighted method of moments
univariate approaches, meta-analytic structural equation modeling (MASEM), 218–21
univariate meta-analysis, 81–116
  categorical predictors, 114–16
  coefficient equality, 113–14
  effect sizes, 50–7
  fixed-effects model, 83–7, 100–1, 107–8, 314–17
  fixed-effects model vs. random-effects model, 93–6
  illustrations using R, 105–16, 295–7

univariate meta-analysis, (*continued*)
- likelihood-based confidence intervals (LBCIs), 110
- mixed-effects model, 96–9, 102–4, 110, 322–5
- moderators,, 110–13, 292–3, 301–4
- M*plus*, 314–27
- vs. multivariate meta-analysis, 122–4
- odds ratio (OR), 105–8
- random-effects model, 87–93, 101–2, 106–7, 109–10, 317–21
- random-effects model vs. fixed-effects model, 93–6
- restricted maximum likelihood estimation (REML), 284–6
- structural equation modeling approach, 100–4

unweighted method of moments (UMM), random-effects model, 87–8

variance component testing
- multivariate meta-analysis, 152–3
- random-effects model, 90–2, 132, 152–3

Wald confidence intervals (Wald CIs), 30–4
Wald test, 28–34
weighted least squares (WLS), 26–7
weighted method of moments (WMM), random-effects model, 87–9
WLS *see* weighted least squares
WMM *see* weighted method of moments